Clashing Views on Controversial Environmental Issues

4th edition

Clashing Views on Controversial Environmental Issues

4th edition

Edited, Selected, and with Introductions by

Theodore D. Goldfarb

State University of New York at Stony Brook

The Dushkin Publishing Group, Inc.

This book is dedicated to my children and all other children for whom the successful resolution of these issues is of great urgency.

Library of Congress Catalog Card Number: 90-84861

Manufactured in the United States of America

Fourth Edition, First Printing

ISBN: 0-87967-937-9

The Dushkin Publishing Group, Inc.
Sluice Dock, Guilford, CT 06437

PREFACE

For the past fourteen years I have been teaching an environmental chemistry course, and my experience has been that the critical and complex relationship we have with our environment is of vital and growing concern to students, regardless of their majors. Consequently, for this fourth edition, I have redoubled my efforts to shape issues and to select articles that do not require a technical background or prerequisite courses in order to be understood. For example, in addition to the sciences, this volume would be appropriate for such disciplines as philosophy, law, sociology, political science, economics, and allied health—any course where environmental topics are addressed.

Faculty are divided about whether it is appropriate to use a classroom to advocate a particular position on a controversial issue. Some believe that the proper role of a teacher is to maintain neutrality in order to present the material in as objective a manner as possible. Others, like myself, find that students rarely fail to recognize their instructors' points of view. Rather than reveal which side I am on through subtle hints, I prefer to be forthright about it, while doing my best to encourage students to develop their own positions, and I do not penalize them if they disagree with my views. No matter whether the goal is to attempt an objective presentation or to encourage advocacy, it is necessary to present both sides of any argument. To be a successful proponent of any position, it is essential to understand your opponents' arguments. The format of this text, with thirty-six essays arranged in pro and con pairs on eighteen environmental controversies, is designed with those objectives in mind.

In the *introduction* to each issue, I present the historical context of the controversy and some of the key questions that divide the disputants. The *postscript* that follows each pair of essays includes comments offered to provoke thought about aspects of the issue that are suitable for classroom discussion. A careful reading of my remarks may reveal the positions I favor, but the essays themselves and the *suggestions for further reading* in each postscript should provide the student with the information needed to construct and support an independent perspective.

Changes to this edition This fourth edition has been considerably revised and updated. There are three completely new issues: *Earth Day 1970 v. 1990: Has the Environmental Movement Been a Success?* (Issue 1); *Can Current Pollution Strategies Improve Air Quality?* (Issue 6); and *Is Brazil Serious About Preserving Its Environment?* (Issue 15). For four of the issues retained from the previous edition, the issue question is similar to what appeared in the previous editions, but both selections have been replaced in order to more sharply focus the debate and bring it up to date: the issue on hazardous waste (Issue 12); the issue on municipal waste (Issue 13); the issue on global warming (Issue 16); and the issue on the ozone depletion problem and the Montreal Protocol (Issue 17). I have replaced one or the other of the YES and NO selections in many of the issues retained from the previous edition. The overall result is that about half of the thirty-six selections are new.

A word to the instructor An *Instructor's Manual with Test Questions* (multiple-choice and essay) is available through the publisher for the instructor using *Taking Sides* in the classroom. Also available is a general guidebook, called *Using Taking Sides in the Classroom,* which has general suggestions for adapting the pro-con approach in any classroom setting.

Acknowledgments I received many helpful comments and suggestions from friends and readers across the United States and Canada. Their suggestions have markedly enhanced the quality of this edition and are reflected in the new issues and the updated selections.

Special thanks go to those who responded to the questionnaire with specific suggestions for the fourth edition:

E. Gene Frankland
Ball State University

Chowdhury Haque
University of Manitoba

Kathryn Hedges
Indiana University Northwest

Arthur Helweg
Western Michigan University

Elmo Law
University of Missouri - Kansas City

Rosemary O'Leary
Indiana University - Bloomington

Peter Pizos
Northwest Community College

Robert L. Vertrees
Ohio State University

Chris White
Community College of the Finger Lakes

I wish to thank my wife, Jane De Young, for her encouragement and for the helpful, intelligent advice she offered. Finally, I am grateful to Mimi Egan, program manager of the Taking Sides series, for her assistance.

Theodore D. Goldfarb
Stony Brook, NY

CONTENTS IN BRIEF

PART 1 GENERAL PHILOSOPHICAL AND POLITICAL ISSUES **1**

Issue 1. Earth Day 1970 *v.* 1990: Has the Environmental Movement Been a Success? **2**

Issue 2. Does Wilderness Have Intrinsic Value? **16**

Issue 3. Do We Need More Stringently Enforced Regulations to Protect Endangered Species? **36**

Issue 4. Does Risk-Benefit Analysis Provide an Objective Method for Making Environmental Decisions? **52**

Issue 5. Is Population Control the Key to Preventing Environmental Deterioration? **70**

PART 2 THE ENVIRONMENT AND TECHNOLOGY **95**

Issue 6. Can Current Pollution Strategies Improve Air Quality? **96**

Issue 7. Is Nuclear Power Safe and Desirable? **112**

Issue 8. Is the Widespread Use of Pesticides Required to Feed the World's People? **130**

Issue 9. Is There a Cancer Epidemic Due to Industrial Chemicals in the Environment? **152**

Issue 10. Is Immediate Legislative Action Needed to Combat the Effects of Acid Rain? **172**

Issue 11. Should Women Be Excluded from Jobs That Could Be Hazardous to a Fetus? **186**

PART 3 DISPOSING OF WASTES **203**

Issue 12. Hazardous Waste: Are Cleanup Efforts Succeeding? **204**

Issue 13. Municipal Waste: Should Incineration Be a Part of Waste Disposal Methods? **216**

Issue 14. Nuclear Waste: Is Yucca Mountain an Appropriate Site for Nuclear Waste Disposal? **228**

PART 4 THE ENVIRONMENT AND THE FUTURE **245**

Issue 15. Is Brazil Serious About Preserving Its Environment? **246**

Issue 16. Does Global Warming Require Immediate Action? **264**

Issue 17. Is the Montreal Protocol Adequate for Solving the Ozone Depletion Problem? **284**

Issue 18. Are Abundant Resources and an Improved Environment Likely Future Prospects for the World's People? **304**

CONTENTS

Preface i

Introduction: The Environmental Movement xii

PART 1 GENERAL PHILOSOPHICAL AND POLITICAL ISSUES **1**

ISSUE 1. Earth Day 1970 *v.* 1990: Has the Environmental Movement Been a Success? **2**

YES: Bil Gilbert, from "Earth Day Plus 20, and Counting," *Smithsonian* **4**

NO: T. Allan Comp, from "Earth Day and Beyond," *American Forests* **11**

Conservation journalist Bil Gilbert documents his claim that environmentalism has produced impressive spiritual and material impacts on society. Forestry scientist T. Allan Comp contends that despite the increase in environmental activity since 1970, most environmental problems have gotten worse.

ISSUE 2. Does Wilderness Have Intrinsic Value? **16**

YES: Robert K. Olson, from "Wilderness International: The New Horizon," *Wilderness* **18**

NO: William Tucker, from "Is Nature Too Good for Us?" *Harper's* **23**

Wilderness activist Robert K. Olson contends that global change and resource depletion have transformed wilderness preservation from a value in its own right to an international imperative. William Tucker, a writer and social critic, asserts that wilderness areas are elitist preserves designed to keep people out.

ISSUE 3. Do We Need More Stringently Enforced Regulations to Protect Endangered Species? **36**

YES: Lewis Regenstein, from "Endangered Species and Human Survival," *USA Today* (a publication of the Society for the Advancement of Education) **38**

NO: Richard Starnes, from "The Sham of Endangered Species," *Outdoor Life* 44

Environmentalist Lewis Regenstein charges that exploiting rather than protecting wildlife, which is a legacy of Reagan administration policies, threatens to accelerate the disappearance of endangered plant and animal species. *Outdoor Life* editor Richard Starnes counters that wildlife management policy should be left to the judgment of professionals in the state and national agencies responsible for carrying out the policies, not to the whims of federal legislators.

ISSUE 4. Does Risk-Benefit Analysis Provide an Objective Method for Making Environmental Decisions? 52

YES: William D. Ruckelshaus, from "Science, Risk, and Public Policy," *Science* 54

NO: Langdon Winner, from "Risk: Another Name for Danger," *Science for the People* 60

Former EPA administrator William D. Ruckelshaus advocates educating the public about risk estimates and separating the scientific process of risk assessment from the management of risks through regulation. Social scientist Langdon Winner asserts that dealing with environmental and health hazards in terms of risk assessment leads to delays and confusion in efforts to regulate pollution and protect the public.

ISSUE 5. Is Population Control the Key to Preventing Environmental Deterioration? 70

YES: Paul R. Ehrlich and John P. Holdren, from "Impact of Population Growth," *Science* 72

NO: Barry Commoner, Michael Corr, and Paul J. Stamler, from "The Causes of Pollution," *Environment* 80

Environmental scientists Paul R. Ehrlich and John P. Holdren argue that population increase is the principal cause of environmental degradation. Environmental scientists Barry Commoner, Michael Corr, and Paul J. Stamler contend that technological change rather than population growth has been the chief cause of environmental stress.

PART 2 THE ENVIRONMENT AND TECHNOLOGY **95**

ISSUE 6. Can Current Pollution Strategies Improve Air Quality? **96**

YES: John G. McDonald, from "Gasoline and Clean Air: Good News, Better News, and a Warning," *Vital Speeches of the Day* **98**

NO: Hilary F. French, from "Communication: A Global Agenda for Clean Air," *Energy Policy* **105**

Oil company president John G. McDonald argues that air pollution from automobiles can be controlled by using cleaner fuels and continuing to improve the automobile engine. Energy policy analyst Hilary F. French claims that adequate control of air pollutants requires major efforts to reorient energy use, transportation, and industrial production toward pollution prevention.

ISSUE 7. Is Nuclear Power Safe and Desirable? **112**

YES: Alvin M. Weinberg, from "Is Nuclear Energy Necessary?" *Bulletin of the Atomic Scientists* **114**

NO: Denis Hayes, from "Nuclear Power: The Fifth Horseman," *Worldwatch Paper 6* **122**

Nuclear physicist Alvin M. Weinberg argues that development of nuclear breeder reactors is needed to assure future energy supplies and that safety can be provided through technical improvements and by locating reactors in remote areas. Energy analyst Denis Hayes contends that developing nuclear power assures nuclear weapons proliferation and that environmentally safer energy sources are technically and economically feasible.

ISSUE 8. Is the Widespread Use of Pesticides Required to Feed the World's People? **130**

YES: William R. Furtick, from "Uncontrolled Pests or Adequate Food," in D. L. Gunn and J. G. R. Stevens, eds., *Pesticides and Human Welfare* **132**

NO: Shirley A. Briggs, from "Silent Spring: The View from 1990," *The Ecologist* **140**

Crop protection specialist William R. Furtick warns that pesticides are essential for the intensive agriculture required to prevent mass starvation.

Environmentalist Shirley A. Briggs counters that pesticides have failed to decrease crop loss while causing widespread ecological harm, as predicted in Rachel Carson's book *Silent Spring*.

ISSUE 9. Is There a Cancer Epidemic Due to Industrial Chemicals in the Environment? **152**

YES: Samuel S. Epstein, from "Losing the War Against Cancer," *The Ecologist* **154**

NO: Elizabeth M. Whelan, from "The Charge of the Cancer Brigade," *National Review* **164**

Physician Samuel S. Epstein claims that preventable environmental exposure is responsible for dramatically rising human cancer rates. Physician Elizabeth M. Whelan asserts that the claim that we are experiencing an environmentally caused cancer epidemic is a misconception.

ISSUE 10. Is Immediate Legislative Action Needed to Combat the Effects of Acid Rain? **172**

YES: Jon R. Luoma, from "Acid Murder No Longer a Mystery," *Audubon* **174**

NO: A. Denny Ellerman, from Testimony before the Subcommittee on Environmental Protection, U.S. Senate **181**

Science writer Jon R. Luoma argues that research has produced convincing evidence that legislation requiring large cutbacks in acid gas emissions is urgently needed to prevent the continued destruction of forest and lake ecosystems. Coal association executive A. Denny Ellerman argues that sulfur oxide emissions are presently declining and further restrictions would produce only marginal benefits at great cost to industry and the public.

ISSUE 11. Should Women Be Excluded from Jobs That Could Be Hazardous to a Fetus? **186**

YES: Hugh M. Finneran, from "Title VII and Restrictions on Employment of Fertile Women," *Labor Law Journal* **188**

NO: Carolyn Marshall, from "Fetal Protection Policies: An Excuse for Workplace Hazard," *The Nation* **196**

Corporate counsel Hugh M. Finneran believes women should be excluded from occupations that threaten the unborn for humanitarian reasons as well as to protect employers from future liability. Health writer Carolyn Marshall counters that reproductive toxins are hazardous to men as well as women and that cleaning up the workplace is the only acceptable response to conditions that threaten the fetus.

PART 3 DISPOSING OF WASTES **203**

ISSUE 12. Hazardous Waste: Are Cleanup Efforts Succeeding? 204

YES: Robert G. Wright, from "Waste Management: A Cooperative Cleanup for Superfund Sites," *Environment* **206**

NO: Robert J. Mentzinger, from "GEMS Landfill: A Superfund Failure," *Public Citizen* **209**

Robert G. Wright, a hazardous waste director for a labor union, describes a partnership among government, labor, and management to produce a training program for waste workers that has proven effective in cleanup efforts. Researcher Robert J. Mentzinger chronicles the problems at one toxic waste site as an introduction to a highly critical assessment of current government cleanup programs.

ISSUE 13. Municipal Waste: Should Incineration Be a Part of Waste Disposal Methods? 216

YES: John Shortsleeve and Robert Roche, from "Analyzing the Integrated Approach," *Waste Age* **218**

NO: Neil Seldman, from "Waste Management: Mass Burn Is Dying," *Environment* **222**

Incineration industry executives John Shortsleeve and Robert Roche argue that an integrated system that incinerates the residue from municipal waste recycling and composting operations is the best disposal option. Waste disposal consultant Neil Seldman claims that intensive recycling and composting can do the job without costly and hazardous mass incineration technology.

ISSUE 14. Nuclear Waste: Is Yucca Mountain an Appropriate Site for Nuclear Waste Disposal? **228**

YES: Luther J. Carter, from "Siting the Nuclear Waste Repository: Last Stand at Yucca Mountain," *Environment* **230**

NO: Charles R. Malone, from "The Yucca Mountain Project: Storage Problems of High-Level Radioactive Wastes," *Environmental Science & Technology* **240**

Science journalist Luther J. Carter describes a variety of technical and political considerations that support the choice of Yucca Mountain as a nuclear waste dump. Nuclear waste specialist Charles R. Malone cites several serious deficiencies in the reliability and accuracy of assessments of the proposed site, which makes it impossible for present efforts to succeed.

PART 4 THE ENVIRONMENT AND THE FUTURE **245**

ISSUE 15. Is Brazil Serious About Preserving Its Environment? **246**

YES: Nira Broner Worcman, from "Brazil's Thriving Environmental Movement," *Technology Review* **248**

NO: Philip M. Fearnside, from "Deforestation in Brazilian Amazonia: The Rates and Causes of Forest Destruction," *The Ecologist* **255**

Brazilian environmental journalist Nira Broner Worcman is cautiously optimistic that new governmental policies in response to national environmental activism and global concern may halt Amazonian rain forest destruction. Brazilian ecologist Philip M. Fearnside cites failure to deal with the economic and social root causes of deforestation as the basis for his fear that the rain forests will be totally destroyed.

ISSUE 16. Does Global Warming Require Immediate Action? **264**

YES: Claudine Schneider, from "Preventing Climate Change," *Issues in Science and Technology* **266**

NO: Ari Patrinos, from "Greenhouse Effect: Should We *Really* Be Concerned?" *USA Today* (a publication of the Society for the Advancement of Education) **274**

Congresswoman Claudine Schneider acknowledges the uncertainties about the extent and consequences of global warming, but she outlines a series of preventative actions that make good sense even if little climate change occurs. Carbon dioxide researcher Ari Patrinos stresses the need for more research, arguing that there is no need for urgent action other than curtailing greenhouse gas emissions.

ISSUE 17. Is the Montreal Protocol Adequate for Solving the Ozone Depletion Problem? **284**

YES: Richard Elliot Benedick, from "Ozone Diplomacy," *Issues in Science and Technology* **286**

NO: Arjun Makhijani, Amanda Bickel, and Annie Makhijani, from "Beyond the Montreal Protocol: Still Working on the Ozone Hole," *Technology Review* **294**

Deputy Assistant Secretary of State Richard Elliot Benedick describes the technical background and political history of the Montreal Protocol, designed to protect the planet from ozone depletion. Energy researchers Arjun Makhijani, Amanda Bickel, and Annie Makhijani acknowledge the significance of the Protocol but argue that it must be strengthened and its outreach broadened if the ozone problem is to be solved.

ISSUE 18. Are Abundant Resources and an Improved Environment Likely Future Prospects for the World's People? **304**

YES: Julian L. Simon, from "Life on Earth Is Getting Better, Not Worse," *The Futurist* **306**

NO: Lindsey Grant, from "The Cornucopian Fallacies: The Myth of Perpetual Growth," *The Futurist* **312**

Economist Julian L. Simon is optimistic about the likelihood that human minds and muscle will overcome resource and environmental problems. Environmental consultant Lindsey Grant fears that unless a "sustainable relationship between people and earth" is developed, the future may bring famine and ecological disaster.

Contributors 322
Index 328

INTRODUCTION

The Environmental Movement

Theodore D. Goldfarb

ENVIRONMENTAL CONSCIOUSNESS

On April 22, 1990, 200 million people in 140 countries around the world participated in a variety of activities to celebrate Earth Day, an event given wide publicity by the media. The date chosen was the anniversary of the first Earth Day (celebrated only in the United States), which many social historians credit with spawning the ongoing global environmental movement. The intervening years have witnessed explosive growth in political, scientific and technical, regulatory, financial, industrial, and educational activities related to an ever-expanding list of environmental problems. We have learned that industrial development has reached such a level that its polluting by-products threaten not only local environments, but also the global ecosystems that control the Earth's climate and the ozone shield that filters out potentially lethal solar radiation. The elevation of environmental concern to a prominent position on the international political agenda has lead to recent speculation by some social commentators that the world is entering "the decade—or even the era— of the environment."

THE HISTORY OF ENVIRONMENTALISM

The current interest in environmental issues in the United States has its historical roots in the conservation movement of the late nineteenth and early twentieth centuries. This earlier, more limited, recognition of the need for environmental preservation was a response to the destruction wrought by uncontrolled industrial exploitation of natural resources in the post–Civil War period. Clear-cutting forests, in addition to producing large devastated areas, resulted in secondary disasters. Bark and branches left in the cutover areas caused several major midwestern forest fires, which leveled villages and killed thousands of people. Severe floods were caused by the loss of trees which previously had helped to reduce surface water runoff. The Sierra Club and the Audubon Society, the two oldest environmental organizations still active today, were founded around the turn of the century and helped to organize public opposition to the destructive practice of exploiting resources. Mining, grazing, and lumbering were brought under government control by such landmark legislation as the Forest Reserve Act of 1891 and the Forest Management Act of 1897. Schools of forestry were established at several of the land grant colleges to help develop the scientific expertise needed for the wise management of forest resources.

Compared to this earlier period of concern about the misuse of natural resources, which developed gradually over several decades, the present environmental movement had an explosive beginning. When Rachel Carson's book *Silent Spring* appeared in 1962, its emotional warning about the inherent dangers in the excessive use of pesticides ignited the imagination of an enormous and disparate audience (e.g., wildlife lovers, health care professionals, hunters, and farmers) who had become uneasy about the proliferation of new synthetic chemicals in agriculture and industry. The atmospheric testing of nuclear weapons began to cause widespread public concern about the effects of nuclear radiation. City dwellers were beginning to recognize the connection between the increasing prevalence of smoky, irritating air and the daily ritual of urban commuter traffic jams. The responses to Carson's book included not only a multitude of scientific and popular debates about the issues she had raised, but also a ground swell of public support for increased environmental controls over all forms of pollution.

The rapid rise in the United States of public concern about environmental issues is apparent from the results of opinion polls. Similar surveys taken in 1965 and 1970 showed an increase from 17 to 53 percent in the number of respondents who rated "reducing pollution of air and water" as one of the three problems they would like the government to pay more attention to. By 1984, pollster Louis Harris was reporting to Congress that 69 percent of the public favored making the Clean Air Act more stringent. A poll taken for the 1988 presidential election revealed that 73 percent of the population consider themselves to be "environmentalists."

The growth of environmental consciousness in the United States swelled the ranks of the older voluntary organizations, such as the national Wildlife Federation, the Sierra Club, the Isaac Walton league, and the Audubon Society, and has led to the establishment of more than 200 new national and regional associations and 3,000 local ones. The newer organizations pursue a more activist agenda and tend to use more aggressive methods than their precursors. Such national and international groups as the Environmental Defense Fund, Friends of the Earth, the National Resources Defense Council, Environmental Action, the League of Conservation Voters, and Zero Population Growth have developed considerable expertise in lobbying for legislation, influencing elections, and litigating in the courts. Critics of the environmental movement have frequently pointed out that the membership of these organizations comes from the upper socioeconomic classes. While acknowledging this is true, environmentalists deny that the causes they champion are elitist, and they cite evidence that most of their goals are supported by majority sentiment among people from all walks of life. It cannot be denied, however, that the effects on health and quality of life that result from pollution tend to be a heavier burden for the poor.

Environmental literature has also grown exponentially since the appearance of *Silent Spring*. Many new popular magazines, technical journals, and organizational newsletters devoted to environmental issues have appeared, as well as hundreds of books, some of which, like Paul Ehrlich's *The Population Bomb* (1968) and Barry Commoner's *The Closing Circle* (1972), have become best-sellers.

CLASHING VIEWS FROM CONFLICTING VALUES

As with all social issues, those on opposite sides of environmental disputes have conflicting personal values. On some level, almost everyone would admit to being concerned about threats to the environment. However, enormous differences exist in individual perceptions about the seriousness of some environmental threats, their origins, their relative importance, and what to do about them. In most instances, very different conclusions can be expressed on these issues, conclusions based on evaluations of the same basic scientific evidence.

What, then, are these different value systems which produce such heated debate? Some are obvious: An executive of a chemical company has a vested interest in placing greater value on the financial security of the company's stockholders than on the possible environmental effects of the company's operation. He or she is likely to interpret the potential health effects of what comes out of the plant's smokestacks or sewer pipes differently than would a resident of the surrounding community. These different interpretations need not involve any conscious dishonesty on anyone's part. There is likely to be sufficient scientific uncertainty about the pathological and ecological consequences of the company's effluents to enable both sides to reach very different conclusions from the available "facts."

Less obvious are the value differences among scientists which can divide them in an environmental dispute. Unfortunately, when questions are raised about the effects of personal value systems on scientific judgments, the twin myths of scientific objectivity and scientific neutrality get in the way. Neither the scientific community nor the general population appear to understand that scientists are very much influenced by subjective, value-laden considerations and will frequently evaluate data in a manner that supports their own interests. For example, a scientist employed by a pesticide manufacturer may be less likely than a similarly trained scientist working for an environmental organization to take data that shows that one of the company's products is a low-level carcinogen in mice and interpret that data to mean that the product therefore poses a threat to human health.

Even self-proclaimed environmentalists frequently argue over environmental issues. Hunters, while supporting the prohibition of lumbering and mining on their favorite hunting grounds, strongly oppose the designation of these regions as wilderness areas because that would result in the prohibition of the vehicles they use to bring home their bounty. Also opposed to wilderness designation are foresters. Although they share many of the environmental goals of preservationists, foresters believe that forest lands should be scientifically managed rather than left alone to evolve naturally.

Political ideology can also have a profound effect on environmental attitudes. Those critical of the prevailing socioeconomic system are likely to attribute environmental problems to the industrial development supported by that system. Others are likelier to blame environmental degradation on more universal factors, such as population growth.

Changes in prevailing social attitudes influence public response to environmental issues. The American pioneers were likely to perceive their natural surroundings as being dominated by hostile forces that needed to be conquered or overcome. This attitude clearly extended to the human inhabitants, as well as the flora and fauna, native to the lands the pioneers were claiming for their own. The notion that humans should conquer nature has only slowly been replaced by the alternative view of living in harmony with the natural environment, but the growing popularity of the environmental movement evidences the public's acceptance of this goal.

PROTECTING THE ENVIRONMENT

There has always been strong resistance to regulatory restraints on industrial and economic activity in the United States. The most ardent supporters of our capitalist economy argue that pollution and other environmental effects have certain costs and that regulation will take place automatically through the marketplace. Despite mounting evidence that the social costs of polluted air and water are usually external to the economic mechanisms affecting prices and profits, prior to the 1960s, Congress imposed very few restrictions on the types of technology and products industry could use or produce.

As noted above, the turn-of-the-century conservation movement did result in legislation restricting the exploitation of lumber and minerals on federal lands. In response to public outrage over numerous incidents of death and illness from adulterated foods, Congress established the Food and Drug Administration (FDA) in 1906, but gave it only limited authority to ban products that were obviously harmful or improperly labeled.

Regulatory Legislation

The environmental movement of the 1960s and 1970s produced a profound and controversial change in the political climate concerning regulatory legislation. Concerns such as the proliferation of new synthetic chemicals in industry and agriculture, the increased use of hundreds of inadequately tested additives in foods, and the effects of automotive emissions were pressed on Congress by increasingly influential environmental organizations. Beginning with the Food Additives Amendment of 1958, which required FDA approval of all new chemicals used in the processing and marketing of foods, a series of federal and state legislative and administrative actions resulted in the creation of numerous regulations and standards aimed at reducing and reversing environmental degradation.

Congress responded to the environmental movement with the National Environmental Policy Act of 1969. This act pronounced a national policy requiring an ecological impact assessment for any major federal action. The legislation called for the establishment of a three-member Council on Environmental Quality, responsible to the president, to initiate studies, make recommendations, and prepare an annual Environmental Quality Report. It also requires all agencies of the federal government to prepare a detailed environmental impact statement (EIS) for any major project or proposed legislation in which they are involved. Despite some initial attempts to evade this requirement, court suits by environmental groups have forced compliance, and now, new facilities like electrical power plants, interstate highways, dams, harbors, and interstate pipelines can only proceed after preparation and review of an EIS.

Another major step in increasing federal anti-pollution efforts was the establishment in 1970 of the Environmental Protection Agency (EPA). Many programs previously administered by a variety of agencies, such as the departments of the Interior, Agriculture, and Health, Education and Welfare, were transferred to this new, central, independent agency. The EPA was granted authority to do research, propose new legislation, and implement and enforce existing laws concerning air and water pollution, pesticide use, radiation exposure, toxic substances, solid waste, and noise abatement.

The year 1970 also marked the establishment of the Occupational Safety and Health Administration (OSHA), the result of a long struggle by organized labor and independent occupational health organizations to focus attention on the special problems of the workplace. A major responsibility of OSHA is the enforcement of legislation regulating the workplace environment. The thousands of synthetic chemicals used in modern industrial activity—most of which have not been tested for effects of chronic exposure—make this task extremely difficult.

The first major legislation to propose the establishment of national standards for pollution control was the Air Quality Act of 1967. The Clean Air Act of 1970 specified that ambient air quality standards were to be achieved by July 1, 1975 (a goal that was not met and remains elusive), and that automotive hydrocarbon, carbon monoxide, and nitrogen oxide emissions were to be reduced by 90 percent within five years—a deadline that has been repeatedly extended. Specific standards to limit the pollution content of effluent wastewater were prescribed in the Water Pollution Control Act of 1970. The Safe Drinking Water Act of 1974 authorized the EPA to establish federal drinking water standards, applicable to all public water supplies. The Occupational Safety and Health Act of 1970 allowed OSHA to establish strict standards for exposure to harmful substances in the workplace. The Environmental Pesticide Control Act of 1972 gave the EPA authority to regulate pesticide use and to control the sale of pesticides in interstate commerce. In

1976, the EPA was authorized to establish specific standards for the disposal of hazardous industrial wastes under the Resource Conservation and Recovery Act—but it wasn't until 1980 that the procedures for implementing this legislative mandate were announced. Finally, in 1976, the Toxic Substance Control Act became law, providing the basis for the regulation of public exposure to toxic materials not covered by any other legislation.

All of this environmental legislation in such a short time span produced a predictable reaction from industrial spokespeople and free-market economists. By the late 1970s, attacks on what critics referred to as over-regulation appeared with increasing frequency in the media. Anti-pollution legislation was criticized as a principal contributor to inflation and a serious impediment to continued industrial development.

One of the principal themes of Ronald Reagan's first presidential campaign was a pledge to get regulators off the backs of entrepreneurs. He interpreted his landslide victory in 1980 to mean that the public supported a sharp reversal of the federal government's role as regulator in all areas, including the environment. Two of Reagan's key appointees were Interior Secretary James Watt and EPA Administrator Ann Gorsuch Burford, both of whom set about to reverse the momentum of their agencies with respect to the regulation of pollution and environmental degradation. It soon became apparent that Reagan and his advisors had misread public attitudes. Sharp staffing and budget cuts at the EPA and OSHA produced a counterattack by environmental organizations whose membership rolls had continued to swell. Mounting public criticism of the neglect of environmental concerns by the Reagan administration was compounded by allegations of misconduct and criminal activity against environmental officials, including Ms. Burford, who was forced to resign. President Reagan attempted to mend fences with environmentalists by recalling William Ruckelshaus, the popular first EPA Administrator, to again head the agency. But throughout Reagan's presidency, few new environmental initiatives were carried out.

During his presidential election campaign, George Bush recognized the public's growing concern with threats to the environment and impatience with the federal government's lack of progress. He entered office pledging to reestablish a strong federal role in dealing with the environmental issues. Recent opinion polls have assessed his record thus far as being heavier on rhetoric than on action.

RECENT DEVELOPMENTS

The differences between the 1970 and 1990 versions of Earth Day in terms of numbers of participants, global involvement, scope of issues considered, and strategies proposed reflect significant changes that have occurred during the environmental movement.

For a comprehensive examination of current environmental developments and a look at what Earth Day 1990 signified and accomplished for the environmental movement, see Issue 1 (*Earth Day 1970 v. 1990: Has the Environmental Movement Been a Success?*). In that issue, the effects of the environmental movement are debated and a critical examination of Earth Day is offered. Many writers and social critics, environmentalists, and policymakers saw in Earth Day 1990 an opportunity to assess the environmental movement and make recommendations for the future direction of environmentalism. The April 30, 1990, issue of *The Nation* contains two critiques of present environmental developments. In the first, "Ending the War Against Earth," Barry Commoner summarizes the principal theme of his recent book *Making Peace with the Planet* (Pantheon, 1990). He proposes that little will be accomplished by merely limiting the pollution produced by existing technology. Instead, he calls for redesigning industrial, agricultural, and transportation systems so that they will be environmentally benign and harmonious with the ecosphere. The other article, "The Trouble with Earth Day," by author and social critic Kirkpatrick Sale, presents four fundamental criticisms of the agenda of Earth Day 1990 organizers. Sale contends that the focus on individual action is misguided because most environmental problems are a result of inappropriate systems of production or policies of governments or institutions that cannot be altered or reversed by each of us acting individually to adopt a more ecological life-style. Second, he complains about the decision to use most of the $3 million and unlimited publicity to put on a "week-long media bash" rather than to organize a long-range campaign with a continuing political thrust. Third, he accuses the organizers of having added support by accepting as partners many of the corporations, politicians, and lobbyists who have helped create existing problems. By doing so, Earth Day organizers have eliminated any possibility of developing a clear analysis of what needs to be done. Finally, Sale points to the narrow anthropocentric focus on human peril rather than a more appropriate ecocentric perspective that would identify the solutions as those that would begin to restore the balance of the Earth's natural systems.

Organizations such as Earth First! and the radical wings of various "green" political movements as well as other proponents of "deep ecology" are even more critical of the strategies linked to Earth Day. This fringe of the environmental movement proposes such radial goals as a return to a much simpler, less technological life-style and a drastic reduction of the present world human population. For an introduction to the policies, philosophies, and recent activities of these "ecorads," read "Radical Ecology on the Rise," by Brian Tokar, "Earth First! and Cointelpro," by Leslie Hemstreet—both in the July/August issue of *Z Magazine*—and "Earth First!ers Wield a Mean Monkey Wrench," by Michael Parfit, *Smithsonian* (April 1990).

Members of the present environmental establishment have offered their own agendas for the future. In the July/August issue of *EPA Journal,* William K. Reilly, administrator of the EPA, presents proposals in "The Greening of EPA" for the greater use of economic incentives to prevent pollution before it occurs. James Gustave Speth, former EPA administrator and current president of the World Resources Institute, calls for a greater emphasis on international problems and a reorganization of the EPA along lines that would enable it to promote far-reaching technological change in his article "EPA and the World Clean-up Puzzle."

Professor of political science and public and environmental affairs Lynton K. Caldwell was one of the principal authors of the National Environmental Policy Act (NEPA) of 1969, which established the basic legislative environmental philosophy and policy under which the EPA and other U.S. governmental agencies have been operating. He now thinks that a constitutional amendment is needed to place environmental protection within the country's fundamental law, as has been done by such other nations as Brazil, China, West Germany, the Netherlands, Sweden, and Switzerland. The December 1989 issue of *Environment* includes Caldwell's article "A Constitutional Law for the Environment—20 Years with NEPA Indicates the Need," as well as several reactions to his proposal and Caldwell's response in the Commentary section.

GLOBAL DEVELOPMENTS

Although initially lagging behind the United States in environmental regulation, other developed industrial countries have been moving rapidly over the past decade to catch up. In a few European countries where "green parties" have become influential participants in the political process, certain pollutant emission standards are now more stringent than their U.S. counterparts. Environmental planning and control are prominent among the controversial issues being debated by the European Economic Community.

While the feeding and clothing of their growing populations continue to be the dominant concerns of developing countries, they too are paying increasing attention to environmental protection. Suggestions that they forgo the use of industrial technologies that have resulted in environmental degradation in developed countries are often viewed as an additional obstacle to the goal of raising their standard of living.

During the past decade, attention has shifted from a focus on local pollution to concern about global environmental degradation. Studies of the potential effects of several gaseous atmospheric pollutants on the Earth's climate and its protective ozone layer have made it apparent that human activity has reached a level that can result in major impacts on the planetary ecosystems. A series of major international conferences of political as well as scientific leaders have been held with the goal of seeking solutions to threatening worldwide environmental problems. Serious discussions are under way about how to guide future development so as to avert or minimize the threats while satisfying the frequently conflicting socioeconomic needs of the developed and developing nations.

New Approaches

An evaluation of the apparent failure to control environmental decay in the past two decades has given rise to demands for new approaches. Environmental policy analysts have proposed that regulatory agencies adopt a more holistic approach to environmental protection, rather than continuing their attempts to impose separate controls on what are actually interconnected problems. The use of economic strategies, such as pollution taxes or the sale of licenses to those who wish to produce limited quantities of pollutants, has received increasing support as potentially more effective than regulatory emission standards. Such suggestions continue to enrage many environmentalists who consider the sale of pollution rights to be unethical. An increasing number of environmental activists are questioning the entire approach of controlling or limiting the amount of pollution produced by existing technologies. They point out that both population growth and increasing worldwide industrial development will result in increasing total quantities of pollutants released despite attempts to reduce the impact of pollution from current, specific sources. Instead, they propose replacing our entire systems of energy production, transportation, and industrial technology with systems that are designed to produce a minimal negative environmental impact.

A new militant wing, Greenpeace, has sprung up within the environmental movement. Greenpeace first received widespread media attention for its actions designed to block the French atmospheric nuclear testing program. As a result of highly successful membership recruiting and fund-raising efforts, it has become the most powerful international grassroots environmental organization. More radical still are the politics and tactics of other "green" organizations such as Earth First! During a 1990 campaign they called Redwood Summer, members chained themselves to trees to prevent the cutting of redwood trees in the ancient forests of northern California. The eco-radicals who constitute the small, but growing, extreme fringe of the environmental movement advocate such policies as a drastic reduction in the world's population and a return to much simpler, less materialistic life-styles.

ECOLOGY AND ENVIRONMENTAL STUDIES

Efforts to protect the environment from the far-reaching effects of human activity require a detailed understanding of the intricate web of interconnected cycles that constitute our natural surroundings. The recent blossoming of ecology and environmental studies into respectable fields of scientific study has provided the basis for such an understanding. Traditional fields of scientific endeavor like geology, chemistry, or physics are too narrowly focused to successfully describe a complex ecosystem. Thus, it is not surprising that chemists and entomologists who helped promote the use of DDT and other pesticides failed to predict the harmful effects that accumulation of these substances in biological food chains had on birds and marine life.

Ecology and environmental studies involve a holistic study of the relationships among living organisms and their environment. It is clearly an ambitious undertaking, and ecologists are only beginning to advance our ability to predict the effects of human intrusions into natural ecosystems.

It has been suggested that our failure to recognize the potentially harmful effects of our activities is related to the way we lead our lives. Industrial development have produced life-styles which separate most of us from direct contact with the natural systems upon which we depend for sustenance. We buy our food in supermarkets and get our water from a kitchen faucet. We tend to take the availability of these essentials for granted until something threatens the supply. It has been claimed that native peoples who lived off the land were more "in touch with nature" and were thus not likely to pollute their environment. This supposition has been discredited by studies showing that the practices of many Native American tribes, despite their generally greater respect for nature, seriously damaged the ecological systems on which they depended. It is unlikely that any people ever set about to intentionally poison their own nests. What clearly distinguishes our society from that of our forebears is the increased capability to employ technology in ways that ultimately result in environmental degradation.

SOME THOUGHTS ON NUCLEAR WEAPONS AND INTERNATIONAL COOPERATION

Most environmental textbooks fail to include any discussion of the enormous ecological devastation that would result from a war involving nuclear weapons. A chilling analysis of the likely environmental effects of a nuclear war is presented in Jonathan Schell's *The Fate of the Earth* (1982). Schell describes the likely destruction of the protective ozone layer, radioactive contamination of the food chain, long-term hazards from radioactive fallout, and a host of other catastrophic environmental consequences of a major nuclear war. In 1983, a nightmarish prediction emerged from a study by a group of eminent scientists that included Carl Sagan and Paul Ehrlich. They theorized that a worldwide "nuclear winter" would result from even a limited nuclear war and that the long-term consequences could be so devastating as to threaten the continued existence of human civilization.

The prevention of nuclear war obviously requires an unprecedented degree of international agreement and cooperation. Many environmental problems such as ozone destruction and climate modification can also result in devastating worldwide effects unless the people of the world can convince their leaders to set aside their usual nationalistic perspectives when facing these problems. Let us hope that the prospect of worldwide environmental catastrophe will help stimulate the political forces and perspectives needed to prevent the ultimate destruction of the Earth and its people.

UN PHOTO 155823/M. Gonzalez

PART 1

General Philosophical and Political Issues

People who regard themselves as environmentalists can be found on both sides of all the issues in this section. But the participants in these debates nonetheless strongly disagree—due to differences in personal values, political beliefs, and what they perceive as their own self-interests—on how best to prevent environmental degeneration.

Understanding the general issues raised in this initial section is useful preparation for examining the more specific controversies that follow in later sections.

Earth Day 1970 *v.* 1990: Has the Environmental Movement Been a Success?

Does Wilderness Have Intrinsic Value?

Do We Need More Stringently Enforced Regulations to Protect Endangered Species?

Does Risk-Benefit Analysis Provide an Objective Method for Making Environmental Decisions?

Is Population Control the Key to Preventing Environmental Deterioration?

ISSUE 1

Earth Day 1970 *v.* 1990: Has the Environmental Movement Been a Success?

YES: Bil Gilbert, from "Earth Day Plus 20, and Counting," *Smithsonian* (April 1990)

NO: T. Allan Comp, from "Earth Day and Beyond," *American Forests* (March/April 1990)

ISSUE SUMMARY

YES: Conservation journalist Bil Gilbert documents his claim that environmentalism has made an impressive and positive impact on society.
NO: Forestry scientist T. Allan Comp contends that despite the large increase in environmental activity since 1970, most environmental problems have gotten worse.

Although Rachel Carson's *Silent Spring* and other popular books about ecological problems written in the early and mid-1960s began an immediate national debate on the environment and swelled the membership ranks of environmental organizations, there was a gestation period of several years before the general public responded with a clear and convincing message to its governmental leaders. That message came in the form of an enormous outpouring of public and political oratory on the first Earth Day, celebrated on April 22, 1970.

Congress and the Nixon administration were quick to react. Although the National Environmental Policy Act had become law on January 1, 1970, and announced the federal government's intention to take a more active role in protecting the environment, responsibility for environmental and natural resource issues was spread among 44 agencies within nine separate federal departments. Recognizing that this would never lead to agreement on a coherent approach to serious public concerns on the environment, the president and Congress created the Environmental Protection Agency (EPA), which came into being on December 2, 1970. Much of the responsibility for protecting and cleaning up the nation's air, land, and water was transferred to this new agency.

There is no dispute about the fact that an unprecedented degree of attention was paid to environmental issues during the two decades since the first Earth Day. Many new national, state, and local agencies were formed, thousands of laws were passed, and budgetary allocations for research, enforcement, and remediation grew.

Far less certain is how much progress has resulted from all of this activity. By some measures it appears that the benefits of environmental protection efforts have outweighed the costs. Many rivers and lakes are less polluted than they were twenty years ago, and the concentrations of some local air pollutants,such as sulfur dioxide and particulate matter, have declined in many large industrial cities.

On the other hand, when overall environmental quality is examined, the conclusion can be much more discouraging. For example, the 1990 annual Environmental Quality Index, published each year by the Wildlife Federation in its journal *National Wildlife,* rated the condition of the air, the water, the soils, wildlife, the forests, energy policy, and the overall quality of life as worse than in 1989. With increasing frequency, reports from international conferences such as the 1989 Forum on Global Change and Our Common Future are warning that fundamental changes in industrial development, agricultural practices, transportation systems, and energy production and use are required to prevent continued, cumulative degradation of the world's ecosystems.

Bil Gilbert, a journalist who has devoted many years to writing about the conservation movement, reviews the progress of environmentalism from Earth Day 1970 to Earth Day 1990 and sees reasons for optimism. He points to some concrete material achievements and, perhaps more importantly, to the enormous increase in public consciousness and sensitivity concerning the environment. T. Allan Comp, who is the American Forestry Association's Director of Program Planning, sees much less to be pleased about from the past twenty years of environmental activity, and he raises questions about what went wrong with the environmental movement. However, as he looks to the future, he commends the organizers of Earth Day 1990 for emphasizing the need for individual action and the necessity of including people from all racial, ethnic, and socioeconomic groups in efforts to solve the diverse set of environmental problems that beset the world today.

YES

Bil Gilbert

EARTH DAY PLUS 20, AND COUNTING

In this country, environmentalists are—with all exceptions granted—commonly Caucasians who possess more than the average means, education and leisure, and are found in relatively attractive and comfortable circumstances. They are distinguished from others who share these attributes by a set of strongly held, fairly homogeneous convictions. They believe:

• The world is divided into two disparate, often conflicting spheres. One comprises humans and their works. Everything else is the "environment," or nature.

• The manner in which humans have encroached upon, despoiled and otherwise abused nature is the cause of virtually all environmental problems.

• Nevertheless, what is left of nature still operates so efficiently, is so beautiful and benign that any major changes in it will probably be for the worse. For example, it is estimated that there are between 10 million and 30 million species of plants and animals now living in this world. Environmentalists are convinced that this, whatever the exact number, is about how many there should be. If the total were to be substantially altered by, say, extinction, mutation or genetic engineering, the consequences would be catastrophic.

• With strong views about what nature and our relationship to it should be, environmentalists feel obliged to instruct the ignorant, inspire the apathetic and confront nonbelievers in regard to these matters. "Clearly, Man has become an unnatural force in the world," says Jay Hair, the president of the National Wildlife Federation, which with 5,600,000 members is the nation's largest environmental organization. "To be an environmentalist now, it is no longer enough to enjoy or study nature passively. We are activists engaged in promoting the wiser use and protection of our global natural resources."

There have been environmentalists among us since at least the Eisenhower era, which is when I began reporting on them and their concerns. However, there was not, or was not generally perceived to be, an environmental movement until about 1970. To the extent the origins of any complicated phenomenon can be traced to a single event, the first Earth Day happenings in the spring of that year were catalysts that transformed a fairly specialized interest into a pervasive popular one. . . .

In the past two decades, environmental concerns have substantially changed, among other things, our habits of work, travel and play, our diets, dress and architecture. Many of these effects (more about some of them later) have been very material and quantifiable. But to my mind the most extraordinary accomplishment of environmentalists—and the one largely responsible for their present economic, social and political influence—has been essentially psychic. In a few decades they have radically altered the image of nature and the opinions that had popularly prevailed for many centuries about how society should treat it. . . .

A NATION OF ENVIRONMENTALISTS?

. . . The pollsters conclude [on the basis of recent public opinion surveys] that a majority—on most questions, a large one—of Americans believe that the poor quality of the environment is one of our most serious national problems, of more magnitude than, for example, homelessness and unemployment. If it will protect the environment, the majority favor, among other things, limiting economic development, changing consumptive habits, increasing government regulations and raising taxes. They say they will support politicians who support such measures.

All the environmentalists I know contend that these sentiments flow inevitably from reality; that they themselves have not been, in the usual manipulative sense, molders of opinion or image-makers. Their role, they insist, has been only that of objective observers who have collected and passed along facts that led rational, public-spirited people to the inescapable conclusion that Nature can no longer protect herself against us, and therefore we must protect her against ourselves. Whatever or whoever is responsible, the defenseless image of the new Nature has, during the past 20 years, come to be generally accepted as a true one. In direct consequence, it seems to me, the environmental movement has prospered, and environmental works have multiplied at an astonishing rate.

With the possible exception of the abolitionists, no reform group in our history has used federal law so effectively—or created more of it—than environmentalists have. Though it was not immediately recognized, the first great legislative leap forward was made in the 1969–70 Congressional sessions. Galvanized by eco-disasters, such as the 1969 Santa Barbara oil spill, and attuned to the on-going Earth Day preparations, the staff of the late Senator Henry (Scoop) Jackson, assisted by Indiana University professor Lynton K. Caldwell, wrote most of what became the National Environmental Protection Act (NEPA). A curious provision of it stated that before leaping into any development projects involving federal lands or funds, public agencies and private entrepreneurs would henceforth have to look at the likely environmental consequences. Environmental Impact Statements (the frequently despised EISes, as they came to be known) would be monitored by a newly created federal advisory body, the Council on Environmental Quality.

Since this council had no enforcement powers, the legislation was generally thought to be innocuous—essentially a statement of good principles on the order of those that speak well of mothers and apple pie. During the next few years, however, environmental activists took NEPA and ran with it to the courts. There, to summarize a number of cases, it was decreed that the EISes had to be substantial documents that considered such things as the effects of the planned works on soil, water, wildlife and recreation.

Once an EIS was completed, public agencies were expected to treat it seriously. An EIS had to be made available for public study and comment. As it turned out, this provision put some very sharp teeth in NEPA, since it gave environmentalists a chance to put intense pressure on agencies to respect EIS findings.

Initially, many public land managers and would-be private developers fiercely resisted the EIS process on the grounds that it was impractical, inconvenient and economically ruinous. After a number of adverse legal rulings, however, most of them concluded that NEPA was a slippery crevice that was best avoided by planning projects and filing EISes that did not bring down the wrath and lawyers of the environmentalists.

NEPA SAID: LOOK BEFORE YOU LEAP

NEPA revolutionized the federal natural-resource business, which is a very considerable one. In addition to overseeing highway construction, irrigation plans, flood control, erosion control, military projects and many other public works, federal agencies own nearly a third of the United States. These holdings include many of the nation's most valuable coal, oil, gas, hard mineral and grazing lands. Since 1971 very little has been done on any of them without some kind of environmental review.

Before the impact of NEPA was fully appreciated, Congress passed, without much technical study or debate, a number of other substantial pieces of pro-environmental legislation. The Endangered Species Act became law during, and generally with the approval of, the Nixon Administration. So did revisions of federal air and water legislation known as the Clean Air Act and the Clean Water Act. As Gaylord Nelson and others have observed, there were then not many true environmental believers in Congress and fewer still in the executive branch.

In 1970 William Ruckelshaus, an assistant attorney general in the Justice Department, was appointed as the first head of the newly created Environmental Protection Agency. Looking back on that time recently, Ruckelshaus told me: "After about 90 days at EPA, I decided the laws we were expected to administer were poorly conceived and in many instances unworkable. Congress mandated that we do the impossible—create a permanently perfect environment."

In Ruckelshaus' opinion, unreasonable expectations about what the EPA—which is slated to receive full Cabinet status this year—could and should do inclined activists to blame the agency for all environmental imperfections. As a result, the environmental community has come to regard the EPA not as an ally as, for example, the Commerce Department is regarded as an ally of the business community—but as a major federal enemy. This antagonism, Ruckelshaus thinks, has often curtailed support and funds for doable projects that could have alleviated, if not eliminated, many environmental problems.

The Endangered Species, Clean Air and Clean Water acts complemented and strengthened NEPA by increasing the number of natural phenomena and federal regulations that had to be considered by anyone preparing an EIS. A number of other statutes—for instance, the Resource Conservation and Recovery Act—were passed to require even more progress. Whatever else it has or has not done, this tough, interlocking mesh of law created what can be described as the environmental-protection industrial establishment, which has come to have a very direct, material impact on the lives of a sizable number of Americans and indirect influences on those of nearly everyone.

This enterprise has become so large and complex that it is difficult to quantify, but the EPA and other sources roughly estimate that public agencies and private firms are now spending about $100 billion a year on pollution control. A decade ago this figure was about $50 billion. All authorities who follow trends in this industry are certain that during the next decades it will grow at least as rapidly as it has in the past.

Environmental protectionists now work on many problems that were ignored or not recognized 20 years ago, and they try to solve them with strategies and devices that did not exist then. In consequence, the industry has created new trades and professions, particularly in the pollution-control and waste-disposal lines. It has also enlarged and upgraded opportunities in a number of traditional vocations. For example, the government agencies that administer EISes, and the private contractors that must abide by them, now employ biologists, ecologists and assorted other terrestrial, aquatic and atmospheric professionals.

The legal profession, which acted somewhat as a midwife at the birth of the environmental movement, has certainly benefited from its growth. Records and recollections are hazy, but there were probably only a few hundred attorneys specializing in this field in the 1960s. Today, there may be as many as 20,000. In 1971 a summary of federal environmental law in the *Environmental Law Reporter* occupied 33 pages. Last year the *ELR* published more than 3,500 pages to cover current developments in the field.

Another group of people whose lives surely would be much different if there were no environmental movement: the cadres who earn their keep and make their mark working for environmental organizations. The National Wildlife Federation annually publishes a directory of conservation-oriented groups. In 1975 there were approximately 200 of them, with a combined membership of 4 million. In 1990 there are 350 or so, with more than 12 million members. They range from Accord Associates ("founded to act as an intermediary in environmental disputes") to Zero Population Growth ("which works to achieve a balance among people, resources and environment by advocating population stabilization"). They vary greatly in size and resources, but I cannot think of any that do not now have more staff, publications and areas of concern than they did. . . .

And so what difference has it all made—the manifestos, legislation, impact statements, money and work? In one of the aforementioned opinion polls, people were asked about the quality of their personal environments (as distinct from those elsewhere that they had read or heard about). Half of them said they thought things were about the same or better than they had been before, and 20 percent thought they might be a bit worse. If the respondents were right, this represents no mean feat of environmental protection. Between 1970 and 1988, the population increased by 40 million; the gross national product, a rough measure of consumptive activity, expanded dramatically. Between 1968 and 1984, the number of cars in the country grew by 60 million vehicles. In short, things from which the environment needs to be protected have so increased that had the effort not been made, its quality would almost certainly have declined drastically.

BY SOME MEASURES, DISCERNIBLE PROGRESS

Here and there, we have even made progress or, more accurately, retaken some ground. In 1970, for example, 200 million tons of sulfur dioxide, carbon monoxide and soot were spewed into the air as gaseous waste. Though we now generate more of these contaminants, their flow into the atmosphere has been reduced by more than 30 percent, with the result that, by some measures, the air is clearer and less noxious. This has been accomplished mainly by adding contaminant-trapping devices to industrial plants and automobiles, or changing the composition of fuels, or both.

The quantity of raw or inadequately treated domestic sewage that is dumped into surface waters has been reduced substantially. Over the past ten years, the number of people served at least by secondary sewage treatment facilities has increased by 72 percent. In the early 1970s it was generally thought that Lake Erie was dying and certainly by this time would be dead—that is, would not support any aquatic flora or fauna. Presently its waters appear to be relatively wholesome and support a thriving commercial and sport fishing industry, though scientists warn that the lake continues to be threatened by toxic chemicals. Similarly a number of other lakes and rivers are now, contrary to earlier predictions, fairly lively places biologically, reasonably safe and attractive for swimming, boating and other recreational uses.

The use of a number of poisonous compounds has been severely regulated or absolutely banned. Like the names of distant battlegrounds where it once seemed the fate of the Republic might be decided, even the designations of many of the most notorious substances are becoming hazy: 1080; asbestos; DDT; Diazinon; Alar. We have been less successful in dealing with many other toxins. Government agencies have yet to regulate—or even study in detail—thousands of chemicals used at work and in the home. But we have become more sophisticated about identifying them, and we are increasingly wary about how they are produced, employed and disposed of.

A NUMBER OF CREATURES ARE BETTER OFF

Since 1970 eight national parks have been established, and the wilderness system has been expanded by 80 million acres. An untotaled amount of land has been set aside for natural preserves by state and local governments, private individuals and organizations. Many highway berms and medians that were formerly seeded and mowed as turf are now planted with species of the native flora. As a result of a 1985 federal law, 34 million acres of farmland particularly vulnerable to erosion have been withdrawn from production and placed in a ten-year "conservation reserve."

It has often been observed that the condition of wildlife is a good measure of general environmental quality. That the status of many creatures was poor and seemed to be worsening rapidly was an argument used with good effect in the 1970s by activists seeking new land and water regulations. Subsequently (and again because the protectionists achieved some successes) the fortunes of a number of species have improved.

. . . Those who are most involved in its affairs tend to be the least impressed with the accomplishments of the environmental movement. Activists point out, for instance, that while we may have cut down a bit on the sulfur dioxide in the air, we are discharging chlorofluorocarbons (chemicals used in such products as refrigerants and aerosol propellants), and generating carbon dioxide as we burn fossil fuels, at a prodigious rate. Unless we reverse this trend, these compounds will probably, within the next 50 years, alter the climate and cause problems that will make all previous ones seem trivial.

Pessimism is a hallmark of the environmental movement. The movement has grown and prospered by being ever ready to point out new crises just around the corner. Consequently, environmentalists are sometimes accused of being merchants of gloom and doom who push the prospect of imminent catastrophes to recruit members, raise money and increase their political clout. Since they deal with the future, most of these indictments remain to be proved. However, in the immediate past and present, environmentalists have been extraordinarily persuasive.

The majority of Americans who look favorably on the environmental movement do not live in circumstances where endangered species, soil erosion, acid rain, the timbering of rain forests, desertification, and the quality of the air, water and land are immediate personal problems. The catastrophic consequences of a hole in the ozone layer, global warming trends and environmental diseases must largely be accepted on faith. As the opinion surveys make clear, they have been.

Jay Hair, of the National Wildlife Federation, thinks the charges that, for reasons of self-interest, environmentalists try to frighten people with predictions of coming disaster are false and cynical. Their role has been more like that of a passer-by who, seeing smoke curl out of a window, warns the occupants that their house may be on fire. Nevertheless, Hair feels that while this approach continues to be necessary, the time may have come to reinforce it with more positive ones. "We are facing enormous problems," he says, ticking off the usual ones. "But I am optimistic that, if we make a global effort, we can create a sustainable environment. Essentially the problems have been created by how we live, and the solutions require that we change how we live. I think we can, while in the long term raising rather than lowering our standards of living. However, I don't think this can be mandated from above by government regulations and treaties. It has to come about because the people themselves *want* to make changes."

Many other environmentalists agree. Their notions about creating a sustainable global environment by greatly altering the global lifestyle may strike some as being wildly impractical, even ludicrous. But so did many of the proposals that were made in 1970.

Since World War II there have been three great national reform movements—those having to do with racial, sexual and environmental relations. For some of the reasons above, a good case can be made for claiming that the latter has had the greatest impact, psychic and material, on society. Whether environmentalists can keep up the pace remains to be seen. But because of their impressive record, the possibility that they will cannot be lightly dismissed.

NO

T. Allan Comp

EARTH DAY AND BEYOND

Earth Day 1970 gained its legendary success from the individual actions of the 20 million Americans who joined the celebration. One mayor helped plant trees. A governor picked up trash. Thousands of university students poured out for teach-ins. But after that glorious day in April 1970, something went wrong with the environmental movement.

Congressman Norm Mineta, deputy whip of the House, recalls that back on April 22, 1970, when he was vice-mayor of San Jose, California, he drove to Booksin Elementary School to plant a grove of trees. Today the trees at the school are alive and well. But Mineta, reflecting on the intervening 20 years, says that instead of getting better, "Many environmental problems have gotten worse."

Air and water pollution, deforestation, the landfill crisis, toxic wastes, and galloping development all demand attention as urgently as they did 20 years ago. Some of the gravest threats we face—global warming, ozone depletion—weren't even recognized in 1970 and thus are nowhere near being solved. So, what went wrong?

Mineta notes that Earth Day inaugurated an era of public awareness of conservation issues. And no doubt that concern is growing—seemingly exponentially in the past year. *Time* magazine reports receiving five times as much mail on environmental issues as it did only a year ago. AFA [American Forestry Association] board member Carl Reidel, head of the University of Vermont's environmental program, relates that 240 students signed up this year for an introductory environmental course that attracted exactly half that number last year. Meanwhile, the national media are racing to produce environmental articles and documentaries to take advantage of the public's current attention.

But the burgeoning public concern doesn't seem to have made environmental leaders complacent. This year's celebration will definitely not be a happy, self-congratulatory commemoration of 20 years of increasing success; rather, Earth Day 1990 is more likely to be a review of how much remains to be done.

From T. Allan Comp, "Earth Day and Beyond," *American Forests* (March/April 1990).

Dennis Hayes, who led the first Earth Day organization and who stepped away from a law practice to take charge of this April's event, has done some serious thinking about that 20-year period and what went wrong. He acknowledges the American conservation movement—including the early role of AFA—as a long and distinguished tradition. But despite the high hopes on Earth Day 1970 and despite the greater willingness of environmentalists to take to the streets since then, Hayes—who once headed the Solar Energy Research Institute and who still remembers marching down Fifth Avenue on Earth Day wearing a gas mask—says, "Those of us who set out to change the world are poised on the threshold of utter failure. Measured on virtually any scale, the world is in worse shape today than it was 20 years ago."

What happened—and did not happen—between 1970 and 1990? For one thing, a long string of federal environmental legislation and court victories and an alphabet soup of federal environmental agencies and armies of environmental lawyers proved unequal to the task of cleaning up our country.

Among those involved in Earth Day 1970 and Earth Day 1990, and in the environmental crusade during the intervening years, the most interesting common conclusion is that Americans rely too heavily on the federal government. Encouraged by the quick response that followed the first Earth Day, we failed to realize that the government is also the largest polluter in the nation and often exempts itself from the same rules it requires industry to follow.

For 1990, the horizon of opportunity is defined by the individual and the changes each of us is willing to make, first in our own life and then in our neighborhood, our state, and our nation. Just as AFA's Global ReLeaf® initiative offers each of us the chance to take a first step toward a better environment, the organizers of Earth Day 1990 are looking first to individual action. They are going back to Earth Day's roots.

The presumption is that only determined individuals willing to act will take the steps that lead to more sustainable national policy. It is now clear that the easy solutions we looked for in 1970 do not exist. Instead, managing one's own life more responsibly is the first and most do-able step in the sequence that leads to change at the national level.

Though there will no doubt be a large number of new participants celebrating on April 22, 1990, also in attendance will be huge numbers of veterans of the first Earth Day, and this time they have an agenda. The organizers of Earth Day 1990 will see that the event marks the beginning of a green era in America—a trend already predicted by various market research firms—in which environmental concerns dominate the public consciousness and drive consumer decisions.

A second distinct break with the older thinking of Earth Day 1970 is a recognition that Americans will have to break a very old cultural habit and begin to plan effectively for distant problems. We have always been good at reacting and rather poor at long-term planning. Launch Sputnik and we will have NASA up and funded in a flash, but tell us the world will eventually run out of oil and nothing much happens. Similarly, we study global warming, ozone depletion, and other doubtlessly catastrophic events foreseen for the future, but we won't do much until the second shoe finally falls.

For example, the current ballyhoo over biodegradable plastics as a solution to the waste problem fails to recognize that mixing biodegradable plastics with other plastics renders them all impossible to recycle. Incinerators that reduce the volume of the waste stream by putting much of it in the air and concentrating the rest in ash that creates its own disposal problem, a nuclear weapons industry with a $150 billion cleanup bill, and agricultural practices that depend on chemicals that concentrate downstream are all aspects of our American willingness to forget about planning for tomorrow.

But perhaps the hardest lesson from Earth Day 1970 is that we did not ask enough of those who took part. The Earth Day 1990 generation of environmentalists must not only preach, they must practice, and we all know that won't come easily.

Research establishes that building more energy-efficient homes could cut energy consumption by two-thirds, that eating lower on the food chain (fish, veggies, etc.) could cut energy use in the food system by two-thirds, and that buying automobiles that are much more efficient relative to their size would cut gasoline consumption by half. We need only to do what we already know how to do.

Similarly, trees are too important for anyone in a sustainable world to use only once. They save energy when planted for shade or as a windbreak, and they are recyclable when used as fuel in efficient, clean woodburning applications.

Asking supporters to wean themselves from a dependence on government, to plan for distant problems before they go beyond the point of no return, and to adopt more responsible lifestyles may not be easy—and it certainly won't be wonderfully popular, but it remains the central plank in the Earth Day 1990 platform.

Equally encouraging, the organizers of this Earth Day have expressed a strong recognition that the environmental movement failed to diversify both in the people who share its goals and in the range of issues it is willing to address. At present, the movement remains sadly narrow. The participation of minorities and women in meaningful roles has been slow to develop.

In the face of this kind of candid assessment, Earth Day 1990 has gathered what the *New York Times* described as the "broadest range of environmental groups ever assembled." Congressman John Lewis, a veteran civil rights activist who spent Earth Day 1970 on a voter registration drive in the South, notes that his Atlanta district is racially and economically mixed—poor, rich, Blacks, Whites, Hispanics, and Asians—but they all unite on one issue: the environment and, more specifically, trees. These diverse people are willing to translate their active interest in the environment into days spent planting trees.

Their involvement is what the Earth Day 1990 organizers want to generate throughout America, and their concern are what will create the green decade of the 1990s. As Congressman Lewis explained recently, "What we do as a nation, as a people, to protect and save not just American society but our whole environment will be the dominant issue of the 1990s. The concerns raised 20 years ago are still with us. The Earth is fragile, and we must continue efforts to protect the environment."

Throughout the U.S., voters are recognizing their greater opportunity to make a difference on a local level, and they are pushing for environmental sensitivities once thought impossible to achieve. Minnesota has passed a law requiring 25 to 35 percent of the state's garbage to be recycled. Vermont has banned CFCs, while Wisconsin, Minnesota, and Massachusetts are all forcing cuts in acid rain. New York, New Jersey, and the six New England states passed auto emission standards equal to the tough regulations already in effect in California. Texas has 90 percent of its public vehicles running on compressed natural gas, and New Jersey has announced a Global Climate Initiative.

These states, and numerous communities as well, are documenting the accuracy of the prediction of a green '90s. The new movement is a popular citizens' crusade requiring no particular credentials or membership—only a determination to act personally and responsibly.

For AFA members and supporters, there is much work to be done for trees, and it is trees that create the first entry step for many Americans into a larger environmental awareness and commitment. AFA sees the greatest need as being in our urban areas, where we are literally deforesting our own communities through neglect.

We need to plant 100 million trees in our towns and cities. The savings in energy, the improvement in air quality, and the increased property-tax income from planted areas could then serve as encouragement and a source of funding for other areas of the country.

This is only a tiny part of the great circle being drawn by Earth Day 1990, but it is one of the few positive steps that you and I can take *right now* and know we are stepping in the right direction, both for ourselves and for our children.

Global ReLeaf® groups all over the nation are planning Earth Day 1990 activities, and Global ReLeaf® and Earth Day 1990 have formed an active partnership. So the prospects for a lot more trees being in the ground on the day after Earth Day are excellent, but there is more to it than that. Twenty years ago, two future congressmen working 3,000 miles apart on the first Earth Day knew nothing of each other, but they knew they were working for something greater than themselves. Today they both serve in the U.S. House of Representatives. The trees one planted are still growing, and the voters the other registered are still voting. Equally important, they are both still doing their best as individuals to be responsible stewards of this planet.

That seems a reasonable model for all of us to follow in 1990 and into the 21st century as well.

POSTSCRIPT

Earth Day 1970 *v.* 1990: Has the Environmental Movement Been a Success?

Gilbert makes the important point that when environmentalists (like Comp) conclude that most of our problems have not been solved, it is necessary to consider how much worse things might be without the past two decades of environmental activism—given the large increases in people, cars, and industrial activity during that period. In presenting his optimistic view of the future, Gilbert mentions the need for changes in life-style, but he does not explore what that might mean. To Comp, the message of Earth Day 1990 is that to secure a healthy future environment, people from all walks of life will have to become more actively involved, less dependent on government, and willing to adopt more environmentally responsible modes of personal behavior. The only specific step in this direction that he cites is the organized campaign to get individuals to participate in planting 100 million trees in towns and cities throughout the world. Comp's inspiration appears to have come from the pronouncements of Denis Hayes, who was the executive director of Earth Day 1970 and the U.S. national chairman of Earth Day 1990. Hayes expands on the theme of relying more on individual commitment and less on government in his article "Earth Day 1990—Threshold of the Green Decade," *Natural History* (April 1990).

If, as proposed by Comp and Hayes, it is important to involve a greater percentage of minorities and people from lower socioeconomic strata in the future environmental movement, definitive action will have to be taken to change many present policies that these groups now view as discriminatory. In "Environmental Racism," *The Amicus Journal* (Spring 1989), Dick Russell documents the charge that facilities that produce pollution are most often built in, or near, communities inhabited by minorities and the poor.

The April 30, 1990, issue of *The Nation* contains two critiques of present environmental initiatives. In the first, "Ending the War Against Earth," Barry Commoner summarizes the principal theme of his recent book *Making Peace With the Planet* (Pantheon, 1990). The other article, "The Trouble With Earth Day," by author and social critic Kirkpatrick Sale, presents four fundamental criticisms of the agenda of Earth Day 1990 organizers. For a complete discussion of these two articles as well as detailed suggestions for further readings on Earth Day and the environmental movement, please turn to pages xvii–xix in the Introduction to this volume. There you will find an in-depth analysis of the concerns raised in this issue and recommendations for other sources of material on this topic.

ISSUE 2

Does Wilderness Have Intrinsic Value?

YES: Robert K. Olson, from "Wilderness International: The New Horizon," *Wilderness* (Fall 1990)

NO: William Tucker, from "Is Nature Too Good for Us?" *Harper's* (March 1982)

ISSUE SUMMARY

YES: Wilderness activist Robert K. Olson contends that global change and resource depletion have transformed wilderness preservation from a value in its own right to an international imperative.
NO: William Tucker, a writer and social critic, asserts that wilderness areas are elitist preserves designed to keep ordinary people out.

The environmental destruction that resulted from the exploitation of natural resources for private profit during the founding of the United States and its early decades gave birth after the Civil War to the progressive conservation movement. Naturalists such as John Muir (1839–1914) and forester and politician Gifford Pinchot (1865–1946) worked to gain the support of powerful people who recognized the need for resource management. Political leaders such as Theodore Roosevelt (1858–1912) promoted legislation during the last quarter of the nineteenth century that led to the establishment of Yellowstone, Yosemite, and Mount Ranier National Parks, and the Adirondack Forest Preserve. This period also witnessed the founding of the Sierra Club and the Audubon Society, whose influential upper-class members worked to promote the conservationist ethic.

Two conflicting positions on resource management emerged. Preservationists, like Muir, argued for the establishment of wilderness areas that would be off-limits to industrial or commercial development. Conservationists, like Roosevelt, supported the concept of "multiple use" of public lands, which permitted limited development and resource consumption to continue. The latter position prevailed and, under the Forest Management Act of 1897, mining, grazing, and lumbering were permitted on U.S. forest lands and were regulated through permits issued by the U.S. Forestry Division.

The first "primitive areas," where all development was prohibited, were designated in the 1920s. Aldo Leopold and Robert Marshall, two officers in the Forest Service, helped establish 70 such areas by administrative fiat.

Leopold and Marshall did this in response to their own concerns about the failure of some of the National Forest Service's management practices. Many preservationists were heartened by this development, and the Wilderness Society was organized in 1935 to press for the preservation of additional undeveloped land.

It became increasingly apparent during the 1940s and 1950s that the administrative mechanism whereby land was designated as either available for development or off-limits was vulnerable. Because of pressure from commercial interests (lumber, mining, etc.), an increasing number of what were then called wilderness areas were lost through reclassification. This set the stage for an eight-year-long campaign that ended in 1964 with the passage of the Federal Wilderness Act. But this was by no means the end of the struggle. The process of implementing this legislation and determining which areas to set aside has been long and tortuous and will probably continue throughout the remainder of this century.

There are more clear-cut differences between values espoused by the opposing factions in the battle over wilderness preservation than in many other environmental conflicts. On one side are the naturalists who see undeveloped "wild" land as a precious resource, where people can go to seek solace and solitude—provided they don't leave their mark. On the opposite extreme are the entrepreneurs whose principal concern is the profit that can be made from utilizing the resources on these lands.

One consequence of the environmental movement has been the proliferation of studies that explore the impact of human developmental activities on remote, isolated regions of the globe. It has become apparent that industrial pollutants move through the air and water and find their way into every nook and cranny of the ecosphere. The notion of totally protecting any area of the Earth from contamination is an ideal that cannot be fully realized. This knowledge has increased the zeal of wilderness advocates who wish to minimize the impact of pollution on the few remaining relatively pristine ecosystems.

Robert K. Olson is a former career officer in the Foreign Service and an active supporter of the Global Tomorrow Coalition. He believes that explosive population growth, the threat of global climate change, and other potential ecological catastrophes have transformed wilderness preservation from an esoteric cause to a pragmatic international priority. William Tucker views the wilderness movement as an elitist enterprise. He sees the notion of excluding most human activity from wilderness areas as an outgrowth of a misguided and romantic ecological ethic.

YES

Robert K. Olson

WILDERNESS INTERNATIONAL: THE NEW HORIZON

Wilderness has surely become one of the most elegant and humane ideas of our time. It is a compliment to the past and to the civilization that has preserved it. Nothing, perhaps, becomes our civilization quite so much as the simplicity and grace with which its limits have been defined by wilderness. But the predicament in which the world finds itself as it moves toward the twenty-first century has added a new and disturbing challenge to the meaning and the preservation of our wilderness heritage.

What of wilderness in an age when exploding populations are destroying the last great forests and invading the remaining scraps of arable land just to stay alive? What of wilderness when human activity is changing the very climate of the earth? What of wilderness when economic growth is expected to quintuple over the next half century? And what of wilderness when, if present trends continue, the future will become even more crowded, more polluted, less stable ecologically, and more vulnerable than the world in which we now live? Most scientists and other authorities on the subject believe that unless we take corrective action soon we are headed for a human and, possibly, natural catastrophe. What then of wilderness? Will it become irrelevant or even more precious than ever?

These are neither vague generalities nor rhetorical questions. We can already see our perspective changing in several ways.

Wilderness hitherto has been principally a cause of its own, a value in its own right to be defended against but not a part of the central dynamics of social and economic development. Today it has to be considered in the light of that complex of questions too obvious to ask, of climate change, food and population, of resource depletion, and of environmental degradation worldwide, not to mention the practices and politics of war and peace. "Politics," as the eminent British ecologist, Sir Frank Frazer Darling, once reminded us, "must never be neglected as a profound ecological factor."

Wilderness hitherto has been a special but relatively marginal concern of society, the experience of some few lovers of the out-of-doors, the crusade of a devoted minority, the residual of economic and social development. But, in the last decade or so, it has been propelled, along with ecological concerns generally, into the mainstream of history—even though society as a whole is still largely unaware of the fact and of its implications.

Wilderness hitherto had been a local, a regional, at best a national issue. Today it has become an integral factor of international life and a vital international issue along with the politics and trade, the prosperity and poverty of nations worldwide.

In short, wilderness, along with practically everything else, is being swept up in a megarevolution in human history that will divide our future as radically from the past as the industrial and agricultural revolutions did in their day. In times to come, what society believed to be of cardinal importance will be secondary or even irrelevant; and what has been seen to be nice but not necessary will become of crucial importance. The prospects for change are daunting and no one really knows what to think. As Worldwatch President Lester Brown has written, "The new era now emerging defies description." This, then, is the context within which we must begin to consider the idea of wilderness both as a value and a resource. I have mentioned three specific ways in which this global predicament can already be perceived to be changing our perspective on wilderness. Let us look at them in more detail, for each includes both challenges and opportunities.

THE GLOBAL QUESTIONS OF FOOD, POPULAtion, resources, and environment are each of them a crisis or a crisis in the making. But more important, we are beginning to realize that they are aspects of a single crisis that includes them all and that is greater than the sum of its parts, a global crisis so vast and so complex that it is almost impossible to describe. As the late Aurelio Peccei, founder of the Club of Rome, has said, the greatest challenge facing us, after all is said and done, is the crisis of complexity. Suddenly, because of the shrinking of the planet and the globalization of life, everything relates to everything else in one huge simultaneous equation. Under the circumstances, there can be no more sovereign nations living either in splendid or in squalid isolation, no more remote and legendary lands waiting to be explored, no more sacred groves dreaming untouched through the centuries. All are being drawn into the web, knitted into the whole, linked together into a complex global system of dubious interdependence.

Yet, behind it lies a very simple truth—that human pressure on the planet may be exceeding the carrying capacity of the biosphere. Again, no one really knows what the limits are; there are too many variables. But it has been estimated by the National Academy of Sciences that, at current rates of economic growth, the limit may be the ten billion we expect to reach by the middle of the next century (or earlier) or, at a maximum, 30 billion with everyone reduced to living at subsistence levels or worse. Whatever they may be, the limits are out there somewhere and we are moving toward them at an increasing rate of speed. The problem we face in its simplest form is how to restrain human growth both in numbers and in demand before we are reduced to universal squalor and before we destroy beyond repair the integrity of the natural world upon which we all depend.

Preserving wilderness for the future simply means coping with these massive and perplexing forces. We can see them already in mineral prospecting here, oil exploration there, and even in recreational overkill. They are all related, mutually reinforcing, and none can be addressed without reference to the others. Failure to cope with one sector, such as population, means failure on all so that environmental protection, by itself, is no longer enough. The world, in other words, is becoming a megasystem of which wilderness is a part. But, to our consternation, it is a system without a master and without a plan.

THE IDEA OF WILDERNESS HAS UNDERGONE many changes over the years and has been much written about so we need not elaborate here. Suffice to say that wilderness, in the Western mind at least, has evolved in modern times from a threat, along with all untamed nature, to a positive good necessary to our spiritual well-being. From the days of Wordsworth we have regretted our increasing separation from nature and mourned its steady destruction by industry and the growth of cities. Back-to-the-land movements have become a periodic impulse in Western life. Somewhere along the way, we have felt, something has gone wrong, a balance has been destroyed, and our materialistic world and way of life has, somehow, to be redeemed.

A new reality now confronts us, and the question of the environment, of conservation, and of wilderness has entered the mainstream of history. Hard values of survival, of the extinction of species, of climate, of the biosphere, of the balance of nature, and of the carrying-capacity of the planet have been added to the soft values of recreation and spiritual renewal. Growth is still our goddess, but it is no exaggeration to say that ecological sanity, of which wilderness is an integral part, is coming to be seen as the ultimate basis for national health, security, and prosperity. We can't get more mainstream than that. It is a sign of the times that the leaders of the superpowers have placed the environment amid the priority items of their political agendas.

UNDER THESE CIRCUMSTANCES, IT IS LITTLE wonder that wilderness has become an international concern and a global issue. America can take pride in being the world's leader in promoting wilderness as a vital human resource and value. The movement for setting aside natural areas like Yellowstone and Yosemite began over a century ago, The Wilderness Society was founded in 1935, and wilderness was given legal status in the 1964 Wilderness Act. But we have tended to see the wilderness movement as a solely American affair, whereas it has, in fact, proliferated all over the world and is becoming an international movement in its own right. . . .

Finland has started a wilderness movement. Italy has its own wilderness society and is working out the details of the designation of the Val Grande area in the Piedmont region as a wilderness area, the first in Europe. Canada has joined the United States in enacting specific wilderness area legislation. Australia, South Africa, and New Zealand, which have administratively designated wilderness areas, are all actively considering national legislation. Zimbabwe established the Maruadona Wilderness Area, the first in Africa, in 1989. France and Australia have teamed up to sponsor an Antarctic Wilderness Park.

The first global inventory of world wilderness resources is being conducted by the Sierra Club with illuminating results. Work to date shows that about one-third of the terrestrial world (most of it uninhabitable) is still wilderness. Antarctica comprises 25 percent of the total, with the rest scattered throughout the other continents. The Soviet Union, for example, is 39 percent wilderness, Canada, 65 percent, Australia 33 percent, Brazil 28 percent, Greenland 90 percent, and China (Tibet) 24 percent wilderness. In general, wilderness outside the antarctic comprises a broad band across the northern latitudes from Alaska to the Soviet Far East. A wilderness of deserts sweeps across the Sahara, through Saudi Arabia and up through Afghanistan and China to the Soviet Union. The United States has more than 90 million acres of designated wilderness lands and the potential for another 100 million, most of it in Alaska and the Western states. This inventory will be invaluable in the future for knowing more about the balance (or imbalance) between human development and lands where nature still predominates, not in detail but in the broadest possible sense.

The international wilderness movement still has a long way to go. It seems to be strongest in countries like the United States and Australia where a wild frontier was a part of national life. Europe has long since forgotten what wilderness was all about. Most of the Third World is inescapably obsessed with the imperatives of development, with devastating results in many places. Yet, if that lonely group of wilderness pioneers could be around today, they would be astonished.

WE HAVE REVIEWED SOME OF THE CHALLENGES facing the wilderness cause, but there are also opportunities. For the first time in history, mankind can see what it is doing. This has never happened before. During the millennia during which humankind was busily destroying the environment of the ancient world, we were almost completely ignorant of what we were doing. Today, what were not too long ago the concerns of a small minority have become the common knowledge of mankind, the subject for UN conferences, of lead articles in national magazines, and of prime-time television programs. We are working it into our foreign policies. There are hundreds of organizations, governmental and private, working on specific problems. Millions of people worldwide are devoting their lives to education and to saving the environment. We may not have a master plan, but the idea of sustainable development has caught on worldwide and provides a central concept to work around and to coordinate our various efforts. Surely this will make a difference.

And never before has wilderness meant so much to so many. From a marginal interest it is moving into the mainstream. We are discovering that wilderness is an integral component of the global scheme of things. To the values of recreation and the preservation of natural beauty for its own sake, there is now added the importance of wilderness as a reservoir of species and a buffer to global warming. There is now an opportunity for the wilderness community to define wilderness in the broadest and most basic terms of mainstream national and international self interest, not as the mere residual after all other interests have been addressed. To the more crowded, more polluted, more vulnerable world of tomorrow, wilderness has a lot to say.

Not least important, the internalization of the wilderness movement has given it a strength and coherence it never had before. Thanks to the wilderness congresses, scattered groups throughout the world, largely unacquainted with one another, have become an international community and hence a global movement. Australians, Americans, and Canadians, Italians, Africans, and Russians, Indians and Inuits, Brazilians and Chinese, scientists, artists, and administrators now talk to one another in global *indabas* and support one another in the councils of the world. In the words of Dr. Roman Zlotin of the USSR Academy of Sciences, "It is impossible to successfully resolve environmental problems in one country, no matter how large or advanced it is, without resolving the problems in the wider world." We are all in it together.

It is also a curious irony that the once unworldly wilderness movement has become a most worldly campaign. From the Thoreauvian impulse to escape the madding crowd, we find ourselves enlisted in what will be the greatest project of social engineering in history. It may very well be, as Thoreau's poetic instinct told us, that the preservation of the world truly depends on wildness.

NO

William Tucker

IS NATURE TOO GOOD FOR US?

Probably nothing has been more central to the environmental movement than the concept of wilderness. "In wildness is the preservation of the world," wrote Thoreau, and environmental writers and speakers have intoned his message repeatedly. Wilderness, in the environmental pantheon, represents a particular kind of sanctuary in which all true values—that is, all nonhuman values—are reposited. Wildernesses are often described as "temples," "churches," and "sacred ground"—refuges for the proposed "new religion" based on environmental consciousness. Carrying the religious metaphor to the extreme, one of the most famous essays of the environmental era holds the Judeo-Christian religion responsible for "ecological crisis."

The wilderness issue also has a political edge. Since 1964, long-standing preservation groups like the Wilderness Society and the Sierra Club have been pressuring conservation agencies like the National Forest Service and the Bureau of Land Management to put large tracts of their holdings into permanent "wilderness designations," countering the "multiple use" concept that was one of the cornerstones of the Conservation Era of the early 1900s.

Preservation and conservation groups have been at odds since the end of the last century, and the rift between them has been a major controversy of environmentalism. The leaders of the Conservation Movement—most notably Theodore Roosevelt, Gifford Pinchot, and John Wesley Powell—called for rational, efficient development of land and other natural resources: multiple use, or reconciling competing uses of land, and also "highest use," or forfeiting more immediate profits from land development for more lasting gains. Preservationists, on the other hand, the followers of California woodsman John Muir, have advocated protecting land in its natural state, setting aside tracts and keeping them inviolate. "Wilderness area" battles have become one of the hottest political issues of the day, especially in western

From William Tucker, "Is Nature Too Good for Us?" *Harper's* (March 1982). Adapted from *Progress and Privilege: America in the Age of Environmentalism* by William Tucker (Doubleday, 1982).

states—the current "Sagebrush Revolt" comes to mind—where large quantities of potentially commercially usable land are at stake.

The term "wilderness" generally connotes mountains, trees, clear streams, rushing waterfalls, grasslands, or parched deserts, but the concept has been institutionalized and has a careful legal definition as well. The one given by the 1964 Wilderness Act, and that most environmentalists favor, is that wilderness is an area "where man is a visitor but does not remain." People do not "leave footprints there," wilderness exponents often say. Wildernesses are, most importantly, areas in which *evidence of human activity is excluded;* they need not have any particular scenic, aesthetic, or recreational value. The values, as environmentalists usually say, are "ecological"—which means, roughly translated, that natural systems are allowed to operate as free from human interference as possible.

The concept of excluding human activity is not to be taken lightly. One of the major issues in wilderness areas has been whether or not federal agencies should fight forest fires. The general decision has been that they should not, except in cases where other lands are threatened. The federal agencies also do not fight the fires with motorized vehicles, which are prohibited in wilderness areas except in extreme emergencies. Thus in recent years both the National Forest Service and the National Park Service have taken to letting forest fires burn unchecked, to the frequent alarm of tourists. The defense is that many forests require periodic leveling by fire in order to make room for new growth. There are some pine trees, for instance, whose cones will break open and scatter their seeds only when burned. This theoretical justification has won some converts, but very few in the timber companies, which bridle at watching millions of board-feet go up in smoke when their own "harvesting" of mature forests has the same effect in clearing the way for new growth and does less damage to forest soils.

The effort to set aside permanent wilderness areas on federal lands began with the National Forest Service in the 1920s. The first permanent reservation was in the Gila National Forest in New Mexico. It was set aside by a young Forest Service officer named Aldo Leopold, who was later to write *A Sand County Almanac,* which has become one of the bibles of the wilderness movement. Robert Marshall, another Forest Service officer, continued the program, and by the 1950s nearly 14 million of the National Forest System's 186 million acres had been administratively designated wilderness preserves.

Leopold and Marshall had been disillusioned by one of the first great efforts at "game management" under the National Forest Service, carried out in the Kaibab Plateau, just north of the Grand Canyon. As early as 1906 federal officials began a program of "predator control" to increase the deer population in the area. Mountain lions, wolves, coyotes, and bobcats were systematically hunted and trapped by game officials. By 1920, the program appeared to be spectacularly successful. The deer population, formerly numbering 4,000, had grown to almost 100,000. But it was realized too late that it was the range's limited food resources that would threaten the deer's existence. During two severe winters, in 1924-26, 60 percent of the herd died, and by 1939 the population had shrunk to only 10,000. Deer populations (unlike human populations) were found to have no way of putting limits on their own reproduction. The case is still cited as the classic example of the "boom and bust" disequilibrium that comes from thoughtless intervention in an ecological system.

The idea of setting aside as wilderness areas larger and larger segments of federally controlled lands began to gain more support from the old preservationists' growing realizations, during the 1950s, that they had not won the battle during the Conservation Era, and that the national forests were not parks that would be protected forever from commercial activity.

Pinchot's plan for practicing "conservation" in the western forests was to encourage a partnership between the government and large industry. In order to discourage overcutting and destructive competition, he formulated a plan that would promote conservation activities among the larger timber companies while placing large segments of the western forests under federal control. It was a classic case of "market restriction," carried out by the joint efforts of larger businesses and government. Only the larger companies, Pinchot reasoned, could generate the profits that would allow them to cut their forest holdings *slowly* so that the trees would have time to grow back. In order to ensure these profit margins, the National Forest Service would hold most of its timber lands out of the market for some time. This would hold up the price of timber and prevent a rampage through the forests by smaller companies trying to beat small profit margins by cutting everything in sight. Then, in later years, the federal lands would gradually be worked into the "sustained yield" cycles, and timber rights put up for sale. It was when the national forests finally came up for cutting in the 1950s that the old preservation groups began to react.

The battle was fought in Congress. The 1960 Multiple Use and Sustained Yield Act tried to reaffirm the principles of the Conservation Movement. But the wilderness groups had their day in 1964 with the passing of the Wilderness Act. The law required all the federal land-management agencies—the National Forest Service, the National Park Service, and the Fish and Wildlife Service—to review all their holdings, keeping in mind that "wilderness" now constituted a valid alternative in the "multiple use" concept—even though the concept of wilderness is essentially a rejection of the idea of multiple use. The Forest Service, with 190 million acres, and the Park Service and Fish and Wildlife Service, each with about 35 million acres, were all given twenty years to start designating wilderness areas. At the time, only 14.5 million acres of National Forest System land were in wilderness designations.

The results have been mixed. The wilderness concept appears valid if it is recognized for what it is—an attempt to create what are essentially "ecological museums" in scenic and biologically significant areas of these lands. But "wilderness," in the hands of environmentalists, has become an all-purpose tool for stopping economic activity as well. This is particularly crucial now because of the many mineral and energy resources available on western lands that environmentalists are trying to push through as wilderness designations. The original legislation specified that lands were to be surveyed for valuable mineral resources before they were put into wilderness preservation. Yet with so much land being reviewed at once, these inventories have been sketchy at best. And once land is locked up as wilderness, it becomes illegal even to explore it for mineral or energy resources.

Thus the situation in western states—where the federal government still owns 68 percent of the land, counting Alaska—has in recent years become a race between mining companies trying to prospect under severely restricted conditions, and environmental groups trying to lock the doors to resource development for good. This kind of permanent preservation—the antithesis of conservation—will probably have enormous effects on our future international trade in energy and mineral resources.

At stake in both the national forests and the Bureau of Land Management holdings are what are called the "roadless areas." Environmentalists call these lands "de facto wilderness," and say that because they have not yet been explored or developed for resources they should not be explored and developed in the future. The Forest Service began its Roadless Area Resources Evaluation (RARE) in 1972, while the Bureau of Land Management began four years later in 1976, after Congress brought its 174 million acres under jurisdiction of the 1964 act. The Forest Service is studying 62 million roadless acres, while the BLM is reviewing 24 million.

In 1974 the Forest Service recommended that 15 million of the 50 million acres then under study be designated as permanent wilderness. Environmental groups, which wanted much more set aside, immediately challenged the decision in court. Naturally, they had no trouble finding flaws in a study intended to cover such a huge amount of land, and in 1977 the Carter administration decided to start over with a "RARE II" study, completed in 1979. This has also been challenged by a consortium of environmental groups that includes the Sierra Club, the Wilderness Society, the National Wildlife Federation, and the Natural Resources Defense Council. The RARE II report also recommended putting about 15 million acres in permanent wilderness, with 36 million released for development and 11 million held for further study. The Bureau of Land Management is not scheduled to complete the study of its 24 million acres until 1991.

The effects of this campaign against resource development have been powerful. From 1972 to 1980, the price of a Douglas fir in Oregon increased 500 percent, largely due to the delays in timber sales from the national forests because of the battles over wilderness areas. Over the decade, timber production from the national forests declined slightly, putting far more pressure on the timber industry's own lands. The nation has now become an importer of logs, despite the vast resources on federal lands. In 1979, environmentalists succeeded in pressuring Congress into setting aside 750,000 acres in Idaho as the Sawtooth Wilderness and National Recreational Area. A resource survey, which was not completed until *after* the congressional action, showed that the area contained an estimated billion dollars' worth of molybdenum, zinc, silver, and gold. The same tract also contained a potential source of cobalt, an important mineral for which we are now dependent on foreign sources for 97 percent of what we use.

Perhaps most fiercely contested are the energy supplies believed to be lying under the geological strata running through Colorado, Wyoming, and Montana just east of the Rockies, called the Overthrust Belt. Much of this land is still administered by the Bureau of Land Management for multiple usage. But with the prospect of energy development, environmental groups have been rushing to try to have these high-plains areas designated as wilderness areas as well (cattle grazing is still allowed in wilderness tracts). On those lands permanently withdrawn from commercial use, mineral exploration will be allowed to continue until 1983. Any mines begun by then can continue on a very restricted basis. But the exploration in "roadless areas" is severely limited, in that in most cases there can be no roads constructed (and no use of off-road vehicles) while exploration is going on. Environmentalists have argued that wells can still be drilled and test mines explored using helicopters. But any such exploration is likely to be extraordinarily expensive and ineffective. Wilderness restrictions are now being drawn so tightly that people on the site are not allowed to leave their excrement in the area.

IMPOSSIBLE PARADISES

What is the purpose of all this? The standard environmental argument is that we have to "preserve these last few wild places before they all disappear." Yet it is obvious that something more is at stake. What is being purveyed is a view of the world in which human activity is defined as "bad" and natural conditions are defined as "good." What is being preserved is evidently much more than "ecosystems." What is being preserved is an *image* of wilderness as a semisacred place beyond humanity's intrusion.

It is instructive to consider how environmentalists themselves define the wilderness. David Brower, former director of the Sierra Club, wrote in his introduction to Paul Ehrlich's *The Population Bomb* (1968):

> Whatever resources the wilderness still held would not sustain (man) in his old habits of growing and reaching without limits. Wilderness could, however, provide answers for questions he had not yet learned how to ask. He could predict that the day of creation was not over, that there would be wiser men, and they would thank him for leaving the source of those answers. Wilderness would remain part of his geography of hope, as Wallace Stegner put it, and could, merely because wilderness endured on the planet, prevent man's world from becoming a cage.

The wilderness, he suggested, is a source of peace and freedom. Yet setting wilderness aside for the purposes of solitude doesn't always work very well. Environmentalists have discovered this over and over again, much to their chagrin. Every time a new "untouched paradise" is discovered, the first thing everyone wants to do is visit it. By their united enthusiasm to find these "sanctuaries," people bring the "cage" of society with them. Very quickly it becomes necessary to erect bars to keep people *out*—which is exactly what most of the "wilderness" legislation has been all about.

In 1964, for example, the Sierra Club published a book on the relatively "undiscovered" paradise of Kauai, the second most westerly island in the Hawaiian chain. It wasn't long before the island had been overrun with tourists. When *Time* magazine ran a feature on Kauai in 1979, one unhappy island resident wrote in to convey this telling sentiment: "We're hoping the shortages of jet fuel will stay around and keep people away from here." The age of environmentalism has also been marked by the near overrunning of popular national parks like Yosemite (which now has a full-time jail), intense pressure on woodland recreational areas, full bookings two and three years in advance for raft trips through the Grand Canyon, and dozens of other spectacles of people crowding into isolated areas to get away from it all. Environmentalists are often critical of these inundations, but they must recognize that they have at least contributed to them.

I am not arguing against wild things, scenic beauty, pristine landscapes, and scenic preservation. What I am questioning is the argument that wilderness is a value against which every other human activity must be judged, and that human beings are somehow unworthy of the landscape. The wilderness has been equated with freedom, but there are many different ideas about what constitutes freedom. In the Middle Ages, the saying was that "city air makes a man free," meaning that the harsh social burdens of medieval feudalism vanished once a person escaped into the heady anonymity of a metropolitan community. When city planner Jane Jacobs, author of *The Death and Life of Great American Cities,* was asked by an interviewer if "overpopulation" and "crowding into large cities" weren't making social prisoners of us all, her simple reply was: "Have you ever lived in a small town?"

It may seem unfair to itemize the personal idiosyncrasies of people who feel comfortable only in wilderness, but it must be remembered that the environmental movement has been shaped by many people who literally spent years of their lives living in isolation. John Muir, the founder of the National Parks movement and the Sierra Club spent almost ten years living alone in the Sierra Mountains while learning to be a trail guide. David Brower, who headed the Sierra Club for over a decade and later broke with it to found the Friends of the Earth, also spent years as a mountaineer. Gary Snyder, the poet laureate of the environmental movement, has lived much of his life in wilderness isolation and has also spent several years in a Zen monastery. All these people far outdid Thoreau in their desire to get a little perspective on the world. There is nothing reprehensible in this, and the literature and philosophy that merge from such experiences are often admirable. But it seems questionable to me that the ethic that comes out of this wilderness isolation—and the sense of ownership of natural landscapes that inevitably follows—can serve as the basis for a useful national philosophy.

THAT FRONTIER SPIRIT

The American frontier is generally agreed to have closed down physically in 1890, the year the last Indian Territory of Oklahoma was opened for the settlement. After that, the Conservation Movement arose quickly to protect the remaining resources and wilderness from heedless stripping and development. Along with this came a significant psychological change in the national character, as the "frontier spirit" diminished and social issues attracted greater attention. The Progressive Movement, the Social Gospel among religious groups, Populism, and Conservation all arose in quick succession immediately after the "closing of the frontier." It seems fair to say that it was only after the frontier had been settled and the sense of endless possibilities that came with open spaces had been constricted in the national consciousness that the country started "growing up."

Does this mean the new environmental consciousness has arisen because we are once again "running out of space"? I doubt it. Anyone taking an airplane across almost any part of the country is inevitably struck by how much greenery and open territory remain, and how little room our towns and cities really occupy. The amount of standing forest in the country, for example, has not diminished appreciably over the last fifty years, and is 75 percent of what it was in 1620. In addition, as environmentalists constantly remind us, trees are "renewable resources." If they continue to be handled intelligently, the forests will always grow back. As farming has moved out to the Great Plains of the Middle West, many eastern areas that were once farmed have reverted back to trees. Though mining operations can permanently scar hillsides and plains, they are usually very limited in scope (and as often as not, it is the roads leading to these mines that environmentalists find most objectionable).

It seems to be that the wilderness ethic has actually represented an attempt psychologically to reopen the American frontier. We have been desperate to maintain belief in unlimited, uncharted vistas within our borders, a preoccupation that has eclipsed the permanent shrinking of the rest of the world outside. Why else would it be so necessary to preserve such huge tracts of "roadless territory" simply because they are now roadless, regardless of their scenic, recreational, or aesthetic values? The environmental movement, among other things, has been a rather backward-looking effort to recapture America's lost innocence.

The central figure in this effort has been the backpacker. The backpacker is a young, unpreopossessing person (inevitably white and upper middle class) who journeys into the wilderness as a passive observer. He or she brings his or her own food, treads softly, leaves no litter, and has no need to make use of any of the resources at hand. Backpackers bring all the necessary accouterments of civilization with them. All their needs have been met by the society from which they seek temporary release. The backpacker is freed from the need to support itself in order to enjoy the aesthetic and spiritual values that are made available by this temporary *removal* from the demands of nature. Many dangers—raging rivers or precipitous cliffs, for instance—become sought-out adventures.

Yet once the backpacker runs out of supplies and starts using resources around him—cutting trees for firewood, putting up a shelter against the rain—he is violating some aspect of the federal Wilderness Act. For example, one of the issues fought in the national forests revolves around tying one's horse to a tree. Purists claim the practice should be forbidden, since it may leave a trodden ring around the tree. They say horses should be hobbled and allowed to graze instead. In recent years, the National Forest Service has come under pressure from environmental groups to enforce this restriction.

Wildernesses, then, are essentially parks for the upper middle class. They are vacation reserves for people who want to rough it—with the assurance that few other people will have the time, energy, or means to follow them into the solitude. This is dramatically highlighted in one Sierra Club book that shows a picture of a professorial sort of individual backpacking off into the woods. The ironic caption is a quote from Julius Viancour, an official of the Western Council of Lumber and Sawmill Workers: "The inaccessible wilderness and primitive areas are off limits to most laboring people. We must have access. . . ." The implication for Sierra Club readers is: "What do these beer-drinking, gun-toting, working people want to do in *our* woods?"

This class-oriented vision of wilderness as an upper-middle-class preserve is further illustrated by the fact that most of the opposition to wilderness designations comes not from industry but from owners of off-road vehicles. In most northern rural areas, snowmobiles are now regarded as the greatest invention since the automobile, and people are ready to fight rather than stay cooped up all winter in their houses. It seems ludicrous to them that snowmobiles (which can't be said even to endanger the ground) should be restricted from vast tracts of land so that the occasional city visitor can have solitude while hiking past on snowshoes.

The recent Boundary Waters Canoe Area controversy in northern Minnesota is an excellent example of the conflict. When the tract was first designated as wilderness in 1964, Congress included a special provision that allowed motorboats into the entire area. By the mid-1970s, outboards and inboards were roaming all over the wilderness, and environmental groups began asking that certain portions of the million-acre preserve be set aside exclusively for canoes. Local residents protested vigorously, arguing that fishing expeditions, via motorboats, contributed to their own recreation. Nevertheless, Congress eventually excluded motorboats from 670,000 acres to the north.

A more even split would seem fairer. It should certainly be possible to accommodate both forms of recreation in the area, and there is as much to be said for canoeing in solitude as there is for making rapid expeditions by powerboat. The natural landscape is not likely to suffer very much from either form of recreation. It is not absolute "ecological" values that are really at stake, but simply different tastes in recreation.

NOT ENTIRELY NATURE

At bottom, then, the mystique of the wilderness has been little more than a revival of Rousseau's Romanticism about the "state of nature." The notion that "only in wilderness are human beings truly free," a credo of environmentalists, is merely a variation on Rousseau's dictum that "man is born free, and everywhere he is in chains." According to Rousseau, only society could enslave people, and only in the "state of nature" was the "noble savage"—the preoccupation of so many early explorers—a fulfilled human being.

The "noble savage" and other indigenous peoples, however, have been carefully excised from the environmentalists' vision. Where environmental efforts have encountered primitive peoples, these indigenous residents have often proved one of the biggest problems. One of the most bitter issues in Alaska is the efforts by environmentalist groups to restrict Indians in their hunting practices.

At the same time, few modern wilderness enthusiasts could imagine, for example, the experience of the nineteenth-century artist J. Ross Browne, who wrote in *Harper's New Monthly Magazine* after visiting the Arizona territories in 1864:

> Sketching in Arizona is . . . rather a ticklish pursuit. . . . I never before traveled through a country in which I was compelled to pursue the fine arts with a revolver strapped around my body, a double-barreled shot-gun lying across my knees, and half a dozen soldiers armed with Sharpe's carbines keeping guard in the distance. Even with all the safeguards . . . I am free to admit that on occasions of this kind I frequently looked behind to see how the country appeared in its rear aspect. An artist with an arrow in his back may be a very picturesque object . . . but I would rather draw him on paper than sit for the portrait myself.

Wilderness today means the land *after* the Indians have been cleared away but *before* the settlers have arrived. It represents an attempt to hold that particular moment forever frozen in time, that moment when the visionary American settler looked out on the land and imagined it as an empty paradise, waiting to be molded to our vision.

In the absence of the noble savage, the environmentalist substitutes himself. The wilderness, while free of human dangers, becomes a kind of basic-training ground for upper-middle-class values. Hence the rise of "survival" groups, where college kids are taken out into the woods for a week or two and let loose to prove their survival instincts. No risks are spared on these expeditions. Several people have died on them, and a string of lawsuits has already been launched by parents and survivors who didn't realize how seriously these survival courses were being taken.

The ultimate aim of these efforts is to test upper-middle-class values against the natural environment. "Survival" candidates cannot hunt, kill, or use much of the natural resources available. The true test is whether their zero-degree sleeping bags and dried-food kits prove equal to the hazards of the tasks. What happens is not necessarily related to nature. One could as easily test survival skills by turning a person loose without money or means in New York City for three days.

I do not mean to imply that these efforts do not require enormous amounts of courage and daring—"survival skills." I am only suggesting that what the backpacker or survival hiker encounters is not entirely "nature," and that the effort to go "back to nature" is one that is carefully circumscribed by the most intensely civilized artifacts. Irving Babbitt, the early twentieth-century critic of Rousseau's Romanticism, is particularly vigorous in his dissent from the idea of civilized people going "back to nature." This type, he says, is actually "the least primitive of all beings":

> We have seen that the special form of unreality encouraged by the aesthetic romanticism of Rousseau is the dream of the simple life, the return to a nature that never existed, and that this dream made its special appeal to an age that was suffering from an excess of artificiality and conventionalism.

Babbitt notes shrewdly that our concept of the "state of nature" is actually one of the most sophisticated productions of civilization. Most primitive peoples, who live much closer to the soil than we do, are repelled by wilderness. The American colonists, when they first encountered the unspoiled landscape, saw nothing but a horrible desert, filled with savages.

What we really encounter when we talk about "wilderness," then, is one of the highest products of civilization. It is a reserve set up to keep people *out*, rather than a "state of nature" in which the inhabitants are "truly free." The only thing that makes people "free" in such a reservation is that they can leave so much behind when they enter. Those who try to stay too long find out how spurious this "freedom" is. After spending a year in a cabin in the north Canadian woods, Elizabeth Arthur wrote in *Island Sojourn:* "I never felt so completely tied to *objects*, resources, and the tools to shape them with."

What we are witnessing in the environmental movement's obsession with purified wilderness is what has often been called the "pastoral impulse." The image of nature as unspoiled, unspotted wilderness where we can go to learn the lessons of ecology is both a product of a complex, technological society and an escape from it. It is this undeniable paradox that forms the real problem of setting up "wildernesses." Only when we have created a society that gives us the leisure to appreciate it can we go out and experience what we imagine to be untrammeled nature. Yet if we lock up too much of our land in these reserves, we are cutting into our resources and endangering the very leisure that allows us to enjoy nature.

The answer is, of course, that we cannot simply let nature "take over" and assume that because we have kept roads and people out of huge tracts of land, then we have absolved ourselves of a national guilt. The concept of stewardship means taking responsibility, not simply letting nature take its course. Where tracts can be set aside from commercialism at no great cost, they should be. Where primitive hiking and recreation areas are appealing, they should be maintained. But if we think we are somehow appeasing the gods by *not* developing resources where they exist, then we are being very shortsighted. Conservation, not preservation, is once again the best guiding principle.

The cult of wilderness leads inevitably in the direction of religion. Once again, Irving Babbitt anticipated this fully.

> When pushed to a certain point the nature cult always tends toward sham spirituality. . . . Those to whom I may seem to be treating the nature cult with undue severity should remember that I am treating it only in its pseudo-religious aspect. . . . My quarrel is only with the asthete who assumes an apocalyptic pose and gives forth as a profound philosophy what is at best only a holiday or weekend view of existence. . . .

It is often said the environmentalism could or should serve as the basis of a new religious consciousness, or a religious "reawakening." This religious trend is usually given an Oriental aura. E. F. Schumacher has a chapter on Buddhist economics in his classic *Small Is Beautiful*. Primitive animisms are also frequently cited as attitudes toward nature that are more "environmentally sound." One book on the environment states baldly that "the American Indian lived in almost perfect harmony with nature." Anthropologist Marvin Harris has even put forth the novel view that primitive man is an environmentalist, and that many cultural habits are unconscious efforts to reduce the population and conserve the environment. He says that the Hindu prohibition against eating cows and the Jewish tradition of not eating pork were both efforts to avoid the ecological destruction that would come with raising these grazing animals intensively.

The implication in these arguments is usually that science and modern technology have somehow dulled our instinctive "environmental" impulses, and that Western "non-spiritual" technology puts us out of harmony with the "balance of nature."

Perhaps the most daring challenge to the environmental soundness of current religious tradition came early in the environmental movement, in a much quoted paper by Lynn White, professor of the history of science at UCLA. Writing in *Science* magazine in 1967, White traced "the historical roots of our ecological crisis" directly to the Western Judeo-Christian tradition in which "man and nature are two things, and man is master." "By destroying pagan animism," he wrote, "Christianity made it possible to exploit nature in a mood of indifference to the feelings of natural objects." He continued:

> Especially in its Western form, Christianity is the most anthropocentric religion the world has seen. . . . Christianity, in absolute contrast to ancient paganism and Asia's religions (except, perhaps, Zoroastrianism), not only established a dualism of man and nature but also insisted that it is God's will that man exploit nature for his proper ends. . . . In antiquity every tree, every spring, every stream, every hill had its own *genius loci*, its guardian spirit. . . . Before one cut a tree, mined a mountain, or dammed a brook, it was important to placate the spirit in charge of that particular situation, and keep it placated.

But the question here is not whether the Judeo-Christian tradition is worth saving in and of itself. It would be more than disappointing if we canceled the accomplishments of Judeo-Christian thought only to find that our treatment of nature had not changed a bit.

There can be no question that White is onto a favorite environmental theme here. What he calls the "Judeo-Christian tradition" is what other writers often term "Western civilization." It is easy to go through environmental books and find long outbursts about the evils that "civilization and progress" have brought us. The long list of Western achievements and advances, the scientific men of genius, are brought to task for creating our "environmental crisis." Sometimes the condemnation is of our brains, pure and simple. Here, for example, is the opening statement from a book about pesticides, written by the late Robert van den Bosch, an outstanding environmental advocate:

> Our problem is that we are too smart for our own good, and for that matter, the good of the biosphere. The basic problem is that our brain enables us to evaluate, plan, and execute. Thus, while all other creatures are programmed by nature and subject to her whims, we have our own gray computer to motivate, for good or evil, our chemical engine. . . . Among living species, we are the only one possessed of arrogance, deliberate stupidity, greed, hate, jealousy, treachery, and the impulse to revenge, all of which may erupt spontaneously or be turned on at will.

At this rate, it can be seen that we don't even need religion to lead us astray. We are doomed from the start because we are not creatures of *instinct*, programmed from the start "by nature."

This type of primitivism has been a very strong, stable undercurrent in the environmental movement. It runs from the kind of fatalistic gibberish quoted above to the Romanticism that names primitive tribes "instinctive environmentalists," from the pessimistic predictions that human beings cannot learn to control their own numbers to the notion that only by remaining innocent children of nature, untouched by progress, can the rural populations of the world hope to feed themselves. At bottom, as many commentators have pointed out, environmentalism is reminiscent of the German Romanticism of the nineteenth century, which sought to shed Christian (and Roman) traditions and revive the Teutonic gods because they were "more in touch with nature."

But are progress, reason, Western civilization, science, and the cerebral cortex really at the root of the "environmental crisis"? Perhaps the best answer comes from an environmentalist himself, Dr. Rene Dubos, a world-renowned microbiologist, author of several prize-winning books on conservation and a founding member of the Natural Resources Defense Council. Dr. Dubos takes exception to the notion that Western Christianity has produced a uniquely exploitative attitude toward nature:

> Erosion of the land, destruction of animal and plant species, excessive exploitation of natural resources, and ecological disasters are not peculiar to the Judeo-Christian tradition and to scientific technology. At all times, and all over the world, man's thoughtless interventions into nature have had a variety of disastrous consequences or at least have changed profoundly the complexity of nature.

Dr. Dubos has catalogued the non-Western or non-Christian cultures that have done environmental damage. Plato observed, for instance, that the hills in Greece had been heedlessly stripped of wood, and erosion had been the result; the ancient Egyptians and Assyrians exterminated large numbers of wild animal species; Indian hunters presumably caused the extinction of many large paleolithic species in North America; Buddhist monks building temples in Asia contributed largely to deforestation. Dubos notes:

> All over the globe and at all times . . . men have pillaged nature and disturbed the ecological equilibrium . . . nor did they have a real choice of alternatives. If men are more destructive now . . . it is because they have at their command more powerful means of destruction, not because they have been influenced by the Bible. In fact, the Judeo-Christian peoples were probably the first to develop on a large scale a pervasive concern for land management and an ethic of nature.

The concern that Dr. Dubos cites is the same one we have rescued out of the perception of environmentalism as a movement based on aristocratic conservatism. That is the legitimate doctrine of *stewardship* of the land. In order to take this responsibility, however, we must recognize the part we play in nature—that "the land is ours." It will not do simply to worship nature, to create a cult of wilderness in which humanity is an eternal intruder and where human activity can only destroy.

"True conservation," writes Dubos, "means not only protecting nature against human misbehavior but also developing human activities which favor a creative, harmonious relationship between man and nature." This is a legitimate goal for the environmental movement.

POSTSCRIPT

Does Wilderness Have Intrinsic Value?

Olson makes a persuasive case that the present state of the world has resulted in a desperate, practical concern about sustainable development, of which wilderness preservation is an intrinsic part. This contradicts Tucker's characterization of the motivation of wilderness enthusiasts as being based on esoteric, romantic notions.

Despite the rapid increase in the popularity of backpacking, Tucker is correct in maintaining that it is still primarily a diversion of the economically privileged. A lack of financial resources and leisure time prevents the majority of U.S. citizens from taking advantage of tax-supported parks and other recreational areas. Is this a potent argument against prohibiting development of some of the total U.S. acreage?

For a more thorough explication of Tucker's analysis of the ecology movement, see his book *Progress and Privilege* (Anchor Press, 1982). Other groups critical of the total exclusion of development in wilderness include hunters, fishermen, and foresters. For a typical hunter's perspective, see Jim Zumbo's article in the June 1979 issue of *Outdoor Life* magazine. Retired forester Fred C. Simmons presents his critique in the May 14, 1982, issue of *National Review*.

The controversy over oil development on Alaska's coastal plain is the most bitterly contested present struggle concerning proposals for exploitation of mineral resources in wilderness regions. The only positive result for environmentalists of the huge oil spill in Prince William Sound by the supertanker *Exxon Valdez* is that it has dealt a political setback to advocates of further oil exploration projects. The battle to preserve Alaskan wilderness is reviewed in articles by Duncan Frazier in *National Parks* (November/December 1987) and by George Laycock in *Audubon* (May 1988).

The Sierra Club periodical *Sierra* and the Wilderness Society's magazine *The Living Wilderness*, as well as the many other publications of those organizations, contain a wealth of writings in support of the wilderness. For a particularly comprehensive collection of essays, see *Voices for the Wilderness* (Ballantine, 1969), edited by William Schwarz.

Those who are sympathetic to the concerns raised by Olson would probably find Bill McKibben's controversial recent book *The End of Nature* (Random House, 1990) both moving and deeply disturbing.

ISSUE 3

Do We Need More Stringently Enforced Regulations to Protect Endangered Species?

YES: Lewis Regenstein, from "Endangered Species and Human Survival," *USA Today*, a publication of the Society for the Advancement of Education (September 1984)

NO: Richard Starnes, from "The Sham of Endangered Species," *Outdoor Life* (August 1980)

ISSUE SUMMARY

YES: Environmentalist Lewis Regenstein charges that the policy of exploiting rather than protecting wildlife, which was the policy of the Reagan administration, threatens to accelerate the disappearance of endangered plant and animal species.
NO: *Outdoor Life* editor Richard Starnes counters that wildlife management policy should be left to the judgment of professionals in the state and national agencies responsible for carrying out the policies, not the whims of federal legislators.

Extinction of biological species is not necessarily a phenomenon initiated by human activity. Although the specific role of extinction in the process of evolution is still being researched and debated, it is generally accepted that the demise of any biological species is inevitable. Opponents of special efforts to protect endangered species invariably point this out. They also suggest that the role of *Homo sapiens* in causing extinction should not be distinguished from that of any other species.

This position is contrary to some well-established facts. Unlike other creatures that have inhabited the Earth, human beings are the first to possess the technological ability to cause wholesale extermination of species, genera, or even entire families of living creatures. This process is accelerating. Between 1600 and 1900, about 75 known species of mammals and birds were hunted to extinction. Wildlife management efforts initiated during the twentieth century have been unsuccessful in stemming the tide, as indicated by the fact that the rate of extinction of species of mammals and birds has jumped to approximately one per year.

In 1973, the Endangered Species Act was adopted, and an international treaty was negotiated, in an effort to combat this worldwide problem. This act united a variety of industrial and business interests with the commercial hunters and trappers who traditionally objected to efforts to restrict their activities. However, opposition developed because the act prohibited construction projects that threatened to cause the extinction of any species. The most celebrated example of a confrontation brought about under this act was an effort by environmentalists to halt construction of the Tellico Dam in Loudon County, Tennessee, on the Little Tennessee River, because it threatened a small fish called the snail darter. Despite the uncompromising language of the law, no major project has been prevented because of its enforcement.

Most public attention given to endangered species has focused on mammals, birds, and a few varieties of trees. Efforts to save timber wolves, whales, whooping cranes, and redwood trees have also received considerable media coverage. But ecologists recognize an even greater threat to the much larger number of species of reptiles, fish, invertebrates, and plants that are being wiped out by human activity. In the past two decades, vast areas in several regions of the world have been cleared to make room for urban development or for food production. Modern agricultural techniques and industries' need for raw material have contributed to the epidemic of extinction. Today it is estimated that a total of 1,000 species of plants and animals may be disappearing annually.

Scientists fear that the vitality of our ecology may be seriously threatened by the reduction of biological diversity resulting from the lost genetic resource contained in the extinct species. They note that the ability of species to evolve and adapt to environmental change depends on the existence of a vast pool of genetic material. This problem joins the issue of endangered species with that of wilderness preservation. Unfortunately, the need to set aside vast undeveloped areas to prevent wholesale extinction is more acute in the poorer, more crowded regions of the world where people are pressured by both their own basic needs and the demand of the industrialized world for their resources.

Lewis Regenstein, who is vice president of the Fund for Animals, bemoans the policies of the Interior Department under the Reagan administration with regard to enforcement of legislation to protect endangered and threatened species. He asserts that commercial interests that wish to exploit rather than nurture species in danger of extinction have been given preferential treatment. Richard Starnes, who is the editor of *Outdoor Life*, claims that it is the bureaucrats empowered by federal legislation that are the real threat to wildlife. He would leave the protection of endangered species in the hands of professionals in agencies like the U.S. Fish and Wildlife Service.

YES

Lewis Regenstein

ENDANGERED SPECIES AND HUMAN SURVIVAL

. . . between 500,000 and 2,000,000 species—15 to 20% of all species on Earth—could be extinct by the year 2000. . . . Extinction of species on this scale is without precedent in human history.

—THE GLOBAL 2000 REPORT TO THE PRESIDENT
(prepared by the Council on Environmental Quality and the U.S. State Department, 1980)

One of the great tragedies of our era—perhaps the greatest—is the massive and unprecedented extinction, at the hands of man, of unique and irreplaceable species of plants and animals. This permanent destruction of numerous fellow life forms with which we share this planet will have grave consequences for humanity, and could eventually threaten man's security and survival.

Perhaps the most damaging losses will occur among the least popular and most obscure creatures, such as plants, insects, and mollusks; those that seem most "expendable" may turn out to be fundamental to man's own well-being. With the Reagan administration's Interior Department refusing to effectively implement the Federal Endangered Species Act and various industry groups attempting to cripple the law, we are now making decisions on protecting rare species that will have profound implications for the future of our own species and society.

Tragically, we do not know, and will probably never learn, the value of the millions of species projected to become extinct in the next 20 years. As pointed out in the 1978 report published by the President's Council on Environmental Quality (CEQ), *The Global Environment and Basic Human Needs,*

> the total number of plant and animal species on Earth may be as high as 10,000,000—only about 15% of which have been identified in scientific literature, let alone been well studied. If current trends continue, a good share of the

> unrecorded majority of species will vanish forever before their existence, or their biological importance, is known.

Over the next decade or two, warns CEQ, "unique ecosystems populated by thousands of unrecorded plant and animal species face rapid destruction—irreversible genetic losses that will profoundly alter the course of evolution."

Because of the ever-increasing exploitation and destruction of the tropical rain forests, most of the forthcoming extinctions of foreign plants will occur in these regions. Rain forests receive extraordinary amounts of warmth, moisture, and sunlight, and thus provide ideal habitat for an amazing variety of species. In the last 30 years alone, half of the world's rain forests have been destroyed, and the remainder are being cut at a rate of 27,000,000 and 50,000,000 acres a year—one or two acres a second!

The Amazon Basin—probably the richest biological community on Earth—alone may house 1,000,000 species (in addition to various unique and endangered tribes of Indians that are also threatened there). More types of woody plant species are found on the slopes of a single Philippine volcano than in the entire U.S.

Worldwide, there are thought to be at least 5,000,000 species of plants and animals that are found nowhere else in the world—many of which are yet to be "discovered" by man—inhabiting tropical forests and dependent on them for survival. At the present rate of deforestation, experts fear that some 500,000 such species are candidates for extinction in the next two decades.

In his book, *The Sinking Ark,* wildlife specialist Norman Myers estimates that at least one species *a day* is already being wiped out in the tropical forests and that, at the increasing rate of timber exploitation, one species *an hour* may soon be lost in the years ahead.

THE LOSS TO HUMANITY

These losses will deprive the world not only of countless beautiful and diverse life forms, but also of future sources of food, drugs, and medicines of incalculably great value. The United Nations' Environment Program, in its *World Conservation Strategy* report of March, 1980, published jointly with the World Wildlife Fund and the International Union for the Conservation of Nature and Natural Resources, described the immense potential value of these plants:

> Penicillin, digitalis, quinine, rubber, pectin, resins, gums, insecticides—these and other medicines and products come from plants. One out of two prescriptions filled in the U.S. each day is for a drug based on an ingredient in a plant. . . . The wheat we know today began as wild plants—and some humans some unknown number of years ago may well have considered those wild plants worthless seeds.

As the 1978 CEQ report observes, the extinction of these species will also entail the loss of many useful products:

> Perhaps the greatest industrial, agricultural, and medical costs of species reduction will stem from future opportunities unknowingly lost. Only about five percent of the world's plant species have yet been screened for pharmacologically active ingredients. Ninety percent of the food that humans eat comes from just 12 crops, but scores of thousands of plants are edible, and some will undoubtedly prove useful in meeting human food needs.

Tropical forests are today the main source of drugs made from plants, and up to half of our prescription drugs come from such flora. In addition, these forests have been the original source of such important food items as bananas, pineapples, rice, millet, sugar cane, cassava, yams, and taro. Such valuable products as rubber and quinine come from plants, and it is possible—even probable—that plants of comparable significance are being wiped out. Thus, in destroying the rain forests, our generation is depriving itself and future ones of plants that could be a possible cure for cancer and other dread diseases, and of which could become important food items to an increasingly hungry and overpopulated planet.

Other types of species of great potential value to mankind are the obscure and little-known snails, clams, scuds, and other mollusks and crustaceans, thousands of which are in serious peril because of water pollution, dredging, stream channelization, the building of highways, dams, housing developments, and other destruction of habitat, carried out mainly by the federal government. As Marc Imlay, formerly a biologist with the Interior Department's Office of Endangered Species, has pointed out,

> Though they seem inconsequential in size, mussels and crustaceans are an indispensable part of the living world. Besides fitting into the food chain, these creatures have recently been recognized as being able to produce poisons, antibiotics, tranquilizers, antispasmodics, and antiseptic chemicals in their systems. Scientists believe these unique abilities can be used as models for the development of synthetic drugs.

Dr. Imlay was known as one of the most determined employees of his office in trying to secure protection for endangered mollusks and crustaceans, some of which, if listed as endangered, could have affected the building of dams and other destructive federal water projects. As a result of his actions, which offended the Interior Department bureaucracy, he was threatened with dismissal in 1977 and later removed from his position and forced to transfer out of Washington, D.C., to Columbia, Mo.

It is not possible to quantify or predict the consequences and magnitude of these losses and the impact they will have on the Earth and future generations. What is clear is that the results will be profound and could be catastrophic. As the President's Council on Environmental Quality concluded in a recent publication, "a decline in the diversity of life forms is of grave concern to all people for a wide range of reasons," especially since "the potential large-scale loss of species is without historical precedent and involves the disruption of ecological systems whose complexity is beyond human grasp." Also, as Eric Eckholm of the Worldwatch Institute has observed,

> Should this biological massacre take place, evolution will no doubt continue, but in a grossly distorted manner. Such a multitude of species losses would constitute a basic and irreversible alteration in the nature of the biosphere even before we understand its workings—an evolutionary Rubicon whose crossing Homo sapiens would do well to avoid. . . . Humans appoint themselves as the ultimate arbiters of evolution and determine its future course on the basis of short-term considerations and a great deal of ignorance. . . . Scientists cannot yet say where the critical thresholds lie, at what level of species extermination the web of life will be seriously disrupted. . . .

Eckholm further notes that, when a plant species is wiped out, some 10 to 30 dependent species can also be jeopardized, such as insects, and even other plants. An example of the complex interrelationship that has evolved between many tropical species is the 40 different species of fig trees native to Central America, each of which has a specific insect pollinator. Other insects, including pollinators for other plants, depend on certain of these fig trees for food. Thus, the extinction of one species can set off a chain reaction, the ultimate effects of which cannot be foreseen. As Eckholm puts it, "Crushed by the march of civilization, one species can take many others with it, and the ecological repercussions and rearrangements that follow may well endanger people."

THE ROLE OF THE REAGAN ADMINISTRATION

With the advent of the Reagan administration, the government's endangered species program has all but ceased to function. Under the direction of Assistant Secretary of the Interior Ray Arnett, an avid big-game trophy hunter, the emphasis has shifted to the exploitation, rather than the protection, of wildlife. Where the U.S. once used to take the lead in international conferences in pushing for protection for rare species, America now fights for increased commercial exploitation of imperiled creatures. The Interior Department has resisted adding more species to the protected lists, instead concentrating on delisting species of interest to hunters and other consumptive interests.

Increasingly large areas of grizzly bear habitat in Montana are being opened up to oil, gas, and mineral exploration, thereby disrupting and destroying some of the last refuges for this threatened species. Montana continues to issue some 1,000 grizzly hunting permits each year, even though this represents several times more grizzlies than are thought to survive in that state and Wyoming, its last two areas of survival in the lower 48 states. (Montana claims that the season is closed after 25 bears have been "taken," but most grizzlies that are shot are either illegally poached or never reported.)

The Administration has also refused to acquire habitats critical to the survival of many endangered species. For example, in 1981, Interior Department plans were halted to acquire several thousand acres in Key Largo, Florida, that provide refuge to several endangered species, including the almost extinct American crocodile, of which only about 50 remain in the U.S. Most of this swampy area is now slated for commercial development. The policies of the administration could spell doom for many other rare species as well.

While the Reagan administration has been surprisingly open about its determination to cater to anti-conservation interests, the records of previous administrations are also disappointing. Conservationists have complained for years that the endangered species listing process has virtually ground to a halt, and that literally thousands of rare species are in danger of becoming extinct because of Interior's inaction. The lynx, the bobcat, the river otter, and thousands of other imperiled species are being excluded from the protected lists and are rapidly being wiped out by hunters, trappers, and the fur industry while Interior Department bureaucrats remain mired in red tape and bureaucratic inertia.

To make matters worse, Interior is still allowing commercial exploitation of species already recognized and listed as threatened or endangered. After The Fund for Animals and other conservationists worked for years to get the African elephant listed and protected, Interior finally listed it in June, 1978, as "threatened," but issued regulations allowing ivory imports to continue. Since the listing, the import into the U.S. of elephant ivory has significantly increased, and the U.S. continues to help provide a major incentive for their massive and illegal slaughter.

One reason why Interior has accomplished so little in the way of listing imperiled species is that it is spending inordinate amounts of time, resources, and manpower removing animals from the endangered list that special interest groups want to exploit. Under pressure from the gun lobby, for example, Interior recently acted to remove from the endangered list the African leopard and the red lechwe (a rare African antelope highly sought by trophy hunters) and to downgrade their status to "threatened," so that American hunters can shoot these animals in Africa and bring them back as "trophies."

The group leading the fight for these delistings—Safari Club, International—in August, 1978, applied to Interior for a permit (which was later withdrawn) to allow its members to kill and import into the U.S. each year some 1,125 animals from various endangered species, including 25 tigers, 100 cheetahs, 5 gorillas, 5 orangutans, 150 African leopards, 5 clouded leopards, 10 snow leopards, 100 mountain zebras, and various other critically endangered species of deer, gazelles, impalas, crocodiles, and others.

Interior has also removed protection from the three large species of Australian kangaroos which are on the U.S. Threatened List by lifting the import ban on and allowing a resumption of imports of kangaroo hides and products. This action has brought about a huge increase in the slaughter of these creatures for the massive U.S. market, which was the world's largest recipient of these hides before the import ban in the mid-1970's. In fact, the U.S., as the world's wealthiest nation, has traditionally played a significant role in the tragedy of disappearing species by supplying the major demand for wildlife products and thus the economic incentive for the slaughter of wildlife in foreign nations.

Before the spotted cats were added to the endangered list in 1972, the U.S. fur industry, between 1968 and 1970, imported 18,456 leopard skins, 31,105 jaguar pelts, and an incredible 349,680 ocelot hides. (These figures are for raw skins and do not include finished products.) Over 2,000 cheetah pelts from Africa were also brought in, a figure representing one and a half times the number of these animals (2,000) estimated to remain in all the parks in Africa in 1971 (and which has declined since then). As the United Nations Environment Program has noted,

> In 1972, some 1,000,000 products manufactured or processed from wildlife were imported into the U.S. In 1978, the number was 91,000,000—an increase of 5,753%! Worldwide trade in tropical birds, aquarium fish, exotic plants, and reptiles has mushroomed.

DESTROYING OUR FELLOW CREATURES AND OURSELVES

As human beings are part of nature, we are bound by its laws. We ignore this fact at great risk to ourselves, for eventually we will certainly destroy a species or ecosystem essential to our own survival. For example, whales, porpoises, and dolphins, slaughtered by the hundreds of thousands each year, play a vital, though little-understood, role in maintaining the health and stability of the world's oceans. By eliminating these intelligent cetaceans, we further upset the delicate balance of life in the seas and imperil oxygen and food-producing ecosystems that are necessary for the survival of all life forms—including man.

Every species plays some role in the environment which may be necessary for the proper function of the ecosystem. Even such "ugly," dangerous, and unpopular creatures as alligators and crocodiles are useful to man. Alligators kill and eat water moccasins and other poisonous snakes and, during times of drought—which occur periodically in the southeastern U.S.—they dig water holes, thus providing water, food, and habitat for fish, birds, and the other creatures of the swamp, allowing them to survive these difficult periods. When crocodiles were eliminated from lakes and river systems in areas of Africa and Australia, many of the food fish also declined or disappeared. It is now thought that this occurred because the crocodiles had been feeding on scavenging or predatory species of fish not eaten by the natives which, left unchecked, multiplied out of control and preyed on or crowded out many of the food fish. Thus, these reptiles serve a much more valuable function than providing hides for shoes, wallets, belts, and other fashion accessories, the demand for which has driven most species and population stocks of them to the verge of extinction.

NO

Richard Starnes

THE SHAM OF ENDANGERED SPECIES

By its very title the Endangered Species Act sounds as righteous as apple pie, home and mother—for what could be more high-minded than seeking by law to protect all of God's creatures from extinction, to end man's vile, age-old practice of wiping out animals that were either very useful or very inconvenient?

When the act became law there was general agreement among conservationists and outdoorsmen that no more idealistic document had emerged from Washington since the pledge of allegiance to the flag. The Endangered Species Act was seen as creating a whole new ethic in man's relationship with the natural world around him. The act, which had its roots in a bill first passed in 1966 and which was substantially rewritten in 1973, was deemed a fitting capstone to the great environmental legislation of the 60s and 70s.

Sad to say, reality has proved to be quite a different thing. Since its inception the act has:

—Become a nightmare of bureaucratic empire building, conflicting authority and interagency gut fighting.

—Given foreign governments vital influence over game management policies in the United States.

—Been used to "protect" species that were never threatened or endangered.

—Actually harmed a number of the species it was designed to protect.

—Hamstrung state game agencies in the management of game and nongame species alike.

—Become a weapon in an international game of chicken that is played by hard-charging Washington bureaucrats to the detriment of American hunters and trappers.

—"The Endangered Species Act," said one Louisiana state game department official who knows a great deal about the big alligator brouhaha (a classic piece of endangered species wrong-headedness which we will examine in some detail) "is probably the worst piece of conservation legislation ever to come out of Washington."

From Richard Starnes, "The Sham of Endangered Species," *Outdoor Life* (August 1980).

Whether that sweeping judgment is justified or not, there is almost universal agreement that the act has worked very badly, has been abused and diverted from its purpose by bureaucrats of a high order of arrogance, and is likely to get no better. Rarely has there been any legislation whose actual workings have been as wide of its intended mark as the Endangered Species Act.

Much of the mischief stemming from the act has come as a consequence of its attempts to impose worldwide the judgment of a handful of Washington bureaucrats. This had led to international horse trading—an art form in which Americans have never excelled—and more often than not it has been our own hunters and trappers who have been skinned. Under the act the United States joined a treaty with a jawbreaker name that few outside the bureaucratic warrens of Washington ever heard of: the Convention on International Trade in Endangered Species of Wild Flora and Fauna. This is shorthanded CITES among the cognoscenti. Like the act of Congress that sired it, CITES has been a mess.

Under the treaty, the 50-odd signatory countries list animals and plants that are either endangered or threatened. Endangered species may neither be exported nor imported without licenses from both importing and exporting countries. Species listed as threatened require licenses only from exporting countries. In practice licenses have been issued or withheld in the United States by the whim of a powerful, independent interagency committee that is called the Endangered Species Scientific Authority (ESSA). In time ESSA became so overbearing and so unresponsible to any of the customary bureaucratic controls that Congress sought to clip its wings. How Congress was frustrated—with the help of the White House—is a diverting tale we will come to in a moment.

How the CITES treaty and ESSA combined to screw up the trade in American bobcat pelts is a story worth telling, for it explains much about how the promise of endangered species protection has been derailed.

"In 1976," a source in the U.S. Department of Interior told *Outdoor Life,* "the United Kingdom succeeded in blanketing all of the family *Felidas* under the CITES treaty. The United States agreed to this without really paying much attention, even though it meant that all the cats on earth—with the exception of pet domestic cats—were now forbidden export without a license. We realized, belatedly, that our own lynx and bobcat were not interdicted, even though the bobcat certainly is not endangered in this country and never has been.

"Still we were stuck with it, and it meant that the U.S. Fish and Wildlife Service, which actually issues the export licenses, could not permit the export of a single bobcat pelt without the authorization of the ESSA committee."

The Fish and Wildlife Service made the point to ESSA that the bobcat was not threatened, that neither bobcat nor lynx was on the U.S. domestic list of endangered or threatened species, and that bobcat pelts were an important item of trade for American trappers.

But, goaded by the Defenders of Wildlife and similar groups of organized handwringers, ESSA insisted there was insufficient data on bobcats to reach any proper determination. ESSA demanded population studies and other information.

"This was really bureaucratic dabbling in state game management affairs," my source said. "The state agencies howled to high heaven that their badly overstrained resources were being diverted from animals that needed and merited management to the bobcat, which did not."

But ESSA was adamant. It ruled there would be no bobcat pelts exported until each state with bobcats to export submitted the studies it wanted. Ultimately ESSA agreed to permit exports, but only from states that had established bobcat management programs of which it approved.

"ESSA went completely out of its tree," one knowledgeable Fish and Wildlife Service source said. "It began making wholesale judgments on state game management programs, although it plainly had no legal nor legislative authority to do so. Ultimately unlimited exports of bobcat pelts were allowed, but not until ESSA had browbeaten the states into adopting bobcat management programs that simply have no justification."

ESSA—the committee is made up of representatives from departments of Interior, Agriculture, Commerce, the National Science Foundation, the Council on Environmental Quality and the Smithsonian Institution—"is actually run by its executive director. The committee rubber stamps his decisions," one insider told me. The executive director is an adroit bureaucrat named William Brown, who combines the unlikely talents of a doctorate in zoology and a law degree.

The Fish and Wildlife Service, which actually must issue import and export licenses for animals listed in the CITES treaty, insists that the law gave it no choice but to follow the "advice" given by ESSA.

Another example of how the international endangered species treaty has been invoked to ride roughshod over U.S. interests is the outrageous tale of the alligator. There is, to begin with, considerable expert opinion that disputes that the alligator was ever endangered. It was first listed by the Department of Interior under the old law in 1966 "when the species was declining rapidly because of habitat loss and commercial exploitation," according to a statement by the Fish and Wildlife Service. But Alan Ensminger, chief of the fur and refuge division of the Louisiana Department of Wildlife and Fisheries, disagrees.

"The alligator was never endangered, in my judgment," he said in an interview. "We probably had no fewer than 50,000 in this state at the lowest point, and we have 500,000 now. Last year we were permitted to have a one-month season in twelve of our parishes [counties] and we took 16,000 alligators. We could probably sustain a yield of 50,000 to 100,000 if we were permitted to manage the animal properly."

The alligator is a prime example of what many thoughtful game managers feel is one of the worst aspects of the endangered species scandals—by robbing state agencies of the right to manage a species, it also robs them of their incentive to protect it and to preserve its habitat. Habitat destruction universally is regarded as the biggest threat to the survival of wildlife, but without viable hunting or pelt interests there is little money, and less will, at the state level to work for healthy conditions for wildlife.

John S. Gottschalk, legislative counsel for the association and a former director of the Fish and Wildlife Service, agreed, noting that "the alligator is the classic example of overapplication of the Endangered Species Act. The American alligator is not now and never has been endangered. When state agencies are deprived of the right to manage their resources, very often state legislatures say, OK, if we can't manage, we won't appropriate any money. The species suffers when that happens.

"The grizzly is another case in point. State authorities can only control marauding grizzlies with the sufferance of the federal government, because the grizzly is on the threatened list and has become a federal ward."

Gottschalk makes the point that "the fundamental problem with the Endangered Species Act is the difficulty in deciding what constitutes an endangered species. Seldom is the actual taking of animals what threatens extinction of a species. It is habitat destruction. Given that, at what point does an animal become endangered? The ultraconservatives say everything is threatened, of course, but it is a very difficult question as to what does constitute an endangered species. We have never had a scientific answer."

One Capitol Hill source intimately associated with attempts to reform the endangered species bureaucracy observed that the alligator hunters in Louisiana had taken it on the chin at least partly because "ESSA was trying to blackjack France into joining the CITES treaty."

It is a persuasive explanation for ESSA's otherwise irrational attitude toward alligators. Almost all alligator hides are exported to France, where they are crafted into ultrachic (and wildly expensive) purses and shoes. "ESSA was using its policy on alligators to punish France for not becoming signatory to the CITES treaty," my Capitol source said. This is a difficult charge to document, but it may be more than coincidence that once France agreed to join CITES, in August 1978, ESSA quickly agreed to license exports of American alligator hides.

The leopard is another notably sore point with American hunters. In 1972 it was added to the CITES endangered species list, although there is ample expert testimony that the animal exists in abundantly huntable numbers in much of sub-Sahara Africa, and that, indeed by listing it as endangered the United States may have struck a mortal blow at its proper management.

Three years ago, biologists James G. Teer and Wendell G. Swank—both authorities in the lore of the African leopard—wrote a report for the Fish and Wildlife Service office of endangered species. In it, with rare forthrightness for a document of this sort, they wrote: "The placement of the leopard on the lists of endangered species . . . is not defensible. The leopard is not endangered under commonly used definitions that endangered status implies, and it has been improperly gazetted to that status."

Swank is back in Africa, but I dug up Teer, and he amplified this judgment, saying, "We simply couldn't find anyone in a responsible position in Africa who would say the leopard was endangered."

By placing the leopard on the endangered list, the United States ensured only that American hunters would quit hunting it, for they could no longer bring home hides or other trophies. But listing the leopard did nothing to curb the poaching and loss of habitat that are the real enemies of the African leopard. Far from it, the action deprived hard-pressed African game departments of revenue they desperately need to manage the species properly.

Smarting badly from the leopard fiasco, the Fish and Wildlife Service says it is rewriting its regulations to permit importation of leopard trophies taken legally. True to the code of the bureaucrats, however, it says it will be "several months" before the new regulations go into effect.

In seeking a rationale for some of the seemingly witless acts of the Fish and Wildlife Service and the ESSA, one need go no further than the obdurate, implacable forces of the anti-hunters. For them, the Endangered Species Act was a godsend, for it enabled them to win by pressure on timid, submissive bureaucrats what they were unable to win on the merits. Unable to stop the hunting of bobcats because the creatures were in such ample supply (up to 1 million at last count), the anti-hunters persuaded the United States to acquiesce in a totally unwarranted international ban on trade in the animal's pelt. Frustrated because African countries still permit the hunting of big cats that were in ample supply, the anti-hunters encouraged a bureaucratic end-run that virtually ended hunting of leopards by Americans, at the same time hamstringing African game agencies in their attempts to put down poaching and loss of habitat.

"The equation is very simple," one Interior Department official who has been scarred by the endangered species wars said wearily. "Remove the economic incentives by prohibiting all hunting and/or trade in hides or pelts, and you have seriously damaged the species—first by depriving the state or foreign game agency of the money it needs to carry out proper management, second by removing any real incentive to nurture and protect the species. If sensible trade is permitted in Louisiana alligators, for example, say 50,000 to 100,000 hides a year at $14 or $15 a foot, you have tapped an abundant natural resource to create a profitable industry. You have also created all the incentive in the world for the state to manage that resource properly."

The long and sleazy record of how the Endangered Species Act was miscarried would not be complete without explaining, or at least trying to explain, what has happened to congressional attempts to bring ESSA to heel. The mills of Washington grind slowly, and sometimes not at all, but the endangered species scandals became so redolent of chicanery and institutionalized stupidity that eventually even Congress became aware of it. As it frequently does when confronted with problems of tiresome complexity, Congress called in the General Accounting Office to investigate the workings of the Endangered Species Act. The result of the investigation was a 112-page report which spelled out some of the problems and made a host of recommendations for corrective action.

The GAO report spawned a number of amendments to the law, the most important of which said ESSA is no longer an independent agency, but becomes a part of the U.S. Fish and Wildlife Service, and its opinions are to be advisory and not mandatory. The Breaux amendment has passed overwhelmingly.

Well, hot dog, the ESSA problem was solved, right?

Wrong. Brace yourself for a short, sharp lesson in how Washington really works. True, ESSA was placed under the authority of the Fish and Wildlife Service and, true, its opinions were to be advisory only. But, lo and behold, there was created yet another bureaucratic organism that was superimposed on ESSA and the Fish and Wildlife Service. It was called the International Convention Advisory Commission (ICAC) and, by one of those wonderful coincidences that reaffirms one's faith in the machinery of government, it was made up of members from the departments of Interior, Agriculture, Commerce and the National Science Foundation, the Smithsonian Institution and the Council on Environmental Quality. Sounds a lot like the old ESSA, doesn't it? It should, for the man temporarily detailed to organize ICAC (and who will likely be appointed to run it) is the former honcho of ESSA, the zoologist-lawyer.

Ah, but the plot thickens to an even syrupier viscosity. No sooner had President Carter signed into law the Breaux amendment than a memorandum arrived in the Fish and Wildlife Service from Stuart Eizenstat, who is President Carter's chief advisor on domestic policy. I talked to an endangered species bureaucrat who had seen the Eizenstat memo, and he was still shaken. Essentially what the memo said was that the ICAC was to make U.S. policy so far as what animals and plants went on the treaty list. What it says, goes.

So, after much travail, much effort on the part of Congress, the GAO and genuinely concerned conservationists, attempts to reform the endangered species mess have apparently come to naught. Like ESSA before it, ICAC is accountable to no one. One final footnote, from an understandably bitter official in the Fish and Wildlife Service: "I'd bet Stuart Eizenstat's ICAC memo was written by the ESSA committee, or more likely by its executive director."

It is, altogether, a disheartening tale, but it does illustrate very neatly where the real power lies in Washington.

POSTSCRIPT

Do We Need More Stringently Enforced Regulations to Protect Endangered Species?

Even the staunchest supporters of the Endangered Species Act (which Congress reauthorized and strengthened in October 1988) and of the CITES treaty would probably agree with Starnes that implementation of these regulations has been far from perfect. Interestingly, many wildlife experts consider the protection of the American alligator to be one of the success stories of endangered species legislative efforts rather than one of its worst scandals, as proposed by Starnes. Clearly Regenstein believes that placing imperiled species on endangered and threatened lists is a key legal step in the effort to protect them. Could these conflicting conclusions be related to a difference in values between the editor of a magazine that caters to hunters and an executive of an organization dedicated to protecting animals?

Starnes's suggestion that removing the restrictions on game hunting would result in better protection of the hunted species, because wildlife management agencies would thereby receive more funding, is not consistent with the history of such activities. It was the failure of these agencies to manage the problem in the first place that resulted in the passage of the legislation and the negotiation of the treaties. Perhaps a more appropriate response would be to chastise those states and nations that have responded to the federal and international endangered species initiatives by withdrawing the funds needed for habitat management. Regenstein agrees with Starnes that habitat destruction is the major cause of wildlife endangerment. Unfortunately, he does not develop this issue, concentrating primarily on the question of overexploitation.

That the National Wildlife Federation filed suit in federal court in August 1990 to ensure enforcement of measures to protect five species of endangered sea turtles is perhaps an indication that lax enforcement of the Endangered Species Act by the Department of Commerce will continue under the Bush administration.

Not all staff members of the U.S. Fish and Wildlife Service are as negative about the Endangered Species Act as are the sources quoted by Starnes. For a very positive discussion of the effects of the legislation, see "Helping to Save Endangered Species," by Gerry Kelly, in the November 1982 issue of *American Forests.*

Attacks on people by grizzly bears is motivating an effort to have them removed from the list of protected threatened species, as explained in Jim Robbins's article "Grizzly and Man—When Species Collide," *National Wildlife* (February/March 1988).

A highly critical analysis of the ecological concern for endangered species is contained in Chapter 7 of *Progress and Privilege* (Anchor Press, 1982), by William Tucker, and in an article by Julian Simon in the May 15, 1986, issue of *New Scientist*.

For a thorough discussion of the relationships among extinction, genetic diversity, and ecological vitality, see *The Sinking Ark*, by Norman Myers (Pergamon Press, 1979), and *Extinction*, by Anne and Paul Ehrlich (Random House, 1981).

A recent, very grim, prognosis on species extinction is given by Jared Diamond in "Playing Dice With Megadeath," *Discover* (April 1990).

ISSUE 4

Does Risk-Benefit Analysis Provide an Objective Method for Making Environmental Decisions?

YES: William D. Ruckelshaus, from "Science, Risk, and Public Policy," *Science* (September 9, 1983)

NO: Langdon Winner, from "Risk: Another Name for Danger," *Science for the People* (May/June 1986)

ISSUE SUMMARY

YES: Former Environmental Protection Agency (EPA) administrator William D. Ruckelshaus advocates educating the public about risk estimates and separating the scientific process of risk assessment from the management of risks through regulation.
NO: Social scientist Langdon Winner asserts that dealing with environmental and health hazards in terms of risk assessment leads to delays and confusion in efforts to regulate pollution and protect the public.

It can be argued that virtually any human activity impacts the ecosystem. However, certain by-products of our technological society are broadly viewed as unacceptable as evidenced by the popularity of efforts to prevent environmental degradation. The first task facing those charged with environmental protection is to determine what constitutes an unacceptable ecological impact.

The field of risk assessment and analysis has grown rapidly in response to legislators' and regulatory bureaucracy's desire for some form of scientific guidance. A variety of methodologies have been proposed. One technique, borrowed from economic analysis, is to attempt to balance the costs and benefits associated with a proposed action—be it the promulgation of a new air pollution regulation or the building of a new dam. Another technique is to compare the relative risks of alternative regulatory actions or alternative technologies.

Proponents of the various methods of risk assessment propose that it is the only scientific, objective means of making environmental decisions. Most environmentalists reject the claim that the methods developed to date are either scientifically precise or objective.

The issue is surrounded by many difficult problems. Given the complex web of interacting cycles in the global ecosystem, is it possible to calculate the likely environmental impacts of any human activity? Once calculated, how can risks, costs, and benefits be weighed one against the other? Should the evaluation of risks and benefits be done by some elite group of scientists or by a public body? Will the risks and benefits be experienced equally by different special-interest groups?

A particularly thorny issue has been the regulation of suspected or known carcinogens. For most other types of hazard, there is some "threshold" exposure below which no effect will be observed. For many, if not all, cancer-causing substances, any level of exposure can be hazardous. This fact has resulted in such controversial legislation as the Delaney Amendment to the Food and Drug Act, which prohibits using any amount of a food additive which has been found to cause cancer in humans or animals. Opponents of this law argue that society is full of risks, and it makes more sense to set some tolerable level of carcinogenic probability. Those who support the law reject the notion of comparing involuntary risks from something added to food during processing with voluntary risks such as those accepted by people who choose to smoke cigarettes. These proponents point out the difficulty of deciding whether an "acceptable" risk should be one or 100 extra cancer deaths per million exposed people.

During the 1980s, the Reagan administration attempted to impose mandatory cost-benefit analyses on environmental regulations. Most of the existing legislation does not require that the societal benefits of a proposed restriction outweigh its economic costs. Courts have gone both ways when asked to decide whether some form of risk- or cost-benefit analysis is an implied part of the regulatory process.

While recognizing the uncertainties inherent in risk estimation, William D. Ruckelshaus, who has twice headed the EPA, advocates increased reliance on this imprecise tool while making its use as uniform and democratic as possible across the federal regulatory agencies. Langdon Winner, a professor of science and technology studies at Rensselaer Polytechnical Institute, objects to the assumptions implicit in the formulation of problems of environmental pollution in terms of assessing acceptable and unacceptable risks. He argues that the process of attempting to make risk assessment more objective leads to waiting for better research results rather than to actions to control environmental degradation.

YES

William D. Ruckelshaus

SCIENCE, RISK, AND PUBLIC POLICY

We are now in a troubled and emotional period for pollution control; many communities are gripped by something approaching panic, and the public discussion is dominated by personalities rather than substance. It is not important to assign blame for this. I appreciate that people are worried about public health and about economic survival, and legitimately so, but we must all reject the emotionalism that surrounds the current discourse and rescue ourselves from the paralysis of honest public policy that it breeds.

I believe that part of the solution to our distress lies with the idea that disciplined minds can grapple with ignorance and sometimes win: the idea of science. We will not recover our equilibrium without a concerted effort to more effectively engage the scientific community. Frankly, we are not going to be able to emerge from our current troubles without a much improved level of public confidence. The polls show that scientists have more credibility than lawyers or businessmen or politicians, and I am all three of those. I need the help of scientists.

This is not a naive plea for science to save us from ourselves. Somehow, our democratic technological society must resolve the dissonance between science and the creation of public policy. Nowhere is this more troublesome than in the formal assessment of risk—the estimation of the association between exposure to a substance and the incidence of some disease, based on scientific data.

SCIENCE AND THE LAW AT EPA

Here is how the problem emerges at the Environmental Protection Agency. EPA is an instrument of public policy, whose mission is to protect the public health and the environment in the manner laid down by its statutes. That manner is to set standards and enforce them, and our enforcement powers are strong and pervasive. But the standards we set, whether technology- or

From William D. Ruckelshaus, "Science, Risk, and Public Policy," *Science*, vol. 221 (September 9, 1983), pp. 54-59.

health-related, must have a sound scientific base.

Science and the law are thus partners at EPA, but uneasy partners. The main reason for the uneasiness lies, I think, in the conflict between the way science really works and the public's thirst for certitude that is written into EPA's laws. Science thrives on uncertainty. The best young scientists flock into fields where great questions have been asked but nothing is known. The greatest triumph of a scientist is the crucial experiment that shatters the certainties of the past and opens up rich new pastures of ignorance.

But EPA's laws often assume, indeed demand, a certainty of protection greater than science can provide with the current state of knowledge. The laws do no more than reflect what the public believes and what it often hears from people with scientific credentials on the 6 o'clock news. The public thinks we know what all the bad pollutants are, precisely what adverse health or environmental effects they cause, how to measure them exactly and control them absolutely. Of course, the public and sometimes the law are wrong, but not all wrong. We do know a great deal about some pollutants, and we have controlled them effectively by using the tools of the Clean Air Act and the Clean Water Act. These are the pollutants for which the scientific community can set safe levels and margins of safety for sensitive populations. If this were the case for all pollutants, we could breathe more easily (in both senses of the phrase); but it is not so.

More than 10 years ago, EPA had the Clean Air Act, the Clean Water Act, a solid waste law, a pesticide law, and laws to control radiation and noise. Yet to come were the myriad of laws to control toxic substances from their manufacture to their disposal—but that they would be passed was obvious even then.

When I departed EPA a decade ago, the struggle over whether the federal government was to have a major role in protecting our health, safety, and environment was ended. The American people had spoken. The laws had been passed; the regulations were being written. The only remaining question was whether the statutory framework we had created made sense or whether, over time, we would adjust it.

SCIENTIFIC REALITIES

Ten years ago I thought I knew the answer to that question as well. I believed it would become apparent to all that we could virtually eliminate the risks we call pollution if we wanted to spend enough money. When it also became apparent that enough money for all the pollutants was a lot of money, I came to believe that we would begin examining the risks very carefully and structure a system that would force us to balance our desire to eliminate pollution against the costs of its control. This would entail some adjustment of the laws, but not all that much, and it would happen by about 1976. I was wrong.

This time around as administrator of EPA, I am determined to improve our country's ability to cope with the risk of pollutants over where I left it 10 years ago. It will not be easy, because we must now deal with a class of pollutants for which it is difficult, if not impossible, to establish a safe level. These pollutants interfere with genetic processes and are associated with the diseases we fear most: cancer and reproductive disorders, including birth defects. The scientific consensus is that any exposure, however small, to a genetically active substance embodies some risk of an effect. Since these substances are widespread in the environment, and since we can detect them down to very low levels, we must assume that life now takes place in a minefield of risks from hundreds, perhaps thousands, of substances. We can no longer tell the public that they have an adequate margin of safety.

This worries all of us, and it should. But when we examine the premises on which such estimates of risk are based, we find a confusing picture. In assessing a suspected carcinogen, for example, there are uncertainties at every point where an assumption must be made: in calculating exposure; in extrapolating from high doses where we have seen an effect to the low doses typical of environmental pollution; in what we may expect when humans are subjected to much lower doses of a substance that, when given in high doses, caused tumors in laboratory animals; and finally, in the very mechanisms by which we suppose the disease to work.

One thing we clearly need to do is ensure that our laws reflect these scientific realities. The administrator of EPA should not be forced to represent that a margin of safety exists for a specific substance at a specific level of exposure where none can be scientifically established. This is particularly true where the inability to so represent forces the cessation of all use of a substance without any further evaluation.

FUNCTIONS OF REGULATORY AGENCIES

It is my strong belief that where EPA, OSHA (the Occupational Safety and Health Administration), or any other social regulatory agency is charged with protecting public health, safety, or the environment, we should be given, to the extent possible, a common statutory formula for accomplishing our tasks. This statutory formula may well weigh public health very heavily, as the American people certainly do.

The formula should be as precise as possible and should include a responsibility for assessing the risk and weighing it, not only against the benefits of continued use of the substance under examination, but against the risks associated with substitute substances and the risks associated with the transfer of the substance from one environmental medium to another through pollution control practices. I recognize that legislative change in the current climate is difficult. It is up to those of us who seek change to make the case for its advisability.

But my purpose here is not to plead for statutory change; it is to speak of risk assessment and risk management and the role of science in both. It is important to distinguish these two essential functions, and I rely here on a recent National Academy of Sciences report on the management of risk in the federal government. Scientists assess a risk to find out what the problems are. The process of deciding what to do about the problems is risk management. The second procedure involves a much broader array of disciplines and is aimed toward a decision about control.

In risk management it is assumed that we have assessed the health risks of a suspect chemical. We must then factor in its benefits, the costs of the various methods available for its control, and the statutory framework for decision. The NAS report recommends that these two functions—risk assessment and risk management—be separated as much as possible within a regulatory agency. This is what we now do at EPA and it makes sense.

RISK ASSESSMENT

We also need to strengthen our risk assessment capabilities. We need more research on the health effects of the substances we regulate. I intend to do everything in my power to make clear the importance of this scientific analysis at EPA. Given the necessity of acting in the face of enormous scientific uncertainties, it is more important than ever that our scientific analysis be rigorous and the quality of our data be high. We must take great pains not to mislead people about the risks to their health. We can help to avoid confusion by ensuring both the quality of our science and the clarity of our language in explaining hazards.

I intend to allocate some of EPA's increased resources to pursuing these ends. Our 1984 request contains significant increases for risk assessment and associated work. We have requested $31 million in supplemental appropriations for research and development, and I expect that risk assessment will be more strongly supported as a result of this increase as well.

I would also like to revitalize our long-term research program to develop a base for more adequately protecting the public health from toxic pollutants. I will be asking the outside scientific community for advice on how best to focus those research efforts.

In the future, this being an imperfect world, the rigor and thoroughness of our risk analyses will undoubtedly be affected by many factors, including the toxicity of the substances examined, the populations exposed, the pressure of the regulatory timetable, and the resources available. Despite these often conflicting pressures, risk assessment at EPA must be based only on scientific evidence and scientific consensus. Nothing will erode public confidence faster than the suspicion that policy considerations have been allowed to influence the assessment of risk.

RISK MANAGEMENT

Although there is an objective way to assess risk, there is, of course, no purely objective way to manage it, nor can we ignore the subjective perception of risk in the ultimate management of a particular substance. To do so would be to place too much credence in our objective data and ignore the possibility that occasionally one's intuition is right. No amount of data is a substitute for judgment.

Further, we must search for ways to describe risk in terms that the average citizen can comprehend. Telling a family that lives close to a manufacturing facility that no further controls on the plant's emissions are needed because, according to our linear model, their risk is only 10^{-6}, is not very reassuring. We need to describe the suspect substances as clearly as possible, tell people what the known or suspected health problems are, and help them compare that risk to those with which they are more familiar.

To effectively manage the risk, we must seek new ways to involve the public in the decision-making process. Whether we believe in participatory democracy or not, it is a part of our social regulatory fabric. Rather than praise or lament it, we should seek more imaginative ways to involve the various segments of the public affected by the substance at issue. They need to become involved early, and they need to be informed if their participation is to be meaningful. We will be searching for ways to make our participatory process work better.

For this to happen, scientists must be willing to take a larger role in explaining the risks to the public—including the uncertainties inherent in any risk assessment. Shouldering this burden is the responsibility of all scientists, not just those with a particular policy end in mind. In fact, all scientists should make clear when they are speaking as scientists, ex cathedra, and when they are recommending policy they believe should flow from scientific information. What we need to hear more of from scientists is science. I am going to try to provide avenues at EPA for scientists to become more involved in the public dialog in which scientific problems are described.

Lest anyone misunderstand, I am not suggesting that all the elements of managing risk can be reduced to a neat mathematical formula. Going through a disciplined approach can help to organize our thoughts so that we include all the elements that should be weighed. We will build up a set of precedents that will be useful for later decision-making and will provide more predictable outcomes for any social regulatory programs we adopt.

In a society in which democratic principles dominate, the perceptions of the public must be weighed. Instead of objective and subjective risks, the experts sometimes refer to "real" and "imaginary" risks. There is a certain arrogance in this—an elitism that has ill served us in the past. Rather than decry the ignorance of the public and seek to ignore their concerns, our governmental processes must accommodate the will of the people and recognize its occasional wisdom. As Thomas Jefferson observed, "If we think [the people] not enlightened enough to exercise their control with a wholesome discretion, the remedy is not to take it from them, but to inform their discretion."

INTERAGENCY AND INTERNATIONAL COORDINATION

Up to this point I have been suggesting how risks should be assessed and managed in EPA. Much needs to be done to coordinate the various EPA programs to ensure a consistent approach. I have established a task force with that charter.

I further believe we should make uniform the way in which we manage risk across the federal regulatory agencies. The public interest is not served by two federal agencies taking diametrically opposed positions on the health risks of a toxic substance and then arguing about it in the press. We should be able to coordinate our risk assessment procedures across all federal agencies. The risk management strategies that flow from that assessment may indeed differ, depending on each agency's statutory mandate or the judgment of the ultimate decision-maker.

But even at the management stage there is no reason why the approaches cannot be coordinated to achieve the goal of risk avoidance or minimization with the least societal disruption possible. I have been exploring with the White House and the Office of Management and Budget the possibility of effecting better intragovernmental coordination of the way in which we assess and manage risk.

To push this one step further, I believe it is in our nation's best interest to share our knowledge of risks and our approach to managing them with the other developed nations of the world. The environmental movement has taught us the interdependence of the world's ecosystems. In coping with the legitimate concerns raised by environmentalists, we must not forget that we cope in a world with interdependent economies. If our approach to the management of risk is not sufficiently in harmony with those of the other developed nations, we could save our health and risk our economy. I do not believe we need to abandon either, but to ensure that it does not happen, we need to work hard to share scientific data and understand how to harmonize our management techniques with those of our sister nations.

In sum, my goal is a government-wide process for assessing and managing environmental risks. Achieving this will take cooperation and goodwill within EPA, among Executive Branch agencies, and between Congress and the Administration, a state of affairs that may partake of the miraculous. Still, it is worth trying, and the effort is worth the wholehearted support of the scientific community. I believe such an effort touches on the maintenance of our current society, in which a democratic polity is grounded in a high-technology industrial civilization. Without a much more successful way of handling the risks associated with the creations of science, I fear we will have set up for ourselves a grim and unnecessary choice between the fruits of advanced technology and the blessings of democracy.

NO

Langdon Winner

RISK: ANOTHER NAME FOR DANGER

The most prevalent way our society explores the possibility of limiting technology is through the study of "risk." Noting how the broader effects of industrial production can damage environmental quality and endanger public health and safety, risk assessment seeks to perfect methods of evaluation that are at once rigorous and morally sound. This approach appears to offer policymakers a way to act upon the best scientific information to protect society from harm. Indeed, if we define "risk" as everything that could conceivably go wrong with the use of science and technology—a definition that many are evidently prepared to accept—then it seems possible that we might arrive at a general understanding of norms to guide the moral aspects of scientific and technical practice.

But the promise of risk assessment is difficult to realize. The arena in which discussions of risk take place is highly politicized and contentious. Specific questions such as those dealing with the safety of nuclear power, as well as more general ones having to do with choosing proper methodologies for studying risks at all, involve high stakes.

Powerful social and economic interests are invested in attempts to answer the question, How safe is safe enough? Expert witnesses are often best identified not by what they know, but rather by whom they represent. Indeed, the very introduction of "risk" as a common way of defining policy issues is itself far from a neutral issue.

At a time in which modern societies are beginning to respond to a wide range of complaints about possible damage various industrial practices have on the environment and public health, the introduction of self-conscious risk assessment adds a distinctly conservative influence. By the term "conservative" here I mean simply a point of view that tends to favor the status quo. Although many of those who have become involved in risk assessment are not conservative in a political sense, it seems to me that the ultimate consequence of this new approach will be to delay, complicate, and befuddle issues in a way that will sustain an industrial status quo relatively free of socially enforced limits. It is the character of this conservatism that I want to explore here.

RISK AND FORTITUDE

If we declare ourselves to be identifying, studying, and remedying hazards, our orientation to the problem is clear. Two assumptions, in particular, appear beyond serious question. First, we can assume that given adequate evidence, the hazards to health and safety are fairly easily demonstrated.

Second, when hazards of this kind are revealed, all reasonable people usually can readily agree on what to do about them. Thus, if we notice that a deep, open pit stands along a path where children walk to school, it seems wise to insist that the responsible party, be it a private person or public agency, either fill the pit or put a fence around it. Similarly, if we have good reason to believe that an industrial polluter is endangering our health or harming the quality of the land, air, or water around us, it seems reasonable to insist that the pollution cease or be strongly curtailed.

Straightforward notions of this kind, it seems to me, lie at the base of a good many social movements concerned with environmental issues, consumer protection, and the control of modern technology. In their own way, of course, such movements are capable of adding elements of complication to policy discussion, for example, notions of complexity from ecological theory. Typically, however, these complications are ones that ultimately reinforce a basic viewpoint that sees "dangers" to human health, other species, and the environment as grave matters that are fairly easy to understand and require urgent remedies.

If, on the other hand, we declare that we are interested in assessing risks, complications of a different sort immediately enter in. Our task now becomes that of studying, weighing, comparing and judging circumstances about which no simple consensus is available.

Both of the common sense assumptions upon which the concern for "hazards" and "dangers" rely are abruptly suspended. Confidence in how much we know and what ought to be done about it vanishes in favor of an excruciatingly detailed inquiry with dozens (if not hundreds) of fascinating dimensions. A new set of challenges presents itself to the scientific and philosophical intellect. Action tends to be postponed indefinitely.

As one shifts the conception of an issue from that of hazard/danger/threat to that of "risk," a number of changes tend to occur in the way one treats that issue. What otherwise might be seen as a fairly obvious link between cause and effect, for example, air pollution and cancer, now becomes something fraught with uncertainty.

What is the relative size of that "risk," the "chance of harm"? And what is the magnitude of the harm when it does take place? What methods are suited to measuring and analyzing these matters in a suitable and rigorous way? Because these are questions that involve scientific knowledge and its present limits, the risk assessor is constrained to acknowledge what are often highly uncertain findings of the best available research.

For example, one must say in all honesty, "We don't know the relationship between this chemical and the harm it may possibly cause." Thus, the norms that regulate the acceptance or rejection of the findings of scientific research become, in effect, moral norms governing judgments about harm and responsibility. A very high premium is placed on not being wrong. Evidence that "the experts disagree" adds further perplexity and a need to be careful before drawing conclusions.

The need to distinguish "facts" from "values" takes on paramount importance. Faced with uncertainty about what is known concerning a particular risk, prudence becomes not a matter of acting effectively to remedy a suspected source of injury, but of waiting for better research findings.[1]

An illustration of this cast of mind can be seen in a study for the Environmental Protection Agency to determine whether or not there were indications that residents of the Love Canal area of New York, an abandoned chemical waste disposal site, showed chromosome damage. The report written by Dante Picciano, a geneticist employed by the Biogenics Corporation of Houston, Texas, drew the following conclusions: "It appears that the chemical exposures at Love Canal may be responsible for much of the apparent increase in the observed cytogenetic aberrations and that the residents are at an increased risk of neoplastic disease, of having spontaneous abortions and of having children with birth defects. However, in absence of a contemporary control population, prudence must be exerted in the interpretation of such results."[2]

Although the chemicals themselves may have been disposed of in reckless fashion, scientific studies on the consequences must be done with scrupulous care. Insofar as law and public policy heed the existing state of scientific knowledge about particular risks, the same variety of caution appears in those domains as well.

Frequently augmenting these uncertainties about cause and effect are the risk assessor's calculations on costs and benefits. To seek practical remedies for man-made risks to health, safety, or environmental quality typically requires an expenditure of public or private money. How much is it reasonable to spend in order to reduce a particular risk? Is the cost warranted as compared to the benefit received?

Even if one is able to set aside troubling issues about equity and "who pays," risk/cost/benefit calculations offer, by their very nature, additional reasons for being hesitant about proposing practical remedies at all. Because it's going to cost us, we must ponder the matter as a budget item. Our budgets, of course, include a wide range of expenditures for things we need, desire, or simply cannot avoid. Informed about how the cost of reducing environmental risks is likely to affect consumer prices, taxes, industrial productivity, and the like, the desire to act decisively with respect to any particular risk has to be weighed against other economic priorities.[3]

A willingness to balance relative costs and benefits is inherent in the very adoption of the concept of "risk" to describe one's situation. In ordinary use the word implies "chance of harm" from the standpoint of one who has weighed that harm against possible gain. What does one do with a risk? Sometimes one decides to *take it*. What, by comparison, does one do with a hazard? Usually one seeks to avoid it or eliminate it.

The use of the concept of "risk" in business dealings, sports, and gambling reveals how closely it is linked to the sense of voluntary undertakings. An investor risks his capital in the hope of making a financial gain. A football team in a close game takes a risk when it decides to run on fourth down and a yard to go. A gambler at a Las Vegas blackjack table risks his or her money on the chance of a big payoff.

In contrast to the concepts of "danger," "hazard," or "peril," the notion of "risk" tends to imply that the chance of harm in question is accepted willingly in the expectation of gain. This connotation makes the distinction between voluntary and involuntary risks outlined in some of the recent literature largely misleading. The word carries a certain baggage, a set of ready associations. The most important of these is the simple recognition that all of us take risks of one kind or another frequently.

Noticing that everyday life is filled with risky situations of various kinds, contemporary risk assessment has focused on a set of psychological complications that further compound the difficulties offered by scientific uncertainty and the calculations of risk/cost/benefit analysis. Do people accurately assess the risk they actually face? How well are they able to compare and evaluate such risks? And why do they decide to focus upon some risks rather than others?

A good deal of interesting and valid psychological research has been devoted to answering such questions. By and large, these studies tend to show that people have a fairly fuzzy comprehension of the relative chance of harm involved in their everyday activities.[4] If one adds to such findings the statistical comparisons of injuries and fatalities suffered in different situations in modern life, then the question of why people become worried about certain kinds of risks and not others becomes genuinely puzzling.[5]

The rhetorical possibilities of this puzzle are often seized upon by writers who assert that people's confusion about risks discredits the claims of those who focus upon the chance of harm from some particular source. Why should a person who drives an automobile, a notorious cause of injury and death, be worried about nuclear power or the level of air pollution? Invidious comparisons of this kind are sometimes employed to show that people's fears about technological hazards are completely irrational.

Hence, one leading proponent of this view argues, "it is not surprising that people with psychological and social problems are unsettled by technological advance. The fears range from the dread of elevators in tall buildings to apprehension about 'radiation' from smoke detectors. Invariably, these fears are evidence of displacing inner anxiety that psychiatrists label as phobic." The same writer explains that normal folk are able to overcome such phobias by reminding themselves of the incalculable good that modern technologies have brought to all of us. "People of sound mind accept the negligible risk and minor inconvenience that often go hand in hand with wondrous material benefits."[6]

Once one has concluded that reports about technological risks are phobia-based, the interesting task becomes that of explaining why people have such fears at all. Tackling this intellectual challenge, anthropologist Mary Douglas and political scientist Aaron Wildavsky have developed a style of analysis based on the assumption that complaints about risk are not to be taken at face value. In their view all reports about environmental risks must be carefully interpreted to reveal the underlying social norms and institutional attachments of those making the complaints.

Different kinds of institutions respond to risk in very different ways. For example, entrepreneurs accept many kinds of economic risk without question. They embrace the invigorating uncertainties of the market, the institutional context that gives their activities meaning. In contrast, public-interest organizations of the environmental movement, organizations that Douglas and Wildavsky describe as "sects," show, in their view, obsessive anxiety about technological risks; the discovery of these risks provides a source of personal commitment and social solidarity the "sects" so desperately need.

Are there any environmental dangers in the world that all reasonable people, regardless of institutional attachment, ought to take seriously? Douglas and Wildavsky find that question impossible to answer. The fact that "the scientists disagree" requires us to be ever skeptical about any claims about particular risks. Instead, Douglas and Wildavsky offer the consolations of social scientific methodology to help us explain (and feel superior to) the strange behavior of our benighted contemporaries.[7]

Entering thickets of scientific uncertainty, wending our way through labyrinths of risk / cost / benefit analysis, balancing skillfully along the fact/value gap, stopping to gaze upon the colorful befuddlement of mass psychology, we finally arrive at an unhappy destination—the realm of invidious comparison and social scorn. This drift in some scholarly writings on risk assessment finds its complement in the public statements and advocacy advertising of corporations in the oil, chemical, and electric power industries.

In the late 1970s the debunking of claims of environmental hazards became a major part of corporate ideology. Closely connected to demands for deregulation and the relaxing of governmental measures to control air pollution, occupational safety and health, and the like, the "risk" theme in the pronouncements of industrial firms assumed major importance.

A typical advertisement from Mobil Oil's "Observations" series illustrates the way in which popularized risk psychology and risk/cost/benefit analysis can work in harmony. "Risky business," the ad announces. "Lawn mowers . . . vacuum cleaners . . . bathtubs . . . stairs . . . all part of everyday life and all hazardous to your health. The Consumer Product Safety Commission says these household necessities caused almost a million accidents last year, yet most people accept the potential risks because of the proven benefits. . . . Risk, in other words, is part of life. Fool's goal. Nothing's safe all the time, yet there are still calls for a "risk-free society."

Although I have read large portions of the recent literature on energy, environment, consumer protection, and the like, I cannot recall having seen even one instance of a demand for a "risk-free" society. The notion appears only as a straw man in advocacy ads like this one. Its text goes on to evoke a string of psychological associations linked to the experience of "risk" in economic enterprise. "Cold feet. What America does need are more companies willing to take business risks, especially on energy, where the risks are high. . . . We're gamblers. . . . Taking risks: it's the best way to keep America rolling . . . and growing."[8] Poker anyone?

There is, then a deep-seated tendency in our culture to appreciate risk-taking in economic activity as a badge of courage. Putting one's money, skill, and reputation on the line in a new venture identifies that person as someone of high moral character. On the other hand, people who have qualms about the occasional side effects of economic wheeling and dealing can easily be portrayed as cowardly and weak-spirited, namby-pambies just not up to the rigors of the marketplace.

Public policies that recognize such qualms can be dismissed as signs that the society lacks fortitude or that the citizenry has grown decadent. Thus, in addition to other difficulties that await those who try to introduce "risk" as a topic for serious political discussion, there is a strong willingness in our culture to embrace risk-taking as one of the warrior virtues. Those who do not possess this virtue should, it would seem, please not stand in the way of those who do.

AVOIDING "RISK"

By calling attention to these features in contemporary discussions about risk, I do not want to suggest, as some have done, that the whole field of study has somehow been corrupted by the influence of selfish economic interests.[9] Neither am I arguing that all or most conversations on this topic show a deliberate, regressive political intent. Indeed, many participants have entered the debate with the most noble of scientific, philosophical, and social goals.

Much of the analytically solid writing now produced on this topic seeks to strengthen intellectual armaments used to defend those parts of society and the environment most likely to experience harm from a variety of technological side effects.[10] And certainly there are many fascinating issues under the rubric of risk assessment that are well worth pursuing. I can only join in wishing that such clearheaded, magnanimous work flourish.

But from the point of view I've described here, the risk debate is one that certain kinds of social interests can expect to lose by the very act of entering. In our times, under most circumstances in which the matter is likely to come up, deliberations about risk are bound to have a strongly conservative drift. The conservatism to which I refer is one that upholds the status quo of production and consumption in our industrial, market-oriented society, a status quo supported by a long history of economic development in which countless new technological applications were introduced with scant regard to the possibility that they might cause harm.

Thus, decades of haphazard use of industrial chemicals provide a background of expectations for today's deliberations on the safety of such chemicals. Pollution of the air, land, and water are not the exception in much of twentieth-century America, but rather the norm. Because industrial practices acceptable in the past have become yardsticks for thinking about what will be acceptable now and in the future, attempts to achieve a cleaner, healthier environment face an uphill battle. The burden of proof rests upon those who seek to change long-existing patterns.

In this context, to define the subject of one's concerns as a "risk" rather than select some other issue skews the subsequent discussion in a particular direction. This choice makes it relatively easy to defend practices associated with high levels of industrial production; at the same time it makes it much more difficult for those who would like to place moral or political limits upon that production to make much headway.

I am not saying that this is a consequence of the way risk assessment is "used," although conservative uses of this sort of analysis are, as we have seen, easily enough concocted. What is more important to recognize is that in a society like ours, discussions centering on risk have an inherent tendency to shape the texture of such inquiries and their outcome as well. The root of this tendency lies, very simply, in the way the concept of "risk" is employed in everyday language. As I have noted, employing this word to talk about any situation declares our willingness to compare expected gain with possible harm.

We generally do not define a practice as a risk unless there is an anticipated advantage somehow associated with that practice. In contrast, this disposition to weigh and compare is not invoked by concepts that might be employed as alternatives to "risk"—"danger," "peril," "hazard," and "threat." Such terms do not presuppose that the source of possible injury is also a source of benefits.

From the outset, then, those who might wish to propose limits upon any particular industrial or technological application are placed at a disadvantage by selecting "risk" as the focus of their concerns. As they adopt risk assessment as a legitimate activity, they tacitly accept assumptions they might otherwise wish to deny (or at least puzzle over): that the object or practice that worries them must be judged in light of some good it brings and that they themselves are recipients of at least some portion of this good.

Once the basic stance and disposition associated with "risk" have defined the field of discourse, all the complications and invidious comparisons I have described begin to enter in. Standards of scientific certainty are applied to the available data to show how little we know about the relationship of cause and effect as regards particular industrial practices and their broader consequences.

Methods of risk/cost/benefit analysis fill out a detailed economic balance sheet useful in deciding how much risk is "acceptable." Statistical analyses show the comparative probability of various kinds of unfortunate events, for example, being injured in a skiing accident as compared to being injured by a nuclear power plant meltdown. Psychological studies reveal peculiarities in the ways people estimate and compare various kinds of risks. Models from social science instruct us about the relationship of institutional structures to particular objects of fear. A vast, intricately specialized division of intellectual labor spreads itself before us.

One path through this mass of issues is to take each one separately, seeking to determine which standards, methods, findings, and models are appropriate to making sound judgments about problems that involve public health, safety, and environmental quality. For example, one might question how reasonable it is to apply the very strict standards of certainty used in scientific research to questions that have a strong social or moral component. Must our judgments on possible harms and the origins of those harms have only a five percent chance of being wrong? Doesn't the use of that significance level mean that possibly dangerous practices are "innocent until proven guilty?"[11]

Similarly, one might reevaluate the role that cost/benefit analysis plays in the assessment of risks, pointing to the strengths and shortcomings of that method. How well are we able to measure the mix of "costs" and "benefits" involved in a given choice? What shall we do when faced with the inadequacy of our measurements? Are criteria of efficiency derived from economic theories sufficient to guide value choices in public policy? In controversies about the status of the intellectual tools used in decision making, such questions are hotly disputed.[12]

But for those who see issues of public health, safety, and environmental quality as fairly straightforward matters requiring urgent action, these exercises in methodological refinement are of dubious value. It is sensible to ask, Why get stuck in such perplexities at all? Should we spend our time working to improve techniques of risk analysis and risk assessment? Or should we spend the same time working more directly to find better ways to secure a beautiful, healthy, well-provided world and to eliminate the spread of harmful residues of industrial life?

The experience of environmentalists and consumer advocates who enter the risk debate will resemble that of a greenhorn who visits Las Vegas and is enticed into a poker game in which the cards are stacked against him. Such players will be asked to wager things very precious to them with little prospect that the gamble will deliver favorable returns. To learn that the stacked deck comes as happenstance rather than by conscious design provides little solace; neither will it be especially comforting to discover that hard work and ingenuity might improve the odds somewhat. For some, it is simply not the right game to enter.

There are some players at the table, however, who stand a much better chance. Proponents of relaxed governmental regulations on nuclear power, industrial pollution, occupational safety and health, environmental protection, and the like will find risk assessment, insofar as they are able to interest others in it, a very fruitful contest. Hence, Chauncey Starr, engineer and advocate of nuclear power, is well advised to take "risk" as the central theme in his repertoire of argument. But the likes of David Brower, Ralph Nader, and other advocates of consumer and environmental interests would do well to think twice before allowing the concept to play an important role in their positions on public issues.

Fortunately, many issues talked about as risks can be legitimately described in other ways. Confronted with any cases of past, present, or obvious future harm, it is possible to discuss that harm directly without pretending that you are playing craps. A toxic waste disposal site placed in your neighborhood need not be defined as a risk; it might appropriately be defined as a problem of toxic waste.

Air polluted by automobiles and industrial smokestacks need not be defined as a "risk"; it might still be called by the old-fashioned name, "pollution." New Englanders who find acid rain falling on them are under no obligation to begin analyzing the "risks of acid rain"; they might retain some Yankee stubbornness and confound the experts by talking about "that destructive acid rain" and what's to be done about it. A treasured natural environment endangered by industrial activity need not be regarded as something at "risk"; one might regard it more positively as an entity that ought to be preserved in its own right.

About all these matters there are rich, detailed forms of discourse that can strengthen our judgments and provide structure for public decisions. A range of theoretical perspectives on environmental protection, public health, and social justice can be drawn upon to clarify the choices that matter. My suggestion is that before "risk" is selected as a focus in any area of policy discussion, other available ways of defining the question be thoroughly investigated.

For example, are health and safety hazards that blue-collar workers encounter on the job properly seen as a matter of "risk" to be analyzed independently of directly related economic and social conditions? Or is it more accurate to consider ways in which these hazards reflect a more general set of social relationships and inequalities characteristic of the free enterprise system?

One's initial definition of the problem helps shape subsequent inquiries into its features. If one identified the issue of worker health and safety as a question of social justice, there would be less need to do all of the weighing of probabilities, comparing of individual psychological responses, and performing of other delicate tasks that risk assessment involves.

It might still be interesting to do research on levels of air pollution in executive offices as compared to those in factories. Of course, one always wants to have the best scientific information on such issues. But, in all likelihood, such studies would reveal little new or surprising. It is common knowledge that our society distributes wealth, income, knowledge, and social opportunities unequally. To establish that it also distributes workplace hazards inequitably merely amplifies the problem. Those concerned with questions of social justice would do well to stick to those questions and not look to risk analysis to shed much light.

NOTES

1. Steven D. Jellinek laments this state of affairs in "On the Inevitability of Being Wrong," *Technology Review* 82(8):8–9, 1980.

2. Dante Picciano, "Pilot Cytogenetic Study of Love Canal, New York," prepared by the Biogenics Corporation for the Environmental Protection Agency, May 14, 1980; quoted in *Hazardous Waste in America*, pp. 113–114

3. See Edmund A. C. Crouch and Richard Wilson, *Risk/Benefit Analysis* (Cambridge: Ballinger, 1982).

4. See Baruch Fischoff et al., *Acceptable Risk* (Cambridge: Cambridge University Press, 1981).

5. Chauncey Starr, Richard Rudman, and Chris Whipple, "Philosophical Basis for Risk Analysis," *Annual Review of Energy* 1:629–662, 1976.

6. Samuel C. Florman, "Technophobia in Modern Times," *Science '82* 3:14, 1982.

7. Mary Douglas and Aaron Wildavsky, *Risk and Culture: The Selection of Technical and Environmental Dangers* (Berkeley: University of California Press, 1982). See my review, "Pollution as Delusion," *New York Times Book Review*, August 8, 1982, 8, 18.

8. From "Observations," an advertisement of the Mobil Oil Corporation, *Parade Magazine*, December 12, 1982, p. 29.

9. David Noble argues this position forcefully in "The Chemistry of Risk: Synthesizing the Corporate Ideology of the 1980s," *Seven Days* 3(7):23–26, 34, 1979. A similar view is offered by Mark Green and Norman Waitzman, "Cost, Benefit and Class," *Working Papers for a New Society* 7(3):39–51, 1980.

10. Intentions of this sort are evident in the research on risk assessment of The Center for Technology, Environment, and Development at Clark University. See Patric Derr, Robert Goble, Roger E. Kasperson, and Robert W. Kates, "Worker/Public Protection: The Double Standard," *Environment* 23(7):6–15, 31–36, 1981; Julie Graham and Don Shakow, "Risk and Reward: Hazard Pay for Workers," *Environment* 23(8):14–20, 44–45, 1981.

11. Talbot Page discusses issues of this kind seeking a balance between "false positives and false negatives" in judgments about risks in "A Generic View of Toxic Chemicals and Similar Risks," *Ecology Law Quarterly* 7:207–244, 1978.

12. Among the helpful criticisms of cost/benefit analysis are Mark Sagoff, "Economic Theory and Environmental Law," *Michigan Law Review* 79:1393–1419, 1981; and Steve H. Hanke, "On the Feasibility of Benefit-Cost Analysis," *Public Policy* 29:147–157, 1981.

POSTSCRIPT

Does Risk-Benefit Analysis Provide an Objective Method for Making Environmental Decisions?

The claim by Ruckelshaus that "there is an objective way to assess risk" puts him clearly at odds with Winner and even with the majority of social scientists in the risk analysis field. The subjectivity that is inherent in all schemes designed to quantify risks and benefits is a principal concern of those who are trying to promote the appropriate use of risk assessment. For a thoughtful discussion of some of the principal approaches to risk analysis, see *Acceptable Risk*, by Baruch Fischhoff, Paul Slovic, and Sarah Lichtenstein (Cambridge University Press, 1981).

Winner's proposal that we focus on curtailing pollution and reducing potential health hazards rather than improving our ability to quantify and compare risks does not solve the practical problems of environmental or health agencies that must decide how much of their limited budgets they should devote to each of many potential environmental problems. While it may be overly simplistic to assume, as Ruckelshaus does, that better public education and a more uniform procedure for risk assessment will provide the basis for resolving hazard management problems, it is clear that our public officials need some method to assess the relative importance of the many potential impacts of technological development on the environment.

"Perceived Risk, Real Risk: Social Science and the Art of Probabilistic Risk Assessment," by William R. Freudenburg, *Science*, October 7, 1988, presents a case for including more social science input in risk assessment.

For an extreme view of the role that culturally determined values and phobias play in the reactions of groups and individuals to environmental degradation, see *Risk and Culture: The Selection of Technical and Environmental Dangers*, by Mary Douglas and Aaron Wildovsky (University of California Press, 1982).

An interesting attempt to develop a quantitative measure of technological "hazardousness," based, in part, on the expressed concerns of lay people, is described by C. Hohenemser, R. W. Kates, and P. Slovic in the April 22, 1983, issue of *Science*.

A scathing attack on the actual methods used by the EPA in assessing and managing risk is the subject of a brief viewpoint piece by environmental scientist Ellen Silbergeld in the November 1987 issue of the *EPA Journal*. She accuses the Agency of failing to assess noncarcinogenic risks, of using unscientific methods, and of inaction and evasion in the face of recognized risks.

ISSUE 5

Is Population Control the Key to Preventing Environmental Deterioration?

YES: Paul R. Ehrlich and John P. Holdren, from "Impact of Population Growth," *Science* (March 26, 1971)

NO: Barry Commoner, Michael Corr, and Paul J. Stamler, from "The Causes of Pollution," *Environment* (April 1971)

ISSUE SUMMARY

YES: Environmental scientists Paul R. Ehrlich and John P. Holdren argue that population increase is the principal cause of environmental degradation.
NO: Environmental scientists Barry Commoner, Michael Corr, and Paul J. Stamler contend that technological change rather than population growth has been the chief cause of environmental stress.

Environmentalists have long debated the extent to which human population growth is a fundamental cause of ecological problems. It is an extremely important issue, since the answer would determine whether population control should be a central strategy of programs designed to protect the environment.

Those seriously concerned about uncontrolled human population growth are often referred to as "Malthusians" after the English parson Thomas Malthus, whose "Essay on the Principle of Population" was first published in 1798. Malthus warned that the human race was doomed because geometric population increases would inexorably outstrip productive capacity, leading to famine and poverty. His predictions were undermined by technological improvements in agriculture and the widespread use of birth control (rejected by Malthus on moral grounds), which brought the rate of population growth in industrialized countries under control during the twentieth century.

The theory of the demographic transition was developed to explain why Malthus's dire predictions had not come true. This theory proposes that the first effect of economic development is to lower death rates. This causes a population boom, but stability is again achieved as economic and social changes lead to lowering of birth rates. This pattern has indeed been followed in Europe, the United States, Canada, and Japan. The less-developed countries of the Third World have more recently experienced rapidly falling death rates. Thus far, the economic and social changes needed to bring down birth rates have not occurred, and many countries in Asia and Latin America suffer from exponential population growth. This fact has given rise to a group of neo-Malthusian theorists who contend that it is unlikely that Third World countries will undergo the transition to lower birth rates required to avoid catastrophe due to overpopulation.

Biologist Paul Ehrlich is the most influential environmentalist to adopt the neo-Malthusian position. He contends that population pressure in both developed and nonindustrial countries is responsible for pollution and other social problems. A rather gloomy prognosis based upon this premise was popularized in his best-seller, *The Population Bomb* (Ballantine, 1968). Ehrlich and those who share his views are so concerned about the national and international consequences of population growth that they believe the United States and other developed countries must take the lead in promoting birth control in order to avert a worldwide ecological disaster. Ehrlich founded the Zero Population Growth organization to promote this goal.

Garrett Hardin, another prominent biologist, has taken the neo-Malthusian environmental position even further. Hardin's "lifeboat ethics" promotes the idea that developed nations (lifeboats) cannot help all of the Third World nations suffering from population-induced resource scarcity. In order to avoid sinking the lifeboat, he proposed that developed nations should share resources only with the select group of Third World nations that are the most likely to survive. The others should be permitted to perish.

Environmental biologist Barry Commoner is the most outspoken critic of the neo-Malthusians. In a series of widely read books, of which *The Closing Circle* (Knopf, 1971) has probably been the most influential, he presents a contrasting analysis of environmental and social problems. In Commoner's view, pollution has been the result of ill-conceived programs of industrial development whose designers and profit-oriented promoters have ignored the ecological consequences of their actions. He rejects Ehrlich's population control strategy as inappropriate and considers Hardin's position both unwarranted and barbaric. Commoner has been accused by some of his critics of opposing industrial development in the Third World. A careful reading of his analysis clearly reveals this to be a false interpretation. In fact, he believes that population control is likely to be achieved in Third World countries only after poverty has been reduced through appropriate development. He optimistically proposes international efforts to promote the restructuring of the forces of production so that resources are used for social good rather than for profit.

The following two articles clearly present the opposing positions on the connections among population growth, industrial development, and environmental degradation.

YES

Paul R. Ehrlich and
John P. Holdren

IMPACT OF POPULATION GROWTH

The interlocking crises in population, resources, and environment have been the focus of countless papers, dozens of prestigious symposia, and a growing avalanche of books. In this wealth of material, several questionable assertions have been appearing with increasing frequency. Perhaps the most serious of these is the notion that the size and growth rate of the U.S. population are only minor contributions to this country's adverse impact on local and global environments. We propose to deal with this and several related misconceptions here, before persistent and unrebutted repetition entrenches them in the public mind—if not the scientific literature. Our discussion centers around five theorems which we believe are demonstrably true and which provide a framework for realistic analysis:

1. Population growth causes a *disproportionate* negative impact on the environment.

2. Problems of population size and growth, resource utilization and depletion, and environmental deterioration must be considered jointly and on a global basis. In this context, population control obviously not a panacea—it is necessary but not alone sufficient to see us through the crisis.

3. Population density is a poor measure of population pressure, and redistributing population would be a dangerous pseudosolution to the population problem.

4. "Environment" must be broadly construed to include such things as the physical environment of urban ghettos, the human behavioral environment, and the epidemiological environment.

5. Theoretical solutions to our problems are often not operational and sometimes are not solutions.

We now examine these theorems in some detail.

POPULATION SIZE AND PER CAPITA IMPACT

In an agricultural or technological society, each human individual has a negative impact on his environment. He is responsible for some of the simplification (and resulting destabilization) of ecological systems which results from the practice of agriculture. He also participates in the utilization of renewable and nonrenewable resources. The total negative impact of such a society on the environment can be expressed, in the simplest terms, by the relation

$$I = P \times F$$

where P is the population, and F is a function which measures the per capita impact. A great deal of complexity is subsumed in this simple relation, however. For example, F increases with per capita consumption if technology is held constant, but may decrease in some cases if more benign technologies are introduced in the provision of a constant level of consumption. (We shall see in connection with theorem 5 that there are limits to the improvements one should anticipate from such "technological fixes.")

Pitfalls abound in the interpretation of manifest increases in the total impact I. For instance, it is easy to mistake changes in the composition of resource demand or environmental impact for absolute per capita increases, and thus to underestimate the role of the population multiplier. Moreover, it is often assumed that population size and per capita impact are independent variables, when in fact they are not. Consider, for example, the recent article by Coale, in which he disparages the role of U.S. population growth in environmental problems by noting that since 1940 "population has increased by 50 percent, but per capita use of electricity has been multiplied several times." This argument contains both the fallacies to which we have just referred.

First, a closer examination of very rapid increases in many kinds of consumption shows that these changes reflect a shift among alternatives within a larger (and much more slowly growing) category. Thus the 760 percent increase in electricity consumption from 1940 to 1969 occurred in large part because the electrical *component* of the energy budget was (and is) increasing much faster than the budget itself. (Electricity comprised 12 percent of the U.S. energy consumption in 1940 versus 22 percent today.) The total energy use, a more important figure than its electrical component in terms of resources and the environment, increased much less dramatically—140 percent from 1940 to 1969. Under the simplest assumption (that is, that a given increase in population size accounts for exactly proportional increase in consumption), this would mean that 38 percent of the increase in energy use during this period is explained by population growth (the actual population increase from 1940 to 1969 was 53 percent). Similar considerations reveal the imprudence of citing, say, aluminum consumption to show that population growth is an "unimportant" factor in resource use. Certainly, aluminum consumption has swelled by over 1400 percent since 1940, but much of the increase has been due to the substitution of aluminum for steel in many applications. Thus a fairer measure is combined consumption of aluminum and steel, which has risen only 117 percent since 1940. Again, under the simplest assumption, population growth accounts for 45 percent of the increase.

The "simplest assumption" is not valid, however, and this is the second flaw in Coale's example (and in his thesis). In short, he has failed to recognize that per capita consumption of energy and resources, and the associated per capita impact on the environment, are themselves functions of the population size. Our previous equation is more accurately written

$$I = P \times F(P)$$

displaying the fact that impact can increase faster than linearly with population. Of course, whether $F(P)$ is an increasing or decreasing function of P depends in part on whether diminishing returns or economies of scale are dominant in the activities of importance. In populous, industrial nations such as the United States, most economies of scale are already being exploited; we are on the diminishing returns part of most of the important curves. . . .

Diminishing returns are also operative in increasing food production to meet the needs of growing populations. Typically, attempts are made both to overproduce on land already farmed and to extend agriculture to marginal land. The former requires disproportionate energy use in obtaining and distributing water, fertilizer, and pesticides. The latter also increases per capita energy use, since the amount of energy invested per unit yield increases as less desirable land is cultivated. Similarly, as the richest fisheries stocks are depleted, the yield per unit effort drops, and more and more energy per capita is required to maintain the supply. Once a stock is depleted it may not recover—it may be nonrenewable.

Population size influences per capita impact in ways other than diminishing returns. As one example, consider the oversimplified but instructive situation in which each person in the population has links with every other person—roads, telephone lines, and so forth. These links involve energy and materials in their construction and use. Since the number of links increases much more rapidly than the number of people, so does the per capita consumption associated with the links. . . .

Not only is there a connection between population size and per capita damage to the environment, but the cost of maintaining environmental quality at a given level escalates disproportionately as population size increases. This effect occurs in part because costs increase very rapidly as one tries to reduce contaminants per unit volume of effluent to lower and lower levels (diminishing returns again!). Consider municipal sewage, for example. The cost of removing 80 to 90 percent of the biochemical and chemical oxygen demand, 90 percent of the suspended solids, and 60 percent of the resistant organic material by means of secondary treatment is about 8 cents per 1000 gallons (3785 liters) in a large plant. But if the volume of sewage is such that its nutrient content creates a serious eutrophication problem (as is the case in the United States today), or if supply considerations dictate the reuse of sewage water for industry, agriculture, or groundwater recharge, advanced treatment is necessary. The cost ranges from two to four times as much as for secondary treatment (17 cents per 1000 gallons for carbon absorption; 34 cents per 1000 gallons for disinfection to yield a potable supply). This dramatic example of diminishing returns in pollution control could be repeated for stack gases, automobile exhausts, and so forth.

Now consider a situation in which the limited capacity of the environment to absorb abuse requires that we hold man's impact in some sector constant as population doubles. This means *per capita effectiveness* of pollution control in this sector must double (that is, effluent per person must be halved). In a typical situation, this would yield doubled per capita costs, or quadrupled total costs (and probably energy consumption) in this sector for a doubling of population. Of course, diminishing returns [effects] may be still more serious: we may easily have an eightfold increase in control costs for a doubling of population. Such arguments leave little ground for the assumption, popularized by Barry Commoner and others, that a 1 percent rate of population growth spawns only 1 percent effects.

It is to be emphasized that the possible existence of "economies of scale" does not invalidate these arguments. Such savings, if available at all, would apply in the case of our sewage example to a change in the amount of effluent to be handled at an installation of a given type. For most technologies, the United States is already more than populous enough to achieve such economies and is doing so. They are accounted for in our example by citing figures for the largest treatment plants of each type. Population growth, on the other hand, forces us into quantitative *and* qualitative changes in how we handle each unit volume of effluent—what fraction and what kinds of material we remove. Here economies of scale do not apply at all, and diminishing returns are the rule. . . .

POPULATION DENSITY AND DISTRIBUTION

Theorem 3 deals with a problem related to the inequitable utilization of world resources. One of the commonest errors made by the uninitiated is to assume that population density (people per square mile) is the critical measure of overpopulation or underpopulation. For instance, Wattenberg states that the United States is not very crowded by "international standards" because Holland has 18 times the population density. We call this notion "the Netherlands fallacy." The Netherlands actually requires large chunks of the earth's resources and vast areas of land not within its borders to maintain itself. For example, it is the second largest per capita importer of protein in the world, and it imports 63 percent of its cereals, including 100 percent of its corn and rice. It also imports all of its cotton, 77 percent of its wool, and all of its iron ore, antimony, bauxite, chromium, copper, gold, lead, magnesite, manganese, mercury, molybdenum, nickel, silver, tin, tungsten, vanadium, zinc, phosphate rock (fertilizer), potash (fertilizer), asbestos, and diamonds. It produces energy equivalent to some 20 million metric tons of coal and consumes the equivalent of over 47 million metric tons.

A certain preoccupation with density as a useful measure of overpopulation is apparent in the article by Coale. He points to the existence of urban problems such as smog in Sydney, Australia, "even though the total population of Australia is about 12 million in an area 80 percent as big as the United States," as evidence that environmental problems are unrelated to population size. His argument would be more persuasive if problems of population *distribution* were the only ones with environmental consequences, and if population distribution was unrelated to resource distribution and population size. Actually, since the carrying capacity of the Australian continent is far below that of the United States, one would *expect* distribution problems—of which Sydney's smog is one symptom—to be encountered at a much lower total population there. Resources, such as water, are in very short supply, and people cluster where resources are available. (Evidently, it cannot be emphasized enough that carrying capacity includes the availability of a wide variety of resources in addition to space itself, and that population pressure is measured relative to the carrying capacity. One would expect water, soils, or the ability of the environment to absorb wastes to be the limiting resource in far more instances than land area.)

In addition, of course, many of the most serious environmental problems are essentially independent of the way in which population is distributed. These include the global problems of weather modification by carbon dioxide and particulate pollution, and the threats to the biosphere posed by man's massive inputs of pesticides, heavy metals, and oil. Similarly, the problems of resource depletion and ecosystem simplification by agriculture depend on how many people there are and their patterns of consumption, but not in any major way on how they are distributed.

Naturally, we do dispute that smog and most other familiar urban ills are serious problems, or that they are related to population distribution. Like many of the difficulties we face, these problems will not be cured simply by stopping population growth; direct and well-conceived assaults on the problems themselves will also be required. Such measures may occasionally include the redistribution of population, but the considerable difficulties and costs of this approach should not be underestimated. . . .

MEANING OF ENVIRONMENT

Theorem 4 emphasizes the comprehensiveness of the environment crisis. All too many people think in terms of national parks and trout streams when they say "environment." For this reason many of the suppressed people of our nation consider ecology to be just one more "racist shuck." They are apathetic or even hostile toward efforts to avert further environmental and sociological deterioration, because they have no reason to believe they will share the fruits of success. Slums, cockroaches, and rats are ecological problems, too. The correction of ghetto conditions in Detroit is neither more nor less important than saving the Great Lakes—both are imperative.

We must pay careful attention to sources of conflict both within the United States and between nations. Conflict within the United States blocks progress toward solving our problems; conflict among nations can easily "solve" them once and for all. Recent laboratory studies on human beings support the anecdotal evidence that crowding may increase aggressiveness in human males. These results underscore long-standing suspicions that population growth, translated into physical crowding, will tend to make the solution of all of our problems more difficult.

As a final example of the need to view "environment" broadly, note that human beings live in an epidemiological environment which deteriorates with crowding and malnutrition—both of which increase with population growth. The hazard posed by the prevalence of these conditions in the world today is compounded by man's unprecedented mobility; potential carriers of diseases of every description move routinely and in substantial numbers from continent to continent in a matter of hours. Nor is there any reason to believe that modern medicine has made widespread plague impossible. The Asian influenza epidemic of 1968 killed relatively few people only because the virus *happened* to be nonfatal to people in otherwise good health, both because of public health measures. Far deadlier viruses, which easily could be scourges without precedent in the population at large, have on more than one occasion been confined to research workers largely by good luck [for example, the Marburgvirus incident of 1967 and the Lassa fever incident of 1970].

SOLUTIONS: THEORETICAL AND PRACTICAL

Theorem 5 states that theoretical solutions to our problems are often not operational, and sometimes are not solutions. In terms of the problem of feeding the world, for example, technological fixes suffer from limitations in scale, lead time, and cost. Thus potentially attractive theoretical approaches—such as desalting seawater for agriculture, new irrigation systems, high-protein diet supplements —prove inadequate in practice. They are too little, too late, and too expensive, or they have sociological costs which hobble their effectiveness. Moreover, many aspects of our technological fixes, such as synthetic organic pesticides and inorganic nitrogen fertilizers, have created vast environmental problems which seem certain to erode global productivity and ecosystem stability. This is not to say that important gains have not been made through the application of technology to agriculture in the poor countries, or that further technological advances are not worth seeking. But it must be stressed that even the most enlightening technology cannot relieve the necessity of grappling forthrightly and promptly with population growth [as Norman Borlaug aptly observed on being notified of his Nobel Prize for development of the new wheats].

Technological attempts to ameliorate the environmental impact of population growth and rising per capita affluence in the developed countries suffer from practical limitations similar to those just mentioned. Not only do such measures tend to be slow, costly, and insufficient in scale, but in addition they most often *shift* our impact rather than remove it. For example, our first generation of smog-control devices increased emissions of oxides of nitrogen while reducing those of hydrocarbons and carbon monoxide. Our unhappiness about eutrophication has led to the replacement of phosphates in detergents with compounds like NTA—nitrilotriacetic acid—which has carcinogenic breakdown products and apparently enhances teratogenic effects of heavy metals. And our distaste for lung diseases apparently induced by sulfur dioxide inclines us to accept the hazards of radioactive waste disposal, fuel reprocessing, routine low-level emissions of radiation, and an apparently small but finite risk of catastrophic accidents associated with nuclear fission power plants. Similarly, electric automobiles would simply shift part of the environmental burden of personal transportation from the vicinity of highways to the vicinity of power plants.

We are not suggesting here that electric cars, or nuclear power plants, or substitutes for phosphates are inherently bad. We argue rather that they, too, pose environmental costs which must be weighed against those they eliminate. In many cases the choice is not obvious, and in *all* cases there will be some environmental impact. The residual per capita impact, after all the best choices have been made, must then be multiplied by the population engaging in the activity. If there are too many people, even the most wisely managed technology will not keep the environment from being overstressed.

In contending that a change in the way we use technology will invalidate these arguments, Commoner claims that our important environmental problems began in the 1940s with the introduction and rapid spread of certain "synthetic" technologies: pesticides and herbicides, inorganic fertilizers, plastics, nuclear energy, and high-compression gasoline engines. In so arguing, he appears to make two unfounded assumptions. The first is that man's pre-1940 environmental impact was innocuous and, without changes for the worse in technology, would have remained innocuous even at a much larger population size. The second assumption is that the advent of the new technologies was independent of the attempt to meet human needs and desires in a growing population. Actually, man's record as a simplifier of ecosystems and plunderer of resources can be traced from his probable role in the extinction of many Pleistocene mammals, through the destruction of the soils of Mesopotamia by salination and erosion, to the deforestation of Europe in the Middle Ages and the American dustbowls of the 1930s, to cite only some highlights. Man's contemporary arsenal of synthetic technological bludgeons indisputably magnifies the potential for disaster, but these were evolved in some measure to *cope* with population pressures, not independently of them. Moreover, it is worth noting that, of the four environmental threats viewed by the prestigious Williamstown study as globally significant, three are associated with pre-1940 technologies which have simply increased in scale [heavy metals, oil in the seas, and carbon dioxide and particulates in the atmosphere, the latter probably due in considerable part to agriculture]. Surely, then, we can anticipate that supplying food, fiber, and metals for a population even larger than today's will have profound (and destabilizing) effect on the global ecosystem under *any* set of technological assumptions.

CONCLUSION

John Platt has aptly described man's present predicament as "a storm of crisis problems." Complacency concerning any component of these problems—sociological, technological, economic, ecological—is unjustified and counter-productive. It is time to admit that there are no monolithic solutions to the problems we face. Indeed, population control, the redirection of technology, the transition from open to closed resource cycles, the equitable distribution of opportunity and the ingredients of prosperity must *all* be accomplished if there is to be a future worth having. Failure in any of these areas will surely sabotage the entire enterprise.

In connection with the five theorems elaborated here, we have dealt at length with the notion that population growth in industrial nations such as the United States is a minor factor, safely ignored. Those who so argue often add that, anyway, population control would be the slowest to take effect of all possible attacks on our various problems, since the inertia in attitudes and in the age structure of the population is so considerable. To conclude that this means population control should be assigned low priority strikes us as curious logic. Precisely because population is the most difficult and slowest to yield among the components of environmental deterioration, we must start on it at once. To ignore population today because the problem is a tough one is to commit ourselves to even gloomier prospects 20 years hence, when most of the "easy" means to reduce per capita impact on the environment will have been exhausted. The desperate and repressive measures of population control which might be contemplated then are reason in themselves to proceed with foresight, alacrity, and compassion today.

NO

Barry Commoner, Michael Corr, and Paul J. Stamler

THE CAUSES OF POLLUTION

Until now most of us in the environmental movement have been chiefly concerned with providing the public with information that shows that there *is* an environmental crisis. In the last year or so, as the existence of the environmental crisis has become more widely recognized, it has become increasingly important to ask: How can we best solve the environmental crisis? To answer this question it is no longer sufficient to recognize only that the crisis exists; it becomes necessary, as well, to consider its causes, so that rational cures can be designed.

Although environmental deterioration involves changes in natural, rather than man-made, realms—the air, water, and soil—it is clear that these changes are due to human action rather than to some natural cataclysm. The search for causes becomes focused, then, on the question: What actions of human society have given rise to environmental deterioration?

Like every living thing on the earth, human beings are part of an ecosystem—a series of interwoven, cyclical events, in which the life of any single organism becomes linked to the life processes of many others. One well-known property of such cyclical systems is that they readily break down if too heavily stressed. Such a stress may result if, for some reason, the population of any one living organism in the cycle becomes too great to be borne by the system as a whole. For example, suppose that in a wooded region the natural predators which attack deer are killed off. The deer population may then become so large that the animals strip the land of most of the available vegetation, reducing its subsequent growth to the point where it can no longer support the deer population; many deer die. Thus, in such a strictly biological situation, overpopulation is self-defeating. Or, looked at another way, the population is self-controlled, since its excessive growth automatically reduces the ability of the ecosystem to support it. In effect, environmental deterioration brought about by an excess in a population which the environment supports is the means of regulating the size of that population.

However, in the case of human beings, matters are very different; such automatic control is undesirable, and, in any case, usually impossible. Clearly, *if* reduced environmental quality were due to excess population, it might be advantageous to take steps to reduce the population size humanely rather than to expose human society to grave dangers, such as epidemics, that would surely accompany any "natural" reduction in population brought about by the environmental decline. Thus, if environmental deterioration were in fact the ecosystem's expected response to human overpopulation, then in order to cure the environmental crisis it would be necessary to relieve the causative stress—that is, to *reduce* actively the population from its present level.

On these grounds it might be argued as well that the stress of a rising human population on the environment is especially intense in a country such as the United States, which has an advanced technology. For it is modern technology which extends man's effects on the environment far beyond his biological requirements for air, food, and water. It is technology which produces smog and smoke; synthetic pesticides, herbicides, detergents, and plastics; rising environmental concentrations of metals such as mercury and lead; radiation; heat; accumulating rubbish and junk. It can be argued that insofar as such technologies are intended to meet human needs—for food, clothing, shelter, transportation, and the amenities of life—the more people there are, and the more active they are, the more pollution.

Against this background it is easy to see why some observers have blamed the environmental crisis on overpopulation. Here are two typical statements:

> The pollution problem is a consequence of population. It did not much matter how a lonely American frontiersman disposed of his waste. "Flowing water purifies itself every ten miles," my grandfather used to say, and the myth was near enough to the truth when he was a boy, for there weren't too many people. But as population became denser the natural chemical and biological recycling processes became overloaded, calling for a redefinition of property rights.
>
> The causal chain of the deterioration [of the environment] is easily followed to its source. Too many cars, too many factories, too much detergent, too much pesticide, multiplying contrails, inadequate sewage treatment plants, too little water, too much carbon dioxide—all can be traced easily to *too many people*.

Some observers, for example, M. P. Miller, chief of census population studies at the U.S. Bureau of the Census, believe that in the U.S. environmental deterioration is only partly due to increasing population, and blame most of the effect on "affluence."

Finally, some of us place the strongest emphasis on the effects of the modern technology that so often violates the basic principles of ecology and generates intense stresses on the environment.

Dr. Paul Ehrlich provides the following statement regarding these several related factors: "Pollution can be said to be the result of multiplying three factors: population size, per capita consumption, and an 'environmental impact' index that measures, in part, how wisely we apply the technology that goes with consumption." As indicated in the previous passage, Dr. Ehrlich appears to consider population size as the predominant factor in this relationship.

Dr. Ehrlich's statement can be paraphrased as an "equation":

$$\text{population size} \times \text{per capita consumption} \times \text{environmental impact per unit of production} = \text{level of pollution}$$

This equation is self-evidently true, as it includes all the main factors and relationships which could possibly influence the environment. The product of population size and per capita consumption gives the total goods consumed; since imports, exports, and storage are relatively slight effects, total consumption can be taken to be approximately equal to total production. When the latter figure is multiplied by the environmental impact (i.e., amount of pollution per unit of production) the final result should be equal to the total environmental effect—the level of pollution.

Precisely because it is so inclusive, however, this equation does not advance our understanding of the causes of environmental problems. All human activities affect the environment to some degree. The equation states this formally, but we are still left with the problem of evaluating the extent to which different activities cause environmental problems, and the extent to which these environmental effects increase with population growth, with increasing per capita consumption, or with changing technologies. If we are to take effective action, we will need a more detailed guide than the equation offers. To begin with, we must know the relative importance of the three factors on the left side of the equation.

Two general approaches suggest themselves. One is to find appropriate numerical values for each of the four factors of the equation. Another way is to examine specific pollution problems and determine to what degree they are caused, explicitly, by a rising population, by increased prosperity, or by the increased environmental impact of new technologies. What follows is an effort to provide some preliminary data relevant to both of these approaches.

To begin with, it is necessary to define the scope of the problem, both in space and time. As to space, we shall restrict the discussion solely to the United States. This decision is based on several factors: (a) The necessary data are available—at least to us—only for the United States. (b) The pollution problem is most intense in a highly developed country such as the United States. (c) In any study involving the comparison of statistical quantities, the more homogeneous the situation, the less likely we are to be misled by averages that combine vastly different situations. In this sense, it might be better to work with a smaller sample of the pollution problem—such as an urban region. Unfortunately, the necessary production statistics are not readily available except on a national scale.

As to time, we have chosen the period 1946-68. There are several reasons for this choice. First, many current environmental problems began with the end of World War II: photochemical smog, radiation from nuclear wastes, pollution from detergents and synthetic pesticides. Another reason for choosing the post-war period is that many changes in production techniques were introduced during this period. The upper limit of the period is a matter of convenience only; statistical data for the two most recent years are often difficult to obtain.

We shall thus be seeking an answer to the following question for the period 1946-68 in the United States: What changes in the levels of specific pollutants, in population size, in environmental impact per unit of production, and in the amounts of goods produced per capita have occurred?

CHANGES IN POLLUTION AND POPULATION LEVELS

Curiously, the first of these questions is the most difficult to answer. Probably the best available data relate to water pollution. These are summarized in a study by Weinberger. For the United States as a whole, in the period 1946-68 the total nitrogen and phosphate discharged into surface waters by municipal sewage increased by 260 percent and 500 percent respectively.

Here are some additional data which, although sparse, are suggestive of the sizes of recent changes in pollution levels. As indicated by glacial deposits, airborne lead has increased by about 400 percent since 1946. Daily nitrogen oxide emissions in Los Angeles County have increased about 530 percent. The average algal population in Lake Erie—one response to, and indicator of, pollution due to nutrients such as nitrate and phosphate—increased about 220 percent. The bacterial count in different sectors of New York harbor increased as much as 890 percent. Such data correspond with general experience. For example, the extent of photochemical smog in the U.S. has surely increased at least ten-fold in the 1946-68 period, for in 1946 it was known only in Los Angeles; it has now been reported in every major city in the country, as well as in smaller areas such as Phoenix, Arizona and Las Vegas, Nevada.

Rough as it is, we can take as an estimate of the change in pollution levels in the United States during 1946-68 increases that range from two- to ten-fold or so, or from 200 to 1,000 percent.

The increase in the U.S. population for the period 1946-68 amounts to about 43 percent. It would appear, then, that the rise in overall U.S. population is insufficient by itself to explain the large increases in overall production levels since 1946. This means that in Ehrlich's equation, the increase in population is too small to bring the left side to approximate equality with the right side unless there have been sufficiently large increases in the per capita production and environmental impact factors.

COMBINED FACTORS OF POPULATION AND PRODUCTION

The equation relates total pollution to three component factors: population size, production per capita, and environmental impact per unit of production. As a second step in evaluating the meaning of this approach, it is useful to determine whether the combined factors of population growth and increased per capita production can account for the changes in pollution levels during the period 1946-68.

A rough measure of overall U.S. production is the Gross National Product (GNP). GNP has increased about 126 percent in that time and GNP per capita has increased about 59 percent. As a first approximation, then, it would appear that the overall increase in total production, as measured by GNP, is also insufficient to account for the considerably larger increases in pollution levels. However, since the GNP is, of course, an average composed of the many separate activities in the total production economy (including not only agricultural and industrial production and transportation, but also various services), a true picture of the relationship between production and environmental pollution requires a breakdown of the GNP into, at the least, some of its main components. . . .

Certain large and basic categories, the necessities of life—food, clothing, and shelter—merit special attention. Data from these categories can be used to determine the degree to which changes in affluence (consumption per capita) or prosperity (production per capita) can account for the large increases in pollution levels for the period 1946-68.

Food production and consumption figures are available from the U.S. Department of Agriculture. Surprisingly, for the period 1910 to 1968, there were very few overall changes in per capita consumption of food materials, especially in the period of interest to us. In 1946-68, total calories consumed dropped from 3,390 per person per day to about 3,250 per person per day, while protein consumption declined slightly from 104 grams per person per day to about 99 grams per person per day.

It should be remembered that these are *consumption* data, whereas the data of interest in connection with environmental stress are those for *total production* (the difference being represented by storage of farm products and the balance of exports and imports). The difference represents, for instance for the year 1968, no more than about 3 percent of the total value of farm production; the consumption data consequently repesent a fairly accurate picture of farm production.

The total production figures do in fact reflect the trends evident in per capita consumption. Thus, total grain production, including grain used for meat production, decreased 6 percent in the 1946-68 period. The figures on declining protein intake tell us that increased meat consumption is more than balanced by declines in other types of protein intake, for instance of eggs, milk, and dry beans. Of course, the increased use of beef and other meat (about 19 percent per capita) does represent some increase in affluence. On the other hand, there has been a corresponding decline in another indicator of affluence, the use of fruit. Taking these various changes into account, then, there is no evidence of any significant change in the overall affluence of the average American with respect to food. And, in general, food production in the U.S. has just about kept up with the 43 percent increase in population in that time.

A similar situation exists with respect to another life necessity—clothing. The following items show either no significant change in per capita production, or a slight decline: shoes, hosiery, shirts, total fibers (i.e., natural plus synthetic), and total fabric production. Again, as in the case of food, the "affluence" or "prosperity" factor in the equation is about 1, so that population increase by itself is not sufficient to explain the large increases in environmental pollution due to production of these items.

In the area of shelter, we find that housing units occupied in 1946 were 0.27 per capita, and in 1968, 0.30 per capita, an increase of 11 percent, although there was some improvement in the quality of units. Again, this change, even with the concurrent 43 percent increase in population, is simply not enough to match the large increases in pollution levels.

Another set of statistics also allows us to arrive at an estimate of the "affluence" factor. These relate to average personal expenditures for food, clothing, and housing (including purchased food and meals, alcoholic beverages, tobacco, rents and mortgage payments, house repairs, wearing apparel, but excluding furniture, household utilities, and domestic service). Such expenditures, adjusted for inflation, increased, per capita, about 27 percent between 1950 and 1968. Again, this increase when multiplied by the concurrent increase in population is insufficient to produce the large increases in pollution levels. It is important to note that these expenditures comprise a sector which represents about one-third of the total United States economy.

All this is evidence, then, that the increases in 1946-68 in two of the factors in Ehrlich's equation—population size and production (or consumption) per capita—are inadequate to account for the concurrent increases in pollution level. This leaves us with the third factor: the nature of the technologies used to produce the various goods, and the impact of these technologies on the environment. We must look to this factor to find the sources of the large increases in pollution.

ENVIRONMENTAL IMPACT

Reference to [figures] enables us to single out the activities which have sharply increased in per capita production in the period 1946-68. They fall into the following general classes of production: synthetic organic chemicals and the products made from them, such as detergents, plastics, synthetic fibers, rubber, pesticides and herbicides; wood pulp and paper products; total production of energy, especially electric power; total horsepower of prime movers, especially petroleum-driven vehicles; cement; aluminum; mercury used for chlorine production; petroleum and petroleum products.

Several remarks about this group of activities are relevant to our problem. First is the fact that the increase in per capita production (and also in total production) in this group of activities is rather high—of the order of 100 to 1,000 percent. This fact, [is] a reminder that the changes in the U.S. production system during the period 1946-68 do not represent an across-the-board increase in affluence or prosperity. That is, the 59 percent increase in per capita GNP in that period obscures the fact that in certain important sectors—for example, those related to basic life necessities—there has been rather little change in production per capita, while in certain other areas of production the increases have been very much larger. The second relevant observation about this group of activities is that their magnitude of increased production per capita begins to approach that of the estimates of concurrent changes in pollution level.

These considerations suggest, as a first approximation, that this particular group of production activities may well be responsible for the observed major changes in pollution levels. This identification is, of course, only suggested by the above considerations as an hypothesis, and is by no means proven by them. However, the isolation of this group provides a valuable starting point for a more detailed examination of the nature of the production activities that comprise it, and of their *specific* relationship to environmental degradation. As we shall see in what follows, this more detailed investigation does, in fact, quite strongly support the hypothesis suggested by the more superficial examination.

Nearly all of the production activities that fall into the class exhibiting striking changes in per capita production turn out to be important causes of pollution. Thus wood pulp production and related paper-making activities are responsible for a very considerable part of the pollution of surface waters with organic wastes, sulphite, and, until several years ago, mercury. Vehicles driven by the internal combustion engine are responsible for a major part of total air pollution, especially in urban areas, and are almost solely responsible for photochemical smog. Much of the remaining air pollution is due to electric power generation, another member of this group. Cement production is a notorious producer of dust pollution and a high consumer of electrical energy. The hazardous effects of mercury released into the environment are just now, belatedly, being recognized.

The new technological changes in agriculture, while yielding no major increase in overall per capita food production, have in fact worsened environmental conditions. Food production in the United States in 1968 caused much more environmental pollution than it did in 1947. Consider, for example, the increased use of nitrogen fertilizer, which rose 534 percent per capita between 1946 and 1968. This striking increase in fertilizer use did not increase total food production, but improved the crop yield per acre (while acreage was reduced) and made up for the loss of nitrogen to the soil due to the increasing use of feedlots to raise animals (with resultant loss of manure to the soil). For reasons which have been described elsewhere, this intensive use of nitrogen fertilizer on limited acreage drives nitrogen out of the soil and into surface waters, where it causes serious pollution problems. Thus, while Americans, on the average, eat about as much food per capita as they used to, it is now grown in ways that cause increased pollution. The new technologies, such as feedlots and fertilizer, have a much more serious effect on pollution than either increases in population or in affluence.

One segment of the group of increasing industrial activities in the period 1946-68, that comprising synthetic organic chemicals, and their products, raises environmental problems of a particularly subtle, but nevertheless important, kind. In the first place, most of them find a place in the economy as substitutes for—some might say, improvements over—older products of a natural, biological origin. Thus synthetic detergents replace soap, which is made from fat—a natural product of animals and plants. Synthetic fibers replace cotton, wool, silk, flax, hemp—all, again, natural products of animals and plants. Synthetic rubber replaces natural rubber. Plastics replace wood and paper products in packaging. In many, but not all, uses plastics replace natural products such as wood and paper. Synthetic pesticides and herbicides replace the natural ecological processes which control pests and unwanted weeds. Both the natural products and their modern replacements are organic substances. In effect, we can regard the products of modern synthetic organic chemistry as man-made variations on a basic scheme of molecular structure which in nature is the exclusive province of living things.

Because they are not *identical* with the natural products which they resemble, these synthetic substances do not fit very well into the chemical schemes which comprise natural ecosystems. Some of the new substances, such as plastics, do not fit into natural biochemical systems at all. Thus, while "nature's plastic," cellulose, is readily degraded by soil microorganisms and thus becomes a source of nutrition for soil organisms, synthetic plastics are not degradable, and therefore accumulate as waste. Automatically they become environmental pollutants. Because there is no natural way to convert them into usable materials, They either accumulate as junk or are disposed of by burning—which, of course, pollutes the air. Nondegradable synthetic detergents, with a branched molecular structure that is incompatible with the requirements of microorganisms which break down natural organic materials, remain in the water and become pollutants. Even degradable synthetic detergents, when broken down, may pollute water with phenol, and add another important water pollutant—phosphate—as well. Thus, these synthetic substitutes for natural products are, inevitably, pollutants. . . .

The point to be emphasized here is that the modern replacements for natural products have become the basis for the new, expansive production activities derived from synthetic organic chemicals, and are, by their very nature, destined to become serious environmental pollutants if they are broadcast into the environment—as, of course, they are.

There is, however, another way in which synthetic organic materials are particularly important as sources of environmental pollution. This relates not to their use but to their production. Let us compare, for example, the implication for environmental pollution of the *production* of, say, a pound of cotton and a pound of a synthetic fiber such as nylon. Both of these materials, consist of long molecular chains, or polymers made by linking together a succession of small units, or monomers. The formation of a polymer from monomers requires energy, part of which is required to form the bond that links the successive monomers. This energy has to be, so to speak, built into the monomer molecules, so that it is available when the intermonomer link is formed. Energy is required to collect together, through cracking and distillation, an assemblage of the particular monomer required for the synthetic process from a mixture such as petroleum. (That is, the process of obtaining a pure collection of the required monomer demands energy.) And it must be remembered that the energy requirement of a production process leads to important environmental consequences, for the combustion required to release energy from a fuel is always a considerable source of pollution.

If we examine cotton production according to these criteria, we find that it comes off with high marks, for relatively little energy capable of environmental pollution is involved. In the first place the energy required to link up the glucose monomers which make up the cotton polymer (cellulose) is built into these molecules from energy provided free in the form of the sunlight absorbed photosynthetically by the cotton plant. Energy derived from sunlight is transformed, by photosynthesis, into a biochemical form, which is then incorporated into glucose molecules in such a way as to provide the energy needed to link them together. At the same time photosynthetic energy synthesizes glucose from carbon dioxide and water. Moreover, glucose so heavily predominates as a major product of photosynthesis that the energy required to "collect" it in pure form is minimal—and of course is also obtained free, from sunlight. And in all these cases the energy is transferred at low temperatures (the cotton plant, after all, does not burn) so that extraneous chemical reactions such as those which occur in high temperature combustion—and which are the source of air pollutants such as nitrogen oxides and sulfur dioxide—do not occur. In fact, the overall photosynthetic process takes carbon dioxide—an animal waste product, and a product of all combustion—out of the air.

Now compare this with the method for producing a synthetic fiber. The raw material for such production is usually petroleum or natural gas. Both of these represent stored forms of photosynthetic energy, just as does cellulose. However, unlike cellulose, these are nonrenewable resources in that they were produced during the early history of the earth in a never-to-be repeated period of very heavy plant growth. Moreover, in order to obtain the desired monomers from the mixture present in petroleum, a series of high-temperature, energy-requiring processes, such as distillation and evaporation, must be used. All this means that the production of synthetic fiber consumes more nonrenewable energy than the production of a natural fiber such as cotton or wool. It also means that the energy-requiring processes involved in synthetic fiber production takes place at high temperatures, which inevitably result in air pollution.

Similar considerations hold for all of the synthetic materials which have replaced natural ones. Thus, the production of synthetic detergents, plastic, and artificial rubber inevitably involves, weight for weight, more environmental pollution than the production of soap, wood, or natural rubber. Of course, the balance sheet is not totally one-sided. For example, at least under present conditions, the production of cotton involves the use of pesticides and herbicides—environmental pollutants that are not needed to produce synthetic fibers. However, we know that the use of pesticides can be considerably reduced in growing a crop such as cotton—and, indeed, must be reduced if the insecticide is not to become useless through the development of insect resistance to the chemicals—by employing modern techniques of biological control. In the same way, it can be argued that wool production involves environmental hazards because sheep can overgraze a pasture and set off erosion. Again, this hazard is not an inevitable accompaniment of sheep-raising, but only evidence of poor ecological management. Similarly, pollution due to combustion could be curtailed and thus reduce the environmental impact of synthetics.

Obviously, much more detailed evaluations of this problem are needed. However, on the basis of these initial considerations it seems evident that the substitution of synthetic organic products for natural ones through the efforts of the modern chemical industry has, until now, considerably intensified environmental pollution. . . .

This brings us to a distinctive and especially important aspect of the environmentally related changes in production which have taken place in the 1946-68 period—electric power production. Electric power production has been noteworthy for its rapid and accelerating rate of growth. Total power production has increased by 662 percent in the period 1946-68; per capita power production has increased by 436 percent. Electric power production from fossil fuels is a major cause of urban air pollution; produced by nuclear reactors it is a source of radioactive pollution. Regardless of the fuel employed, power production introduces heat into the environment, some of it in the form of waste heat released at the power plant into either cooling waters or the air. Ultimately all electric power, when used, is converted to heat, causing increasingly serious heat pollution problems in cities in the summer. One of the striking features of the present U.S. production system is its accelerating demand for more and more power—with the resultant exacerbation of pollution problems.

AFFLUENCE AND INCREASED PRODUCTION

What is striking in the data discussed above is that so many of the new and expanding production activities are highly power-consumptive and have replaced less power-consumptive activities. This is true of the synthetic chemical industry, of cement, and of the introduction of domestic electrical appliances.

It is useful to return at this point to the question of affluence. To what extent do these increased uses of electric power, which surely contribute greatly to environmental deterioration, arise from the increased affluence, or well-being of the American public? Certainly the introduction of an appliance such as a washing machine is, indeed, a valuable contribution to a family's well-being; a family with a washing machine is without question more affluent than a family without one. And, equally clear, this increased affluence adds to the total consumption of electric power and thereby adds to the burden of environmental pollutants. Such new uses of electricity therefore do support the view that affluence leads to pollution.

On the other hand, what is the contribution to public affluence of substituting a power-consumptive aluminum beer can for a less power-demanding steel can? After all, what contributes to human welfare is not the can, but the beer (it is interesting to note in passing that beer consumption per capita has remained essentially unchanged in the 1946-68 period). In this instance the extra power consumption due to the increased use of aluminum cans—and the resulting environmental pollution—cannot be charged to improved affluence. The same is true of the increased use of the nonreturnable bottle, which pollutes the environment during the production process (glass products require considerable fuel combustion) and pollutes it further when it is discarded after use. The extra power involved in producing aluminum beer cans, the extra power and other production costs involved in using nonreturnable bottles instead of reusable ones, contribute both to environmental pollution and GNP. But they add nothing to the affluence or well-being of the people who use these products.

Thus, in evaluating the meaning of increased productive activity as it relates to the matters at issue here, a sharp distinction needs to be made between those activities which actually contribute to improved well-being and those which do not, or do so minimally (as does the self-opening aluminum beer can). Power production is an important area in which this distinction needs to be made. Thus, the chemical industry and the production of cement and aluminum, taken together, account for 18 percent of the present consumption of power in the United States. For reasons already given, some significant fraction of the power used for these purposes involves the production of a product which replaces a less power consumptive one. Hence this category of power consumption—and its attendant environmental pollution—ought not to be charged to increased affluence. It seems likely to us that when all the appropriate calculations have been made, a very considerable part of the recent increases in demand for electric power will turn out to involve just such changes in which well-being, or affluence, is not improved, but the environment and the people who live in it suffer.

Transportation is another unique interesting area for such considerations. At first glance changes in the transportation scene in the United States in 1950-68 do seem to bear out the notion that pollution is due to increased affluence. In that period of time the total horsepower of automotive vehicles increased by 260 percent, the number of car registrations per capita by 110 percent, the vehicle miles traveled per capita by 100 percent, the motor fuel used per capita for transportation by 90 percent. All this gives the appearance of increased affluence—at the expense of worsened pollution.

However, looked at a little more closely the picture becomes quite different. It turns out that while the use of individual vehicles has increased sharply, the use of railroads has declined—thus replacing a less polluting means of transportation with a more polluting one. One can argue, of course, that it is more affluent to drive one's own car than to ride in a railroad car along with a number of strangers. Accepting the validity of that argument, it is still relevant to point out that it does *not* hold for comparison between freight hauled in a truck or in a railroad train. The fuel expenditure for hauling a ton of freight one mile by truck is 5.6 times as great as for a ton mile hauled by rail. In addition, the energy outlay for cement and steel for a four-lane expressway suitable for carrying heavy-truck traffic is 3.5 times as much as that required for a single track line designed to carry express trains. Rights-of-way account for a considerable proportion of the environmental impact of transportation systems. In unobstructed country requiring no cuts or fills, a 400-foot right-of-way is desirable for an expressway, while a 100-foot right-of-way is desirable for an express rail line (in both cases, allowing for future expansion to more lanes or two rail lines respectively). Then, too, motor-vehicle-related accidents might be included in environmental considerations; in 1968 there were 55,000 deaths and 4.4 million injuries due to motor vehicle accidents, while there were only about 1,000 railway deaths, almost none of them passenger deaths. Aside from the loss of life and health, motor vehicle accidents are responsible for the expenditure of $12 billion a year on automobile insurance, which is equivalent to 16 percent of total person consumption expenditures for transportation. In the case of urban travel it is very clear that efficient mass transit would be not only a more desirable means of travel than private cars, but also far less polluting. The lack of mass transit systems in American cities, and the resulting use of an increasing number of private cars is, again, a cause of increased pollution that does not stem from increased affluence.

It seems to us that the foregoing data provide significant evidence that the rapid intensification of pollution in the United States in the period 1946-68 cannot be accounted for solely by concurrent increases either in population or in affluence. What seems to be far more important than these factors in generating intense pollution is the *nature* of the *production* process; that is, its impact on the environment. The new technologies introduced following World War II have by and large provided Americans with about the same degree of affluence with respect to basic life necessities (food, clothing, and shelter); with certain increased amenities, such as private automobiles, and with certain real improvements such as household appliances. Most of these changes have involved a much greater stress on the environment than the activities which they have replaced. Thus, the most powerful cause of environmental pollution in the United States appears to be the introduction of such changes in technology, without due regard to their untoward effects on the environment.

Of course, the more people that are supported by *ecologically faulty technologies*—whether old ones, such as coal-burning power plants, or new ones, such as those which have replaced natural products with synthetic ones—the more pollution will be produced. But if the new, ecologically faulty technologies had *not* been introduced, the increase in U.S. population in the last 25 years would have had a much smaller effect on the environment. And, on the other hand, had the production system of the U.S. been based *wholly* on sound ecological practice (for example, sewage disposal systems which return organic matter to the soil; vehicle engines which operate at low pressure and temperature and therefore do not produce smog-triggering nitrogen oxides; reliance on natural products rather than energy-consumptive synthetic substitutes; closed production systems that prevent environmental release of toxic substances) pollution levels would not have risen as much as they have, despite the rise in population size and in certain kinds of affluence. . . .

THE PRIMARY CAUSE

These considerations and the ones discussed earlier in connection with agriculture, synthetics, power, and transportation, allow us, we believe, to draw the conclusion that the predominant factor in our industrial society's increased environmental degradation is neither population nor affluence, but the increasing environmental impact per unit of production due to technological changes.

Thus, in seeking public policies to alleviate environmental degradation it must be recognized that a stable population with stable consumption patterns would still face increasing environmental problems if the environmental impact of production continues to increase. The environmental impact per unit of production is increasing now, partially due to a process of displacement in which new, high-impact technologies are becoming predominant. Hence, social choices with regard to productive technology are inescapable in resolving the environmental crisis.

POSTSCRIPT

Is Population Control the Key to Preventing Environmental Deterioration?

Recent U.S. policy does not support the positions espoused in either of the two preceding articles. Contributions to worldwide birth control efforts have been curtailed by both the Reagan and Bush administrations. They have also opposed direct governmental involvement in constraining industry to adopt ecologically sound development strategies.

Anyone with a serious interest in environmental issues should certainly read Ehrlich's *The Population Bomb* (Ballantine, 1968) and Commoner's *The Closing Circle* (Knopf, 1971). Ehrlich and Holdren were so distressed by the arguments contained in Commoner's popular book that they wrote a detailed critique, which Commoner answered with a lengthy response. These two no-holds-barred pieces were published as a "Dialogue" in the May 1972 issue of the *Bulletin of the Atomic Scientists*. They are interesting reading, not only for their technical content, but as a rare example of respected scientists airing their professional and personal antagonisms in public.

Another frequently cited controversial essay in support of the neo-Malthusian analysis is "The Tragedy of the Commons," by Garrett Hardin, which first appeared in the December 13, 1968, issue of *Science*. For a thorough attempt to justify his authoritarian response to the world population problem, see Hardin's book *Exploring New Ethics for Survival* (Viking Press, 1972).

An economic and political analyst who is concerned about the connections among population growth, resource depletion, and pollution—but who rejects Hardin's proposed solutions—is Lester Brown, director of the Worldwatch Institute. His world view is detailed in *The Twenty-Ninth Day* (Norton, 1978).

Anyone willing to entertain the propositions that pollution has not been increasing, natural resources are not becoming scarce, the world food situation is improving, and population growth is actually beneficial, might find economist Julian Simon's *The Ultimate Resource* (Princeton University Press, 1982) amusing, if not convincing.

Frances Moore Lappé and Joseph Collins, directors of the Institute for Food and Development Policy, share Commoner's view that world hunger, overpopulation, and related problems need to be solved through economic and political change initiated by Third World people rather than by enforced population control or other interventionist strategies initiated by developed countries. The results of their research are presented in *Food First* (Houghton Mifflin, 1977).

For a recent assessment of the need to control population growth by several international authorities, including Barry Commoner, see "A Forum: How Big is the Population Factor?" in the July/August 1990 issue of *EPA Journal*. Paul Ehrlich presents his current views on the issue in an article he coauthored with Anne Ehrlich entitled "The Population Explosion," *The Amicus Journal* (Winter 1990).

UNITED NATIONS/RICK GRUNBAUM NJ.

PART 2

The Environment and Technology

Most of the environmental concerns that are the focus of current regulatory debates are directly related to modern industrial development—the pace of which has been accelerating dramatically since World War II. Thousands of new synthetic chemicals have been introduced into manufacturing processes and agricultural pursuits. New technology and its byproducts, and the exponential increases in the production and use of energy, have all contributed to the release of environmental pollutants. How to continue to improve the standard of living for the world's people without increasing ecological stress and exposure to toxins is the key question that underlies the issues debated in this section.

Can Current Pollution Strategies Improve Air Quality?

Is Nuclear Power Safe and Desirable?

Is the Widespread Use of Pesticides Required to Feed the World's People?

Is There a Cancer Epidemic Due to Industrial Chemicals in the Environment?

Is Immediate Legislative Action Needed to Combat the Effects of Acid Rain?

Should Women Be Excluded from Jobs That Could Be Hazardous to a Fetus?

ISSUE 6

Can Current Pollution Strategies Improve Air Quality?

YES: John G. McDonald, from "Gasoline and Clean Air: Good News, Better News, and a Warning," *Vital Speeches of the Day* (August 15, 1990)

NO: Hilary F. French, from "Communication: A Global Agenda for Clean Air," *Energy Policy* (July/August 1990)

ISSUE SUMMARY

YES: Oil company executive John G. McDonald argues that air pollution from automobiles can be controlled by using cleaner fuels and continuing to improve the automobile engine.
NO: Energy policy analyst Hilary F. French claims that adequate control of air pollutants requires major efforts to reorient energy use, transportation, and industrial production toward pollution prevention.

The fouling of air due to the burning of fuels has long plagued the inhabitants of populated areas. After the industrial revolution, the emissions from factory smokestacks were added to the pollution resulting from cooking and household heating. The increased use of coal combined with local meteorological conditions in London produced the dense, smoky, foggy condition first referred to as "smog" by Dr. H. A. Des Vouex in the early 1900s.

Dr. Des Vouex organized British smoke abatement societies. Despite the efforts of these organizations, the problem grew worse and spread to other industrial centers during the first half of the twentieth century. In December 1930, a dense smog in Belgium's Meuse Valley resulted in 60 deaths and an estimated 6,000 illnesses due to the combined effect of the dust and sulfur oxides from coal combustion. This event and a similar one in Donora, Pennsylvania, received public attention but little response. Serious smog control efforts did not begin until a disastrous "killer smog" in London in December 1952 resulted in approximately 4,000 deaths.

The first major air pollution control victory was the regulation of high-sulfur coal burning in populated areas. The last life-threatening London smog incidents were recorded in the mid-1960s. Unfortunately, a new smog problem, linked to the rush hour traffic in large cities like Los Angeles, was fast developing. This highly irritating, smelly, smoky haze—now referred to as photochemical smog—is caused by sunlight on air laden with nitrogen oxides and unburned hydrocarbons from automotive exhausts and other sources. The resulting chemical reactions produce ozone, and a variety of more exotic chemicals, that are very irritating to lungs and nasal passages—even at very low levels.

The first comprehensive legislation aimed at controlling air pollution from stationary sources (such as factories or electric power stations) and mobile sources (such as cars and trucks) was the Clean Air Act of 1963. Subsequent amendments to this law established National Ambient Air Quality Standards for sulfur dioxide, suspended particulates (dust), carbon monoxide, nitrogen dioxide, ozone, and airborne lead. Federal emission standards were mandated for automobiles, and states were required to establish and implement industrial air pollution control programs—subject to Environmental Protection Agency (EPA) approval—for meeting the air quality standards. In 1977, an additional provision was enacted, aimed at the prevention of significant deterioration of air quality in areas, previously unpolluted, which were now threatened by new pollution sources.

The most significant improvement that has resulted from efforts to comply with this complex legislation has been moderate decreases in carbon monoxide and sulfur dioxide concentrations in some highly polluted urban areas. Congress and the EPA have had to repeatedly extend deadlines for meeting the standards, and many areas in the United States still exceed the limit for one or more pollutants.

After nine years of frustrated effort, undermined by a lack of White House leadership during the Reagan administration, a comprehensive revision and reauthorization of the Clean Air Act has recently been signed into law by President Bush. This legislation includes an attempt to improve control of the major regional and local air pollutants as well as to attack such international and global problems as acid rain, global warming, and ozone depletion—see Issues 10, 16, and 17. The strategies for achieving these ends are along the lines recommended in the essay by John G. McDonald, executive of British Petroleum (BP) Oil Company. They involve a continuation of efforts to reduce the emission of pollutants by modifying rather than eliminating the sources. Hilary F. French, an energy policy analyst for the Worldwatch Institute, fears that this approach will fail to stem the worldwide degradation of air quality. He advocates a comprehensive restructuring of energy and transportation systems with pollution elimination as the driving force.

YES

John G. McDonald

GASOLINE AND CLEAN AIR: GOOD NEWS, BETTER NEWS, AND A WARNING

Today my job is to talk to you about America's air quality problem and what the oil industry proposes to do to clean it up. Let me start with a categorical statement:

BP [*British Petroleum Oil Company. McDonald is executive vice president of BP America.*—ED.] supports the clean air initiative and compliance with the current Federal standards. Air is a natural resource, of which we are the temporary custodians, and I believe we have a duty to provide future generations with clean air and a healthy environment in which to live. I therefore applaud President Bush's efforts to improve air quality in current non-attainment areas. What I wish to address today is a major contribution that the oil industry can make to this initiative.

I am aware that there may be some of you in the City Club's large audience who are skeptical about the oil industry's intent to make a meaningful contribution to the clean air initiative. Immediately following the Prince William Sound oil spill by the Exxon Valdez, the reputation of the oil industry and its perceived concern for the environment were at low ebbs. Despite the skepticism, and long before the Prince William Sound spill, most oil companies have operated to a very high level of compliance with environmental regulations. Post the Alaskan spill, there is an even heightened sensitivity to the environment.

It is clear that the public is calling on the industry not only to provide the fuels for personal transportation, but to help reduce air pollution. There is considerable debate about how to reduce pollution from automobiles, but there is no mistaking the public's demand for air quality improvements and there is no question that transportation fuels are a part of the solution. The industry simply must respond to the need. I believe that we as an industry and BP in particular recognize that need and we will be part of the solution. I hope to describe for you a few ways in which we can hope to alleviate the very difficult air quality problem facing us all.

Getting ready for this talk, I jotted down an informal description for what I want to say. It's: "Good news, better news, and a warning." I think I have

From John G. McDonald, "Gasoline and Clean Air: Good News, Better News, and a Warning," *Vital Speeches of the Day* (August 15, 1990).

some good news to share with you. I think I have some news that is even better. But I do have a warning to add. Let me get to work.

First of all, when we talk about air quality, please realize that we're talking about an issue with many dimensions.

Air quality concerns range from acid rain to global warming to smog and health issues. Some of the concerns are very tangible. You can see and smell smog. Others are the subject of ongoing debate. Is global warming even real? How many parts per million of a particular air toxic pose a health threat, and so on.

Air contaminants which contribute to this variety of air quality problems include sulfur dioxide, particulate emissions, hydrocarbons, nitrogen oxides, carbon monoxide, carbon dioxide, and toxic traces. The contaminants stem from a variety of sources. Power plants, factories, and automobiles are contributors along with a host of less often considered sources ranging from dry cleaners to paint fumes and even natural causes.

To keep the topic today to manageable proportions, I intend to focus my comments only on one important element of the debate, the impact of automobile emissions on air quality. There are three main areas of current concern arising from conventionally fueled automobiles and they are air toxics, carbon monoxide, and smog.

The first area of concern is air toxics, and in terms of automobile emissions, benzene is the most frequently discussed contaminant. Benzene is considered to be an air toxic because it is a known carcinogen. Motor vehicles account for approximately 85 percent of the nation's total benzene emissions. Automobiles release benzene into the atmosphere primarily through tail pipe emissions, and to a much lesser degree, evaporation from gasoline itself. How serious is the question of air toxics from automobiles? The Environmental Protection Agency has released figures which indicate that at the extreme, those emissions account for less than a tenth of one percent of all cases of cancer.

The second area of concern is carbon monoxide. Carbon monoxide causes dizziness and fatigue and can impair central nervous system functions. Gasoline-powered vehicles release about half of the nation's carbon monoxide emissions, with pre-1985 vehicles accounting for a disproportionate share. Automobiles emit carbon monoxide as a result of incomplete combustion in the engine chamber, caused by fuel burning without quite enough air. There is a federal ambient air quality standard for carbon monoxide which forty-four metropolitan areas in the United States fail to meet today.

The third area of concern is smog, with which we are all familiar. Hydrocarbons and nitrogen oxides react in the presence of sunlight to form ground-level ozone, which is a major constituent of smog. Gasoline-powered vehicles account for about one-fourth of the nitrogen oxide and hydrocarbon emissions in the United States. Elevated ozone concentrations result in reduced lung function, especially during vigorous physical activity. This health problem is particularly acute in children and the elderly.

A federal ambient air quality standard exists for ground-level ozone, and one hundred and one metropolitan areas in the United States fail to meet the standard today. The degree of the problem varies significantly. Some non-attainment areas are out of compliance only a few days each year whereas Los Angeles, with the most severe air quality problems, fails to meet the federal standard 145 days each year.

The good news is that the auto/fuel package we buy today is a far better environmental system than just six years ago. Lead has been eliminated from the gasoline, and gasoline vapor pressure (its tendency to evaporate) is significantly reduced. Nitrogen oxide emissions from the tail pipe have been reduced by 75 percent, and hydrocarbon emissions, and carbon monoxide have been slashed by over 95 percent. This represents tremendous progress, but more must be done!

Unfortunately, identifying the types of air pollution created by the automobile is the easy part. The combustion and atmospheric chemistry underlying the production of these emissions is nowhere near as well understood. This is somewhat surprising since the automobile has been with us for a century, but nonetheless a fact. That is why BP together with 13 other oil companies have combined with the three major U.S. auto manufacturers to carry out research on how changes in gasoline components and automobile technology will further reduce vehicle emissions and improve air quality.

Gasoline is not a single chemical, but rather a blend of over 200 different hydrocarbons. Automobile emissions, both from the evaporation of gasoline and from the tail pipe contain an array of hydrocarbons and other components, each of which have a different propensity to react and form ozone.

The results from the first phase of the Auto/Oil Industry Research program are expected to go beyond an understanding of the quantitative relationships between fuel composition, automobile engines and control technology, and the associated emissions to an understanding of the relative tendency of those various emissions to form ozone. In other words, the research is underway to determine how much reduction in ozone (I.E., smog) can be expected from various gasoline reformulations. This first phase of the research will be completed later this year and will cost some $15 million.

I believe strongly that the solution to the air quality problem must first and foremost be based on sound science which at this point we've been lacking. The degree of cooperation between the oil and auto industries in this cooperative effort is completely unprecedented, and there is no question that it is advancing our understanding. At the moment however, this platform of sound science is just being built.

One alternative in the air quality issue that has a number of supporters is to substitute methanol for gasoline in automobiles. Methanol exhaust emissions promise advantages over *today's* gasoline, especially with regard to ozone formation, but we don't know how it compares to reformulated gasoline. We will have a better quantification of this against the benefits of *reformulated* gasoline when the Auto/Oil Industry Research program is completed, and I would not wish to prejudge these results.

People who are less enthusiastic about methanol have raised concerns about its toxicity and solubility in water (therefore the potential difficulty to detect and clean up a spill). They have raised concerns about formaldehyde emissions from methanol-powered vehicles. Formaldehyde, like benzene, is a toxin and a probable human carcinogen. Like some hydrocarbons, formaldehyde reacts with nitrogen oxides in the presence of sunlight to form ozone. To some extent, there is an element of "the devil you know versus the devil you don't." There are no easy answers.

Without entering the methanol debate, there *is* an important point to be considered. Namely, regardless of how extensive a role methanol plays in the air quality solution it cannot be done as quickly as can changes to gasoline.

Air quality improvements from methanol-powered vehicles take time; time to establish a fleet of methanol vehicles, time to develop distribution networks, time to expand methanol production capabilities. Even in a fairly localized area with severe air quality problems, like Los Angeles for example, it will take a long time for methanol to make a significant air quality contribution. The average age of an automobile today is 8 years. That typical car cannot burn methanol. With such a slow vehicle turnover, it still takes years for methanol-powered vehicles to become a significant proportion of the fleet even after the vehicles are generally available and accepted by the public.

Reformulated gasoline doesn't require a new fleet of vehicles or a new distribution network. Therefore, the air quality benefits can be achieved sooner. Reformulated gasoline will provide a way to reduce the emissions of the older, worse-polluting vehicles as well as new automobiles. The impact is immediate and multiplied by a tremendously larger number of cars.

If there is such a large potential prize to be gained by optimizing the fuel to the current fleet, why isn't the petroleum industry supplying this new gasoline in quantity now? The main reason is the lack of a scientific platform that I referred to earlier. The kind of fuels being talked about cannot be made in quantity today and will require huge investments by the oil industry. These huge investments, which are obviously going to translate into higher costs for the industry and ultimately for the consumer, should not be made without understanding the air quality improvements to be achieved. Nobody wants to buy a pig in a poke. The investments should be made where the greatest air quality improvements may be realized.

Ideally, we would like to eliminate *all* vehicle emissions that are detrimental to our air quality. But this would be enormously costly.

Fortunately, due to competition and today's oversupply of crude oil, gasoline now costs less, after adjusting for inflation, than it did 20 years ago before the first large oil price hikes. Nonetheless, the consumer is very sensitive to gasoline prices and, in particular, to price increases. This is because we all enjoy the freedom, flexibility, and pleasure provided by the automobile. Cars are an integral part of our culture. They are perhaps the foremost symbol of our high standard of living in the United States. The infrastructure of many of our cities is dependent on use of the private automobile.

The key dilemma we face is how to make the greatest air quality improvements without making the personal transportation freedom we have come to value either prohibitively expensive or significantly restricted. This trade-off lies at the heart of the issue. These tough choices will not be made by the oil industry. They are choices that you and our elected officials will make. It is my hope that our industry can help in three ways.

First, through research we can define the air quality benefits to be gained from various fuel alternatives and provide an understanding of the costs involved. Basically, we can help provide the information needed to make a rational choice.

Second, we should encourage legislation to take a form that achieves air quality objectives in the most cost-effective manner. Ideally, legislation should identify emission performance targets to be reached. The industry could then determine the most efficient formulations and investments to achieve compliance. Emission performance targets rather than fuel specifications would guarantee air quality objectives are met.

Unfortunately, in my view, the legislative momentum has swung from specifying emission performance targets and allowing market forces to come up with the best technical solution, to legislating a specific gasoline composition. Without the scientific base which is now just being developed, there is no guarantee that the desired level of emissions will be achieved nor that the investments which will be required will result in the greatest air quality benefit. Our legislators are playing doctor. If they write a prescription, I hope the laws retain flexibility to change when the real doctors find a better formula!

The third and most important area where our industry can help is to meet the dual demand now facing us. Provide affordable fuel for personal transportation and meet the need with fuel that is more friendly to our environment.

As alluded to earlier, another dimension of the issue is that not all areas of the country *have* an air quality problem, nor do all non-attainment areas have the same magnitude of problems. Los Angeles, due to its climate and density of population, has by far the worst air quality in the nation. In fact, some projections suggest that even if all the automobiles were removed from LA's roads, it still would not comply with the national ozone standard. During the period 1986 through 1988, LA averaged 145 days each year above the ozone standard whereas Cleveland averaged only 6 days per year.

As I've said, air quality improvements do not come without costs. To the extent that an area like Los Angeles has an extreme problem, the solutions in Los Angeles will likely be quite costly. It doesn't make sense to impose what may be very expensive solutions for a problem as large as that in Los Angeles on all non-attainment areas. And we don't need any changes in those areas that have attained the legal standards. On the other hand, it is uneconomical to develop dozens of different alternatives for various degrees of non-attainment. Some complex choices lie ahead.

As you are probably aware, the Clean Air Act is currently being debated in the House, having passed through the Senate. From my understanding of the current status of the fuel proposals, I would make the following comments. Allow me to address the three areas of concern regarding automobile emissions; air toxics, carbon monoxide, and ozone (smog).

First, air toxics. I would support the reduction in benzene content of gasoline down to the vicinity of 1 percent from the current industry average of around 2.5 percent. Reducing benzene much below 1 percent, however, would mean a very steep rise in costs with extremely small cancer reduction benefits.

Second, carbon monoxide. I believe that carbon monoxide pollution will soon fade as an issue. This is some of the "good news" which I promised. Because of the advances in automotive technology, older cars now account for 90 percent of the mobile source carbon monoxide emissions. Post 1985 cars all have an oxygen sensor which adjusts the fuel to air ratio to avoid incomplete combustion and the creation of excess carbon monoxide. Therefore, as the car fleet turns over, with new cars replacing old, carbon monoxide emissions will naturally diminish. We may, however, continue to need good vehicle inspection programs to assure that the newer vehicles continue to run as designed by the manufacturer.

To accelerate the reduction of carbon monoxide pollution, a number of Western cities and the current version of the Clean Air Bill being debated in the House mandate the use of oxygenates in gasoline. When oxygenates are added to the gasoline consumed in older cars, especially those without oxygen sensors, combustion is more complete and carbon monoxide emissions are reduced. Examples of oxygenates are ethanol and other alcohols and various hydrocarbon ethers, which contain oxygen within their molecular structure.

Since the use of oxygenates in the fuel can reduce the carbon monoxide produced by older cars, I would support their use as short-term expedients in the most severe carbon monoxide non-attainment areas until these areas come into compliance. However, use of oxygenates is unnecessary in new cars since their carbon monoxide emissions are very low and oxygenates are expensive gasoline components. Further, the supply of oxygenates is very limited and cost substantially more per gallon than today's gasoline. Mandating their use will make them even more expensive.

Having covered the Clean Air Act issues relative to benzene and carbon monoxide, this brings me finally to ozone, which is probably the most difficult of the automobile air quality issues. Recall that hydrocarbons and nitrogen oxides react in sunlight to form ozone (or smog). While the hydrocarbon tail pipe emissions of new cars have been reduced to below 5 percent of their uncontrolled level and nitrogen oxide to below 25 percent, the automobile fleet has grown substantially. Turnover of the car fleet alone won't solve the ozone problem in the worst areas. It is also the most complicated issue because differing hydrocarbons have differing reactivities, and different geographic areas have different tendencies to form ozone depending upon different hydrocarbon to nitrogen oxide ratios that exist. It is also important to realize that although cars emit both hydrocarbons and nitrogen oxides, they constitute only about one-fourth of the overall emissions that contribute to ozone. Large stationary sources such as power stations and factories as well as dry cleaners, and household items like charcoal lighter fluids contribute to the problem.

A certain class of hydrocarbons in gasoline called aromatics and another called olefins are believed to have a greater tendency to form ozone, than other hydrocarbons. This is one of the key areas being addressed by the Auto/Oil Industry Research program. It may be that some, but not all aromatics and olefins are bad. It may be that some of those compounds are created in the engine combustion chamber regardless of their content in the gasoline. We need the scientific research to understand the relationships.

However, before the research is completed, the current version of the Clean Air Act mandates a progressive reduction in maximum aromatics content and mandate levels of oxygenates for the nine worst ozone non-compliance areas in the country. If fuel specification mandates are inevitable, the legislation should be written to leave wiggle room for the U.S. Environmental Protection Agency to adjust the requirements, based on sound research, to the most effective specifications.

The reduction sought in aromatics content is probably directionally right. Mandating of oxygenates, on the other hand, may actually make the ozone air quality problem worse in some areas. Mother nature can sometimes be perverse. Those same oxygenates which reduce carbon monoxide emissions in carbon monoxide non-attainment areas lead to increased quantities of nitrogen oxide emissions, thereby aggravating the ozone situation in ozone non-attainment areas. Only two of the nine worst ozone areas are also serious carbon monoxide non-attainment areas, yet Congress is on the verge of requiring oxygenates in all of the nine worst ozone non-attainment areas. Furthermore, the current supply of oxygenates falls short of the amount required to satisfy the proposed oxygenate mandate even in those nine areas.

As I mentioned before, we would prefer legislation that sets a tail pipe performance *standard* and allows the petroleum industry to mix and match the gasoline components to meet the emission standards in the most effective way. At the end of the day the consumer will pick up at least part of the cost and an efficient solution is in everyone's best interest.

Let me make a categorical statement here: If research shows that aromatics or olefins turn out to be bad, then the industry will reduce them. It's as simple as that. And I intend to have BP America among the leaders.

So, what have I said in this talk about clean air?

Some good news: Automobile emissions today are lower than ever before and the carbon monoxide problem from cars is disappearing.

Some better news: The oil industry recognizes the need for further substantial changes in transportation fuels and is actively working to improve air quality.

And a warning: Cleaner transportation fuels mean higher cost for those fuels, and mandating fuel specifications will likely mean higher costs than mandating emission standards.

Please understand clearly: I agree completely that cleaning up the air is a serious problem. No one in the oil industry should quibble with that in today's world. Please understand, too, that we see very clearly that gasoline is still part of the problem, and needs further improvement.

I'm trying to share facts today, not pass the buck.

We at BP are working hard to create new fuels that provide the greatest air quality improvement at a reasonable cost for our customers. I can't promise you when or where BP's own new green gasoline will appear, but I assure you it will come.

NO

Hilary F. French

COMMUNICATION: A GLOBAL AGENDA FOR CLEAN AIR

Keywords: **Pollution; Energy; Transportation**

The world's first clean air act may have been a 1306 proclamation by King Edward I of England banning the burning of 'sea coles' (chunks of coal found along the seashore) by London craftsmen. Almost seven centuries later the problem is far from solved all over the world. Indeed, air pollution has proven so intractable a phenomenon that a book could be written about the history of efforts to combat it.

As one air pollution challenge has been met, a new one has frequently emerged to take its place. Even some of the solutions have become part of the problem: the tall smokestacks built in the 1960s and 1970s to disperse emissions from huge coal-burning power plants became conduits to the upper atmosphere for the pollutants that form acid rain.

Recently, global warming has arisen as a main focus of environmental concern, sometimes conveying the misleading impression that air pollution is yesterday's problem. But traditional air pollutants and greenhouse gases stem largely from the burning of fossil fuels in energy, transportation and industrial systems. Having common roots, air pollution and global warming can also have common solutions. Unfortunately, policy makers persist in tackling them in isolation, which runs the risk of lessening one set of problems while exacerbating the other.

The reason efforts to clear the air have only been marginally successful at best is because they have focused on specific measures to combat individual pollutants rather than addressing the underlying societal structures that give rise to the problem in the first place. Turning the corner on air pollution requires moving beyond patchwork, end-of-the-pipe approaches to confront pollution at its sources. This will mean reorienting energy, transportation

From Hilary F. French, "Communication: A Global Agenda for Clean Air," *Energy Policy*, vol. 18, no. 6 (July/August 1990), pp. 583-585. Notes omitted.

and industrial structures towards prevention.

REDUCING EMISSIONS

Scrubbers, nitrogen oxides control technologies, and new cleaner-burning coal technologies can all reduce emissions dramatically, but they are not the ultimate solutions. For one, they can create environmental problems of their own, such as the need to dispose of scrubber ash, a hazardous waste. Second, they do little if anything to reduce carbon dioxide (CO_2) emissions, so make no significant contribution to slowing global warming.

For these reasons, technologies of this kind are best viewed as a bridge to the day when energy-efficient societies are the norm and pollution-free sources such as solar, wind and water power provide the bulk of the world's electricity.

Improving energy efficiency is a clean air priority. Such measures as more efficient refrigerators and lighting can markedly, and cost effectively, reduce electricity consumption, which will in turn reduce emissions. Equally important, the savings that result from not building power plants because demand has been cut by efficiency can more than offset the additional cost of installing scrubbers at existing plants.

Similar rethinking can help reduce car emissions. To date, modifying car engines and installing catalytic converters have been the primary strategies employed to lower harmful emissions. Although catalytic converters are sorely needed in countries that do not require them—they reduce hydrocarbon emissions by an average of 87%, carbon monoxide by an average of 85%, and nitrogen oxides by 62% over the life of a vehicle—they alone are not sufficient. Expanding car fleets are overwhelming the good they do, even in countries that have mandated their use.

Reducing air pollution in cities is likely to require a major shift away from automobiles as the cornerstone of urban transportation systems. As congestion slows traffic to a crawl in many cities, driving to work is becoming unattractive anyway. Convenient public transportation, car pooling and measures that facilitate bicycle commuting are the cheapest, most effective, ways for metropolitan areas to proceed.

Driving restrictions already exist in many of the world's cities. For example, Florence has turned its downtown area into a pedestrian mall during daylight hours. Budapest bans motor traffic from all but two streets in the downtown area during particularly polluted spells. In Mexico City and Santiago, one-fifth of all vehicles are kept off the streets each weekday based on their licence-plate numbers.

Improving the fuel efficiency of automobile fleets would also help reduce air pollution. An analysis of 781 car models by Chris Calwell of the Natural Resources Defense Council shows that the 50 most efficient emitted one-third less hydrocarbons than the average car and roughly half as much as the 50 least efficient ones.

As with power plant and auto emissions, efforts to control airborne toxic chemicals will be most successful if they focus on minimizing waste rather than simply on controlling emissions. The Congressional Office of Technology Assessment has concluded that it is technically and economically feasible for US industries to lower the production of toxic wastes and pollutants by up to 50%

within the next few years. Similar possibilities exist in other countries.

A POLITICAL PROGRESS REPORT

In much of the world, air pollution is now squarely on the public policy agenda. This is a promising sign. Unfortunately, the public's desire for clean air has not yet been matched with the political leadership necessary to provide it. Recent developments at the national and international levels, though constituting steps forward, remain inadequate to the task. In the US, for example, Congress is on the brink of adopting major amendments to the Clean Air Act of 1970 that will cut in half the emissions that cause acid rain, tighten emissions standards for automobiles significantly, and require much stricter control of toxic air pollutants.

Almost any legislation would be an improvement. Twenty years after the Act became law, 487 countries still are not in compliance. But, the proposed legislation fails to address the problem at a fundamental level by not encouraging energy efficiency, waste reduction, and a revamping of transportation systems and urban designs.

Los Angeles—with the worst air quality in the US—has been one of the first regions in the world to really understand that lasting change will not come through mere tinkering. Under a bold new air quality plan embracing the entire region, the city government will discourage automobile use, boost public transportation, and control household and industrial activities that contribute to smog.

Because air pollution crosses borders with impunity, international cooperation on reducing emissions is essential. The past few years have seen important incremental advances in international forums. Under the auspices of the UN Economic Commission for Europe, agreements to reduce emissions of sulphur dioxide (SO_2) and nitrogen oxides have been reached. The SO_2 protocol, adopted in July 1985, calls for a 30% reduction in emissions or their transboundary flows from 1980 levels by 1993. The nitrogen oxides accord, signed in November 1988, calls for a freeze on emissions at 1987 levels in 1994, as well as further discussions beginning in 1996 aimed at actual reductions.

A November 1988 directive by the European Economic Community (EEC) represents a binding commitment by the members to reduce acid rain-causing emissions significantly. The directive will lower community-wide emissions of SO_2 from existing power plants by a total of 57% from 1980 levels by 2003, and of nitrogen oxides by 30% by 1998.

The European Community, though quicker than the US to cut back sharply on the emissions that cause acid rain, has been slower to tackle urban air quality. This finally changed in June 1989, when the EEC Council of Environmental Ministers ended a nearly four-year debate and approved new standards for small cars. These will be as tough as those now in effect in US.

AN AGENDA FOR CLEAN AIR

As we enter the 1990s, air pollution policy around the world is in flux. Though numerous technologies and strategies are available to reduce emissions, their piecemeal adoption has led to overall failure in the battle for clean air. To clear the air, it will be necessary to supplement traditional air pollution strategies with innovative policies that encourage energy efficiency and non-polluting, renewable energy sources, minimize dependence on the automobile and reduce the production of toxic wastes.

Developing a more environmentally benign energy system depends critically on correcting market imperfections. The first step is to ensure that subsidies are removed so that the cost of energy reflects its true value, providing an incentive to use less. In PR China, for example, energy efficiency has improved an average of 3.7% annually since the country's economic reform programme began in 1979.

Incorporating the environmental costs of burning fossil fuels into the power planning process is a crucial second step. In the US the State of New York is pioneering this approach. Under an innovative programme, a competitive market is being set up whereby independent power producers must bid against each other for power supply contracts. Suppliers planning to burn fossil fuels are required to add nearly one cent/KWh to their bids to account for air pollution, and an additional half cent/KWh to account for other environmental costs.

Taxing emissions also can serve as a strong disincentive to pollute. Swedish officials have drafted proposals that would tax sulphur, nitrogen oxides and CO_2 emissions. If the proposed plan is approved, by the end of the decade it would reduce SO_2 emissions by 7% from their 1987 level, nitrogen oxides by 8%, and CO_2 from 9–17%.

New transportation policies can help to lessen society's dependence on automobiles—especially the most polluting and fuel-inefficient varieties. In addition to restricting the entry of cars into downtown areas, municipal governments can invest more in efficient public transportation systems and help commuters organize car pools to keep vehicles off the road. Developing policies to ensure that drivers pay the full cost of using their cars could also curb automobile dependence. Taxing or eliminating free parking benefits, collecting tolls or fees for road use, and imposing hefty gasoline and automobile sales taxes are all means to this end.

To encourage the manufacture and sale of less-polluting cars, governments could consider an emissions-pegged sales tax. Analysts at the Lawrence Berkeley Laboratory in Berkeley, CA, have suggested a revolving fund that would tax new high-emission cars, and subsidize cleaner ones. Consumer demand would then spur the industry to produce less-polluting cars.

Because emissions of toxic chemicals into the air are commonly less strongly regulated than those onto land or into water, stricter controls on them are sorely needed. Over the long term, this will serve as an incentive to reduce waste: as disposal becomes more difficult and expensive, waste reduction and recycling become increasingly attractive. Certain other economic incentives could also help. For example, a 'deposit-refund' system might be implemented whereby industries are taxed on the purchase of hazardous inputs, but given a refund for wastes produced from them that are recovered or recycled.

While the means are available to clean the air, it will be a difficult task. In the West, powerful businesses such as car manufacturers and electric utilities, will strongly resist measures that they fear will cost them money. In Eastern Europe, the USSR and the developing world, extreme economic problems, coupled with shortages of hard currency, mean that money for pollution prevention and control is scarce.

On the promising side, faced with ever-mounting costs to human health and the environment, people on every continent are beginning to look at pollution prevention through a different economic lens. Rather than a financial burden, they are seeing that it is a sound investment. The old notion that pollution is the price of progress seems finally to be becoming a relic of the past.

POSTSCRIPT

Can Current Pollution Strategies Improve Air Quality?

Given his position as the president of a major international petroleum company, it is not surprising that McDonald's proposals include reformulation of petroleum-based fuels and continued improvements in the efficiency of vehicles powered by internal combustion engines rather than replacing fossil fuels with nonpolluting energy sources.

Although he mentions the fact that the developing world faces severe economic constraints that impede pollution prevention efforts, French fails to elaborate on this very serious problem. An obvious priority of leaders of Third World nations is to reduce the enormous disparity between the standards of living of their people and those in developed industrial nations. If the leaders of the industrial nations that have made the largest contributions to global air pollution problems wish to prevent the developing nations from contributing to the problem, they may have to be willing to pay a significant part of the cost. The recently revealed severe pollution problems in eastern European countries illustrate the fact that poorly planned, underfinanced, rapid industrial development can be environmentally disastrous.

To understand the compromises that finally permitted passage of the 1990 Clean Air Act revisions, it is necessary to learn about the heated controversies that evolved during the long and troubled history of this complex legislation. For a synopsis of the environmentalist and industry positions on the need for new air pollution legislation near the start of the debate, see "Clean Air, The Public Speaks—Will Congress Listen?" *Sierra Club Bulletin* (January/February 1982) and "Why Business Wants the Clean Air Act Changed," *Nation's Business* (June 1982). For a critique of air regulation enforcement under the Reagan administration, see "The Elusive Bubble," by Rochelle Stanfield, *National Journal* (April 5, 1986). One of the key controversies is how to reduce ground level ozone concentrations. Three alternative strategies for achieving this goal are the subject of "Innovative Approaches for Revising the Clean Air Act," by Robert W. Hahn, *Natural Resources Journal* (Winter 1988).

While bemoaning the compromises with industry that resulted in less stringent requirements and delays in implementation of some of its key provisions, environmentalists are in general agreement that the new Act will result in significant progress, especially the acid rain reduction initiatives.

A growing number of environmental analysts share French's view that global improvement in air quality can only be achieved through the replacement of present polluting technologies by environmentally benign alternatives. This is certainly the message of environmental scientist Barry Commoner's new book, *Making Peace With the Planet* (Pantheon, 1990). A key element in most plans for environmentally sound, sustainable development is a greater reliance on clean, renewable energy sources. "Renewing Renewable Energy," by Susan Williams, Scott Fenn, and Terry Clausen in the Spring 1990 issue of *Issues in Science and Technology* presents an optimistic assessment of the potential for increased reliance on renewable energy.

ISSUE 7

Is Nuclear Power Safe and Desirable?

YES: Alvin M. Weinberg, from "Is Nuclear Energy Necessary?" *Bulletin of the Atomic Scientists* (March 1980)

NO: Denis Hayes, from "Nuclear Power: The Fifth Horseman," *Worldwatch, Paper 6* (May 1976)

ISSUE SUMMARY

YES: Nuclear physicist Alvin M. Weinberg argues that development of nuclear breeder reactors is needed to assure future energy supplies and that safety can be provided through technical improvements and siting in remote areas.
NO: Energy analyst Denis Hayes contends that development of nuclear power assures nuclear weapons proliferation and that environmentally safer energy sources are technically and economically feasible.

Nuclear fission provides a technologically feasible means of producing energy to power steam turbines for large-scale electric power generation. This basic fact is about the only relevant issue on which the proponents and opponents of nuclear power agree.

Ever since former president Eisenhower's announcement of the "atoms for peace" program in the early 1950s there has been a growing controversy surrounding the development of commercial nuclear power. Some social scientists contend that the U.S. government promoted utility-operated nuclear plants to undermine the developing public opposition to atmospheric testing of nuclear weapons. If so, this hope has not been realized.

Initially, there were sharply conflicting estimates of the potential hazards. The central issue now is whether the public can be protected from exposure to the lethal radioactive isotopes produced in the core of a reactor. The safety issue includes all steps in the fuel cycle—from the mining of the ore to the necessary long-term storage of the wastes. The difficulty in finding an acceptable means of dealing with radioactive waste (see the issue on nuclear waste disposal) illustrates an essential ecological fact: When something is discharged into the environment, it doesn't simply disappear. No matter where something is buried or stored, it will ultimately resurface.

The possibility of a major release of radiation from an accident at a nuclear plant has been a constant theme in the nuclear debate. The near disaster at the Three Mile Island plant in 1979 resulted in widespread loss of confidence in the pro-nuclear experts who had offered assurances that such an event was extremely unlikely. The deaths, injuries, and international radioactive contamination resulting from the explosion and fire at the Soviet Chernobyl reactor in Spring of 1986 has renewed and heightened opposition to nuclear power.

Economic factors have recently dealt a possible death blow to the efforts of the nuclear industry to prosper in the face of growing public fear. High interest rates and expensive new safety systems have driven up the price of nuclear-generated electricity. Estimates of the total cost of building a nuclear plant rose by more than a factor of 15 between 1974 and 1986. Faced with the staggering task of raising billions of dollars for a single new facility and the declining demand for electricity, utility planners have rapidly retreated from the nuclear option. No new nuclear plants were ordered in the United States from 1979 through 1988. Numerous orders that were placed prior to that period have since been cancelled.

Present efforts to revive the nuclear industry are motivated by dwindling supplies of oil, concern about "greenhouse" atmospheric warming, and the general need to reduce air pollution. As explained in the issue on global warming, the coal and petroleum that are currently the principal sources of energy in the industrialized world are the major causes of the increased atmospheric carbon dioxide that threatens to cause global warming over the next several decades. Nuclear power, despite its other drawbacks, produces none of the major air pollutants. Nuclear opponents contend that conservation and renewable solar energy sources are more cost-effective, environmentally safe alternatives and have a wider variety of applications than does nuclear power.

Alvin M. Weinberg has long been a supporter of nuclear power. He contends that we must keep the nuclear option open and develop the breeder reactors that can assure a long-term source of energy for continued worldwide economic development. Unlike some nuclear proponents, he accepts the likelihood of major nuclear accidents but contends that the risk can be reduced to a tolerable level by technical improvements and by placing nuclear plants in remote locations.

Denis Hayes supports the solar alternative. He argues that proliferation of nuclear weapon capability is an inevitable consequence of nuclear power development. Thus a commitment to the nuclear option precludes the worldwide stability and security which he views as essential if the potential hazards of this technology are to be contained.

YES

Alvin M. Weinberg

IS NUCLEAR ENERGY NECESSARY?

Two questions dominate the nuclear issue: Is nuclear energy necessary? Can nuclear energy be made acceptable? Unless the answer to the first is affirmative, there is little incentive to devise the improvements in nuclear energy necessary to rescue it from its present malaise. In my view, nuclear energy is highly desirable, if not absolutely necessary. It can be made acceptable. And rather than continue the bitter confrontations between proponents of renewable sources and proponents of nuclear sources, the energy community ought to put its efforts into achieving an acceptable nuclear system. We shall need all sources of energy; we cannot afford to reject nuclear energy because its current embodiments are faulted.

The fission chain reaction is a bit of a scientific fluke: for example, there is no theoretical reason why the number of neutrons emitted per fission must be sufficient to maintain a chain reaction. One cannot therefore argue that mankind would perish had fission not been discovered. Indeed, before 1939 energy futurologists, recognizing that fossil fuels were finite, speculated on the possibilities of drawing energy in the long run from the various solar sources—including wind, waves, ocean thermal gradients, and biomass—as well as geothermal sources. Nevertheless, the outlook at the time was rather pessimistic. Most pessimistic was the assessment of Charles G. Darwin, descendant of the biologist. In his book, *The Next Million Years,* published in 1953, he predicted a brutish, Malthusian future for man unless he developed an inexhaustible energy source other than the sun. Darwin's candidate for such a source was controlled fusion.[1]

Many of us in the fission community have recognized that in the fission breeder man had another path to salvation from Darwin's ultimate Malthusian disequilibrium. It is therefore understandable that we addressed the development of the breeder with such enthusiasm: here was an embodiment of Darwin's inexhaustible energy source that would, to use H. G. Wells' words of 1914, "Set Man Free."[2]

From Alvin M. Weinberg, "Is Nuclear Energy Necessary?" *Bulletin of the Atomic Scientists* (March 1980).

Given this noble, perhaps noblest, of all technological dreams, it is almost incomprehensible to us why the world is now asking: Is fission necessary? Is not the relevant question: What can be done to eliminate the deficiencies of fission, rather than eliminating fission itself?

The short range: to 2000. What can we say about the necessity of fission in the near term, roughly to the year 2000? Obviously the need for fission depends upon the availability of alternative fuels or sources of energy, and upon the future demand for energy. These are not matters that can readily be settled for the whole world. In the United States, with its abundant coal, for example, a moratorium on fission would be less serious than it would be in Japan or France. The Institute for Energy Analysis (IEA), in its study, "Economic and Environmental Implications of a U.S. Nuclear Moratorium, 1985-2010," concluded that the United States could weather a limited moratorium with a loss of 0.5 percent in GNP. The moratorium would allow completion of all reactors under construction by 1985; no new reactors would be built after 1985.

The moratorium would place great pressure on coal or imported oil or both. It was estimated that between 18 and 27 billion tons of additional coal would be needed by 2010 to fuel those stations that would serve in place of nuclear plants not built; alternatively, the additional imported oil would amount to from 6 to 9 billion tons. Re-examining these estimates three years after they were first made, I would say the impact might be overstated, particularly because we expected electricity to capture 46 percent of the energy market by 2000, compared to its present market share of 30 percent. While electricity's market share will probably increase, it seems unlikely that it will increase this much. Nevertheless, in view of the great environmental problems associated with coal mining, let alone the political tensions created by expanded import of oil, I conclude that even this limited nuclear moratorium is very undesirable.

Most other countries do not have coal. For them, rejection of nuclear energy would certainly entail costly importation of coal and oil. A one-gigawatt (electric) oil-fired power plant uses 1.4 million tons of oil per year. Throughout the world 800 million tons of oil are burned annually in central electric power stations; this represents 25 percent of the world's consumption of oil in the year 1978. Should all the oil-fired plants be replaced over the next decade with nuclear plants, the pressure on oil would be reduced significantly.

In recent years it has become fashionable to fault this line of argument as simplistic. Rather than replace oil in electric power stations with coal or uranium, we are asked to believe we can so reduce our energy demand as to make many existing, let alone future, electric power plants superfluous. In any case, the use of electricity for such purposes as heating of houses or water is deemed inelegant and wasteful and ought to be discouraged.

It goes without saying that conservation must be central to any energy policy; indeed, much has already been accomplished. For example, D. Reister of the Institute for Energy Analysis points out that in the United States, the ratio of energy to gross national product has decreased by 10 percent between 1970 and 1978.[3] But as Stobaugh and Yergin point out in the Harvard Energy Futures study, conservation requires decisions by innumerable consumers; by contrast, increased supply requires far fewer decisions. Thus the prediction of how much conservation is actually achieved—as contrasted with how much is theoretically achievable—is intrinsically less certain than is the prediction of how much supply can be increased. (To be more accurate, since energy supply and demand must balance, at issue is the relative freedom of choice afforded by policies that depend on conservation rather than on increased supply). The difficulty has recently been analyzed by P. C. Roberts of the United Kingdom.[4] Roberts gave evidence that, in the United Kingdom, the amount a family is likely to spend to retrofit its house with energy-conserving devices goes as the square of the family income. Roberts thus estimates that, even if fuel costs double, the amount of conservation induced by market forces in the next 25 years is about 35 percent of the theoretical. Conservation in houses, at least in the United Kingdom, does not happen automatically; subsidies of various sorts seem to be necessary. This is not to say that conservation is unimportant; it is simply that conservation in the residential sector is probably more difficult to achieve than in the industrial sector.

The current mood of rejection of electricity seems irrational. If oil is scarce and coal and uranium are abundant, it makes sense to replace oil with coal and uranium, even if in so doing one must resort to inefficient resistive heating or other devices that use electricity. After all, we are not driven by thermodynamic imperatives: economic or political considerations, such as reducing our dependence on foreign oil, certainly take precedence over the much discussed, but often irrelevant, structure to improve the second law efficiency.

Two technical developments, the electric car and the heat pump, could swing the balance toward an electrical future dominated by large central stations. The recent announcement by General Motors of a car powered by a zinc battery whose lifetime is 50,000 kilometers, with a cruising speed of 80 kilometers per hour and a range between recharges of 165 kilometers could, if realized, alter our attitude toward the electric future. In addition, the electrically driven heat pump is a proven device. In the United States 560,000 heat pumps were installed last year, and this number has been increasing by 40 percent each year.

If one concedes that:

- a predominately electric future is at least as plausible as a nonelectric one,
- oil will continue to be scarce,
- the solar technologies will not penetrate on a large scale (cost, intermittency, storage?),
- coal is not generally available except in countries that possess indigenous deposits (would the United States be prepared, in 50 years, to mine an additional billion tons of coal, to be shipped overseas?),

then it seems inescapable that fission is necessary, at least in some part of the world. Beyond this, if the cost per joule of coal reaches that of oil—that is $80 per ton of coal at present, possibly $140 per ton with oil at $40 per barrel—then even though the cost of nuclear reactors is high (say, $1,5000 per kilowatt), electricity from current reactors probably will still be cheaper than electricity from coal-fired stations.

There are uncertainties in these assumptions. But it is because we cannot know with certainty even the next 20 years that we ought to preserve all our energy options, including nuclear. In this sense, I would judge nuclear energy in the short term to be a necessity.

The long term—beyond 2030. We must look at the time, perhaps 50 to 100 years from now, when oil and gas have become scarce. Fission, if it survives, eventually would be based on breeders, though not necessarily fast breeders. Three issues dominate the long-term outlook for fission: the availability and acceptability of alternatives, the long-term energy demand, and carbon dioxide.

On the scale under discussion, only fusion or the various solar sources are large enough to compete with fission. Here I must admit to being agnostic. Fusion *may* turn out to be technically and economically feasible, but there is no way of knowing. And in good measure this is true of solar energy. Despite the many claims of technological and economic breakthroughs, the uncertainties remain. Such figures as 50 cents per peak watt, the aim of the Department of Energy for solar cells, are still no more than a hope.[5] Given these uncertainties about alternatives to the fission breeder, prudence requires us to develop the breeder, and to deploy it if the alternatives prove too costly or turn out to be unfeasible. In short, it is far too early to reject any of the long-term energy options, particularly the breeder.

Long-term scenarios. Here I can do no better than to borrow extensively from recent studies completed at the International Institute for Applied Systems Analysis (ILASA).[6] These scenarios, which are close to those developed by R. Rotty, at the IEA, contemplate a world population of 8 billion in 2030. Because China, Latin America and possibly India are likely to increase both their per-capita energy expenditure and their populations faster than the developed world, it is plausible to expect these countries to use a much larger fraction of the world's energy than they do now. This, of course, could change somewhat if China's announced program of one child per family succeeds.

At present, the primary per-capita energy demand averaged over the entire world comes to about 2.1 kilowatts per year. The total primary energy demand comes to 8.2 terrawatts (1 terrawatt = 1 trillion watts). If the average per capita demand grows to about 4.3 kilowatts per capita, the total primary energy demand might reach around 35 terrawatts; should it grow to only about 2.8 kilowatts per capita, then the total might be 22 terrawatts. These projections are designated by ILASA as "high" and "low"; the IEA scenario falls between these ILASA projections.

Nuclear energy provides about one fourth the world's energy in each of the scenarios, whereas hydro, solar, and others provide about 8 percent in the high scenario, 10 percent in the low. Can the renewables replace the 25 percent provided by nuclear energy? Here there is a sharp divergence, with nuclear opponents, such as A. Lovins, insisting that nuclear is unnecessary, that the various solar sources will suffice. To be sure, the total primary energy demand contemplated by these authors is around 17-terrawatts—an amount available from fossil sources according to ILASA, and a little more than twice the world's present energy budget. Whether a 17 terrawatt world, especially one that is increasingly urbanized, can be socially stable, no one can say. My own view is that in planning the future we do best to prepare for a higher rather than a lower energy demand. It almost goes without saying that the less energy shortages become the focus for social tensions, the better for all.

One must recognize that even the 5.2 terrawatts supplied by fission in the ILASA low scenario is formidable: some 1,500 reactors, each producing about 1,000 megawatts (electric) or 3,300 megawatts of heat. And if, in the very long run, fission takes over the role now played by oil and natural gas, the number of required reactors, even in ILASA's low scenario, would rise to over 4,000! In the high scenario these numbers would be roughly doubled.

Carbon dioxide—a sword of Damocles? The possible constraint on fossil fuel, imposed by accumulation of carbon dioxide in the atmosphere, may be demonstrated by a few numbers. The world's total fossil fuel reserve is estimated to contain about 10 trillion tons of carbon. Should all of this be burned, and if 50 percent of the resulting carbon dioxide remains airborne, the total carbon dioxide in the atmosphere might increase sevenfold. The temperature of the lower atmosphere is estimated by Manabe and Weatherald to increase by from 2.5 to 3 degrees centigrade for every doubling of carbon dioxide in the atmosphere.[7] Thus a sevenfold increase could raise the temperature of the lower atmosphere by as much as 8 degrees centigrade.

H. Flohn has estimated the climatic regimes that might prevail if various amounts of carbon dioxide were added to the atmosphere. He characterizes each regime by comparing it with similar regimes in the geologic past.[8] Flohn's estimates of the temperature rise caused by a given addition of carbon dioxide are somewhat higher than those of other authors, largely because he includes the effect of other "greenhouse" gases.

Perhaps the most alarming aspect of Flohn's estimates is that the burning of only 20 percent of the world's fossil fuel would lead to an ice-free Arctic; and he argues that an ice-free Arctic would profoundly change the entire world's climate—to a regime that last occurred two million years ago. Flohn also offers evidence, from the very rapid disappearance of forests at the end of previous interglacials, that once a climate change begins, its ecological consequences could be manifested over a few decades.

Whether the climatic changes induced by carbon dioxide accumulation produced by fossil fuel would constitute the enormous catastrophe envisaged by Flohn is controversial. Nevertheless, we must ask: What is a prudent course in the face of such predictions, even granting their inherent uncertainty? The answer seems obvious. Reduce consumption—a feat far easier to achieve in the energy-rich parts of the world than in the energy-poor parts, and be prepared to shift to energy sources that produce no carbon dioxide. In the latter category are the solar sources, the nuclear sources (fission and fusion), and geothermal.

If the geothermal source is as small as it now appears, and if fusion remains far from technical realization, the sun and uranium are the only remaining alternatives. And if one further concedes that the solar sources may remain expensive, especially if one tries to provide reliable power from inherently stochastic sources like the wind and the sun, then caution dictates that we be prepared to use the only other technically feasible, very large energy source that does not add carbon dioxide to the atmosphere—nuclear fission. To be sure, in making this judgment I am implicitly calculating the risk of a carbon dioxide catastrophe. Moreover, I am assuming that if and when (perhaps in a decade or two) the reality of carbon dioxide emerges as a clearly defined political issue, then the move to nuclear and solar, coupled with conservation, can be made fast enough to forestall the catastrophe envisaged by Flohn.

We cannot as yet invoke the carbon dioxide catastrophe as justification for a nuclear future because there is still so much scientific controversy surrounding it. But if we are, as Palmer Putnam put it in 1953, "prudent custodians of man's future," then we would be acting irresponsibly if we reject any energy source, including fission, that produces no carbon dioxide.

There is a likelihood that the world may see many thousands of nuclear reactors within, say, the next 100 years. Is such a world feasible? In short, can man live with fission?

I set aside the issues of proliferation, of waste disposal, and of low-level radiation: the first because it is a political, not a technical question; the second because, as the recent Interagency Review Group established by President Carter asserted, satisfactory confinement of wastes for periods of several thousand years is technically feasible (after about 1,000 years the ingestion toxicity is comparable to that of the original uranium from which the wastes were derived); the third, because the hazard of low-level radiation has been grossly overplayed. But the matter of reactor accidents is not so easily disposed of. As the incident at Three Mile Island demonstrated, an accident in a nuclear plant is a real possibility; nuclear energy may not survive many incidents like Three Mile Island, even though no one was harmed there.

The *a priori* probability of an accident that releases sizable amounts of radioactivity was estimated by Rasmussen to be 50 millionths per reactor year. This number has since been criticized as being too low. In any event, the Three Mile Island accident—which released about a dozen curies of iodine-131 and according to the Kemeny commission probably caused no bodily harm—occurred after a few hundred reactor-years of operation. Its *a priori* probability has been estimated at around one in 400 reactor-years.

For any given reactor, a satisfactory accident probability would appear to be 50 millionths per reactor-year. A particular reactor, during its 40 years of operation, would have a likelihood of one in 500 of suffering an accident. But if the world energy system involved as many as 5,000 reactors—that is, 10 times as many as are now either in operation or under construction—one might expect an accident that released sizable amounts of radioactivity every four years. Considering that a nuclear accident anywhere is a nuclear accident everywhere, I believe this accident probability is unacceptable. If man is to live with fission over the long term, he must reduce the *a priori* probability of accident by a large factor—say 100.

A relevant comparison is today's volume of air travel, which would have been impossible had not the accident probability per passenger mile been reduced drastically as the number of passenger miles increased. This was accomplished in air transport by a combination of technical and organizational improvements, and the same kinds of improvements will be needed in nuclear energy. Can we visualize the needed changes?

Perhaps the easiest are the technical improvements. Certainly the deficiencies that led to Three Mile Island will be corrected: German light water reactors, for example, have two relief valves in series so that if one fails to close, the other is available. Control panels will be provided with positive indication of valve position, something that was lacking at Three Mile Island. And other technical improvements will be forthcoming.

But beyond the technical fixes, various institutional changes are needed. Among these I mention first the establishment by the utilities of the Institute for Nuclear Power Operations and the Nuclear Safety Analysis Center. These are very significant developments. But I believe, in the long run, even more important is adherence to the principle of confined, permanent siting in relatively remote areas, which is already characteristic of most of the sites in the United States. We have estimated that a nuclear system of 615 gigawatts (electric) could be accommodated in the United States on 80 of the present 100 sites, augmented by 20 new sites to replace those not suited for expansion. My proposal, at least for countries well embarked on nuclear energy, is essentially a moratorium on new sites, not on new reactors. The ultimate nuclear system would consist of large centers, located at those existing sites that are adequately remote, plus a few new sites that are also remote.

The advantages of clustered, permanent locations seem compelling to me. They include a larger cadre of experts available on site; better organizational memory, and therefore better operation; more effective security and easier control of fissile material; handling of low-level wastes and spent fuel elements on site; easier surveillance of decommissioned reactors; and, as at existing nuclear sites like Oak Ridge, a surrounding population that understands radiation and is prepared to respond in case of accident. Beyond this, the entities operating large clustered sites are likely to be stronger than are entities operating, say, a single reactor. This certainly appears to be the case in Canada, where Ontario Hydro operates Pickering, Bruce and Darlington, each with four or more reactors; and the Tennessee Valley Authority in the United States, which intends to confine all its reactors to the seven sites now under construction. Soviet nuclear specialists N. Dollezhal and Y. I. Koryakin (*Bulletin*, Jan. 1980) have proposed a similar confined siting policy for the Soviet Union.

The measures I suggest may not be sufficient to rescue nuclear energy from its present disaffection. But in view of the strong incentive to maintain the nuclear option, it seems important to devise the fixes that will make nuclear energy acceptable. To do less might impose on those who follow us a future much bleaker than one that uses all energy sources—including an acceptable nuclear energy.

NOTES

1. Charles G. Darwin, *The Next Million Years* (Garden City, N.Y.: Doubleday, 1953).
2. H. G. Wells, *The World Set Free: A Story of Mankind* (New York: Dutton and Company, 1914).
3. D. Reister to Alvin Weinberg, Nov. 7, 1979.
4. P. C. Roberts, "Energy and Society," presented at the Commission of the European Communities Energy Systems Analysis International Conference, Dublin, Ireland, Oct. 11, 1979.
5. "Study on Solar Photovoltaic Energy Conversion," prepared for the Office of Technology Policy and the U.S. Department of Energy by the American Physical Society, New York, 1979.

6. World Häfele, "World-Regional Energy Modelling," presented at the Commission of the European Communities Energy Systems Analysis International Conference, Dublin, Ireland, Oct. 9, 1979.

7. S. Manabe and R. T. Wetherald, "The Effects of Doubling the CO_2 Concentration on the Climate of a General Circulation Model," *Journal of Atmospheric Science* 32, (1975, 3–15).

8. H. Flohn, "Possible Climatic Consequences of a Man-Made Global Warming," presented at the Commission of the European Communities Energy Systems Analysis International Conference, Dublin, Ireland, Oct. 11, 1979.

NO

Denis Hayes

NUCLEAR POWER: THE FIFTH HORSEMAN

Arguments against nuclear power are rooted in a simple paradox. Commercial nuclear power is viable only under social conditions of absolute stability and predictability. Yet the mere existence of fissile materials undermines the security that nuclear technology requires.

A commitment to nuclear fission is uncompromising and unending. This power source cannot brook natural disasters or serious mechanical failures, human mistakes or willful malevolence. It demands an unprecedented vigilance of our social institutions and demands it for a quarter million years. At the same time, the use of commercial nuclear power dramatically increases the fragility of human civilization. Acceptance of nuclear technology amounts to acceptance of the inevitable spread of nuclear weapons from nation to nation, and the near-certainty that some nuclear bombs will end up in terrorist hands. The debate is not whether nuclear power will lead to nuclear weapons; that is beyond question. What is unknown is who will control these bombs, how they will be used, and what their use will portend for even the most stable institutions.

The weapons proliferation question ranks about all others. But a swarm of auxilliary problems, many of which have received more scrutiny in some countries than in others, deserves attention. If the world is indeed to "go nuclear," all will be legitimate matters of international concern. The entire case against commercial nuclear power deserves broad scrutiny *before* we become irreversibly committed to a nuclear future. . . .

Global nuclear development was initially spurred by the belief that fission would provide a cheap, clean, safe source of power for rich and poor alike. However, the dream of "electricity too cheap to meter" has foundered under a heavy burden of technical, economic, and moral problems—some of which appear to be inherently unsolvable. A growing body of analysis suggests that the total costs of nuclear power far outweigh the total benefits. . . .

Early nuclear critics tended to be gadflies, pointing out flaws in reactor designs and calling for immediate remedies. With the passage of years, many of these reformers became outright opponents, convinced that the problems with nuclear technology were so intractable that commercial fission should be bypassed as a major energy source. In early 1976, three high-level officials resigned from the U.S. General Electric Company in order to work full-time on behalf of the Nuclear Safeguards Initiative in California. Antinuclear sentiment is now coalescing into an international movement with a rapidly growing base of political support. . . .

ENVIRONMENTAL IMPACT

Although fission was at one time viewed as a "clean" source of energy, it is now opposed by almost every major environmental organization in the world. Nonetheless, nuclear proponents continue to argue that nuclear power is ecologically benign.

The environmental argument is almost always framed in terms of the relative environmental costs of nuclear power and coal. Moreover, the environmental effects of a perfectly functioning nuclear fuel cycle are usually measured against the American experience with coal *before* the passage of the landmark Coal Mine Safety Act and *before* the implementation of the Clean Air Act of 1970. Coal mine accidents in the United States have decreased steadily over the last few years, and improvements in mining conditions will result in a dramatic decrease in the incidence of black lung disease. When electric utilities are finally compelled to meet the terms of the Clean Air Act, airborne emissions (which in the past have been a major health menace, causing widespread premature death among the elderly and people suffering from respiratory ailments) will be significantly reduced.

Moreover, comparisons between coal and nuclear power plants are irrelevant to the great majority of nations without significant coal reserves. Thus, if environmental comparisons are to be made, they ought to focus on the relative impacts of nuclear power and renewable sources (solar power, wind power, wood and other organic sources, etc.) that are available to all countries.

At present, thermal pollution and radiation are the principal environmental dangers spawned by the nuclear fuel cycle. Other threats may be unleashed. Should nuclear power grow to the point where massive amounts of low-grade ore were being processed for fuel, environmental repercussions would surely be felt. If nuclear power were to survive until mankind stumbled into a nuclear war, the environmental consequences would be of an altogether different magnitude.

All mechanical processes that generate electricity also generate thermal pollution. But nuclear power plants cast off more waste heat per unit of electricity produced than does any other commercial technology. This heat must be dissipated, either through the use of cooling towers or through direct discharge into a body of water. Since cooling towers are exceedingly expensive, most reactors planned around the world will inject heat directly into lakes or streams. If a power plant operates constantly, the local habitat would undergo a massive transformation and a new ecosystem that could thrive at the higher temperature would develop. However, reactors must be shut down regularly for fuel changes, and irregularly for various other reasons. Thus, the consequent erratic temperature changes make it difficult for any stable ecological community to survive.

Many of the problems associated with a nuclear reactor would be multiplied manyfold with the coming of a proposed pattern of intensive nuclear development misleadingly called a "nuclear park." The park would contain uranium enrichment facilities, a fuel fabrication plant, a large number (10 to 40) of individual reactors, and a fuel reprocessing plant. The thermal burden associated with such a development could be sufficient to alter the local climate and, possibly, to generate a continuous cloud cover. Yet Dr. Chauncey Starr, President of the Electric Power Research Institute, considers such parks the "inevitable result of the growth of the nuclear power industry."

The environmental threats posed by the nuclear power cycle cannot be fully measured without an understanding of the effects of radiation on life at the molecular level—an understanding that is at present far from complete. The radiation associated with nuclear power is emitted through the spontaneous decay of reactor-produced radioactive materials. (In addition to its 100 tons of uranium oxide fuel, one large modern reactor contains about two tons of various radioactive isotopes—one thousand times as much radioactive material as the Hiroshima bomb produced.) Each radioactive isotope has a precisely measurable half-life (time during which half the atoms in any piece of that radioactive material will decay) that ranges from less than a second to more than a million years. The half-life of plutonium-239, the most controversial isotope associated with nuclear power, is 24,400 years. . . .

The effects of "ionizing radiation" have been studied in many experiments performed on hamsters, guinea pigs, and beagles. Some clear statistical correlations have emerged—especially in experiments involving high dosages of radiation. But we have almost no knowledge of the cancer-causing and mutation-causing mechanisms at the molecular level. We know that radiation causes cancer, but we don't know *how* it causes cancer.

Information on the effects of radiation on human beings is sketchier than that on radiation-exposed laboratory animals. The debate over the effects of atmospheric nuclear bomb tests, sparked by two-time Nobel Prize winner Linus Pauling, has never been resolved (although a large and growing body of evidence seems to support Pauling's claims). A U.S. medical data bank, established in 1968 to monitor nuclear workers, has been handicapped by the refusal of some private nuclear companies to cooperate. Moreover, this "Transuranium Registry" has been unable to track down many exposed workers from the early years of the nuclear era—a severe research handicap since many radiation-induced effects have very long lag times before symptoms appear.

Because we lack a scientific understanding of how ionizing radiation actually affects discrete biological processes, the link between cause and effect is necessarily speculative (much as is the case with cigarette smoking). This inductive leap leaves the experts divided. The Natural Resources Defense Council, an environmental group, is petitioning U.S. authorities to lower the permissible concentration of plutonium aerosols by 115,000 times.

Many radiation disputes revolve around linear concepts, around the probability of an ill-effect increasing in direct proportion to the amount of radiation received. Will even very low dosages cause some cancers, or is there a threshold below which exposure to radiation is harmless? If a miniscule amount of lung tissue is subjected to a very high dosage of radiation, should the likelihood of cancer be derived from the high intensity in the physically affected area or from the low intensity in relation to the whole lung?

We don't understand the molecular biology of radiation-induced cancer, so our policies are necessarily based upon statistical inferences. But radiation statistics are particularly ambiguous because (1) routine radiation from nuclear power is in addition to inescapable radiation from other sources and considerable variation exists in the amount of natural background radiation and medical radiation that people are subjected to; (2)interrelationships between different types of cancers and different types of radiation are extremely confusing; (3) the public is exposed to many carcinogens other than radiation that cannot be controlled in large-scale epidemiological surveys; and (4) because of the lag time, many potential radiation victims die from other causes before radiation-induced effects appear.

The low levels of radioactive isotopes routinely emitted through a perfectly functioning nuclear fuel cycle, or from a leaking low-level waste repository, have not been "proven" either safe or unsafe. In several developed countries, reactor emission standards have been dramatically tightened in recent years, although not enough to satisfy many critics. Other countries have no standards whatsoever—even though radionuclides can become concentrated up to several thousand times as they move up a food chain, so dilute emissions may be in a more dangerous form when ingested by people at the top of the food chain. . . .

NUCLEAR POWER AND SOCIETY

Every major energy transition brings with it profound social change. The substitution of coal for wood and wind ushered in the industrial revolution. The petroleum era revolutionized mankind's approach to movement—restructuring our cities and shrinking our world. Now, at the twilight of the petroleum age, we face another energy transition in the certain knowledge that it will radically alter tomorrow's society. Each of the many energy options available to us today carries with it far-reaching social implications.

Nuclear power is highly centralized, technically complex, capital intensive, and fraught with long-term dangers. It produces electricity, a form of energy that is difficult to store and that can be transported only along expensive, vulnerable corridors. Some of the consequences of the widespread use of nuclear power can be easily anticipated.

Increased deployment of nuclear power must lead to a more authoritarian society. Reliance upon nuclear power as the principal source of energy is probably possible only in a totalitarian state. Nobel Prize winning physicist Hannes Alfven has described the requirements of a stable nuclear state in striking terms:

> Fission energy is safe only if a number of critical devices work as they should, if a number of people in key positions follow all of their instructions, if there is no sabotage, no hijacking of transports, if no reactor fuel processing plant or waste repository anywhere in the world is situated in a region of riots or guerrilla activity, and no revolution or war—even a "conventional one"—takes place

in these regions. The enormous quantities of extremely dangerous material must not get into the hands of ignorant people or desperados. No acts of God can be permitted.

The existence of highly centralized facilities and their frail transmission tendrils will foster a garrison mentality in those responsible for their security. Such systems are vulnerable to sabotage, and a coordinated attack could immobilize even a large country, since storing a substantial volume of "reserve" electricity is so difficult. Moreover, 100,000 shipments of plutonium each year would saddle societies with risks that have no peacetime parallel.

Nuclear power is viable only under conditions of absolute stability. The nuclear option requires guaranteed quiescence—internationally and in perpetuity. Widespread surveillance and police infiltration of all dissident organizations will become social imperatives, as will deployment of a paramilitary nuclear police force to safeguard every facet of the massive and labyrinthine fissile fuel cycle.

Broad nuclear development could, of course, be attempted with precautions no more elaborate or oppressive than those that have characterized nuclear efforts to date. Such a course would assure a nuclear tragedy, after which public opinion would demand authoritarian measures of great severity. Orwellian abrogations of civil liberties would be tolerated if they were deemed necessary to prevent nuclear terrorism.

Guarding long-lived toxic radioactive waste will require not just the sworn vigilance of centuries; it will require an eternal commitment. Thoughtful nuclear supporters are suggesting the creation of a nuclear "priesthood" to assume the burden of perpetual surveillance. Since the nuclear wastes now being created will remain toxic for 100 times longer than all recorded human history, an approach with quasi-religious overtones is only fitting.

The capital-intensive nature of nuclear development will foreclose other options. As governments channel massive streams of capital into directions in which they would not naturally flow, investment opportunities in industry, agriculture, transportation, and housing—not to mention those investments in more energy-efficient technologies and alternative energy sources—will be bypassed. The U.S. Project Independence effort would require one trillion dollars by 1985, four-fifths of which would be earmarked for new rather than replacement facilities. Under such a scenario, new energy plants would require two-thirds of all net capital investment during that period. If Project Independence were more exclusively nuclear, the figure would be even higher.

With such a large portion of its capital tied up in nuclear investments, a nation will have no option but to continue to use this power source, come what may. Already it has become extremely difficult for many countries to turn away from their nuclear commitments. If current nuclear projections hold true for the next few years, it will be too late. Already there have been frightening examples of falsified reports filed by nuclear owners seeking to avoid expensive shutdowns. When vast sums are tied up in initial capital investments, every idle moment is extremely costly. After some level of investment, the abandonment of a technology becomes unthinkable.

In a world where money is power, these same large investments will cause inordinate power to accrue to the managers of nuclear energy.

These managers will be a highly trained, remote, technocratic elite, making decisions for an alienated society on technical grounds beyond the public ken. C. S. Lewis has written that, "what we call Man's power over Nature turns out to be a power exercised by some men over others with nature as its instrument." As nations grow increasingly reliant upon exotic technologies, the authority of the technological bureaucracies will necessarily become more complete. Energy planners now project that by the year 2000 most countries will be building the equivalent of their total 1975 energy facilities *every three years.* Although central planners may have no difficulty locating such a mass of energy facilities on their maps, they will face tremendous difficulties siting them in the actual countryside of a democratic state. . . .

The nations of the world must together make an end-of-an-epoch decision. As the finite remaining supply of petroleum fuels continues to shrink, the need for a fundamental transition becomes increasingly urgent. The nuclear Siren is presently attracting much interest, but hopefully her appeal will prove short-lived. Alternatives are abundant.

Coal can play an important role in the immediate future. The energy content of the world's remaining coal far exceeds that of the remaining oil, and recent advances in mining and combustion will allow much of this resource to be tapped without imposing unacceptable environmental costs.

A wide range of solar devices is becoming available. Systems to capture low-grade heat to warm buildings and water—uses which constitute more than 25 percent of current energy needed in all countries—are now on the market at competitive prices. Photovoltaics, thermalelectric systems, bioconversion processes, windpower, and other benign, renewable options promise large amounts of relatively low-cost, reliable, high-quality energy with less effort than would be required by the new generation of breeder reactors. Solar options can be decentralized, simple, adapted to indigenous materials, and dependent only upon a country's energy "income" from the sun. They produce no toxic wastes and no potential bomb materials.

Finally, it is of critical importance that greater attention be paid to opportunities for energy conservation. The United States could, according to several analyses, eliminate about half of its fuel budget without significant alterations in its economic system or its way of life. Even greater reductions might be accompanied by improvements in public health and the general quality of life. Sweden and West Germany manage to achieve an excellent standard of living on about half the U.S. per capita level of fuel consumption. Enormous savings can be made throughout much of the industrialized world, where for the last several decades cheap energy has been systematically substituted for labor, capital, and materials. Even in poor countries, the replacement of open fires with cheap, efficient stoves, the use of inexpensive pressure cookers instead of pots, etc., would allow significant energy savings. Moreover, such countries should employ anticipatory conservation measures in their development plans, taking care to avoid the sloppiness and wastefulness that characterize those nations that industrialized in the era of cheap oil.

It is already too late to halt entirely the widespread dissemination of the scientific principles underlying nuclear power. What *can* still be sought, however, is the international renunciation of this technology and all the grave threats it entails. Although the nuclear debate has been dominated by technical issues, the real points of controversy fall in the realm of values and ethics. No person, regardless of technical skill, has a right to impose a personal moral judgment on society. No country, regardless of strength, will be able to make the nuclear decision for the world. But if increasing numbers of people and countries begin independently and actively to oppose nuclear power, the world may follow.

POSTSCRIPT

Is Nuclear Power Safe and Desirable?

Weinberg was among the first to propose that new, "ultrasafe," reactor designs are the key to regaining public support for nuclear power. Although other energy analysts supported this view, spokespeople for the nuclear industry were initially unreceptive. They were concerned about the cost-effectiveness of the new designs and feared that public doubts about already operating plants would be exacerbated by the admission that new reactors must be much safer. There were indications that opinions in the pro-nuclear community were shifting in favor of a fundamental change in plant design before the Chernobyl disaster. This shocking event has accelerated this change of attitude.

It is clear that Hayes places much more value on the issue of nuclear proliferation than Weinberg, who dismisses this concern because it is a political rather than a technical problem. The growing international movement against nuclear weapons has redirected the nuclear power debate. A persuasive, detailed argument for the inevitability of the connection between bombs and power plants is given in Amory and Hunter Lovins's *Energy/War: Breaking the Nuclear Link* (Harper Colophon, 1981).

Hayes's concern about the effects of low-level radiation and the confinement of radioactive waste is shared by most nuclear power critics. For a treatise on the radiation issue see, *Radiation and Human Health* (Sierra Club Books, 1981), by John Gofman, who has credentials both as a nuclear scientist and a medical doctor.

For a discussion and descriptions of the new reactor designs favored by advocates of reviving the nuclear option, see articles by Richard K. Lester in the March 1986 *Scientific American*; by David Rose in the August 1985 *Bulletin of the Atomic Scientists*; by Russ Manning in the May 1985 *Environment*; and by Joseph Eppinger in the February 1985 *Science Digest*.

Among the more informative antinuclear books are *Cult of the Atom* (Simon & Schuster, 1982), by Daniel Ford, and *The Menace of Atomic Energy* (Norton, 1977), by Ralph Nader and John Abbotts.

Nuclear power's potential contribution to decreasing the global warming threat is critically discussed in articles by William Lanouette, *Bulletin of the Atomic Scientists* (April 1990), and by Alan Miller and Irving Mintzer in the June 1990 issue of the same journal. The Chernobyl accident and its aftermath are examined in articles by Christoph Hohenemser and Ortwin Renn, *Environment* (April 1988), and by William Sweet, *Technology Review* (July 1989).

ISSUE 8

Is the Widespread Use of Pesticides Required to Feed the World's People?

YES: William R. Furtick, from "Uncontrolled Pests or Adequate Food," in D. L. Gunn and J. G. R. Stevens, eds., *Pesticides and Human Welfare* (Oxford University Press, 1976)

NO: Shirley A. Briggs, from "Silent Spring: The View From 1990," *The Ecologist* (March/April 1990)

ISSUE SUMMARY

YES: Crop protection specialist William R. Furtick warns that pesticides are essential for the intensive agriculture required to prevent mass starvation.
NO: Environmentalist Shirley A. Briggs counters that pesticides have failed to decrease crop loss while causing the widespread ecological harm predicted in Rachel Carson's book *Silent Spring.*

The use of naturally occurring chemicals in agriculture has a history that dates back many centuries. After World War II, however, the application of synthetic chemical toxins to croplands became sufficiently intensive to cause widespread environmental problems. DDT, used during the war to control malaria and other insect-borne diseases, was promoted by agribusiness as the choice solution for a wide variety of agricultural pest problems. As insect resistance to DDT increased, other chlorinated organic toxins such as heptachlor, aldrin, dieldrin, mirex, and chlordane were introduced by the burgeoning chemical pesticide industry. Environmental scientists became concerned about the effects of these fat-soluble, persistent toxins whose concentrations became magnified in carnivorous species at the top of the ecological food chain. The first serious problem to be documented was reproductive failure resulting from DDT ingestion in such birds of prey as falcons, pelicans, osprey, and eagles. Chlorinated pesticides were also found to be poisoning marine life.

Marine scientist Rachel Carson's best-seller *Silent Spring* (Houghton Mifflin, 1962) raised public and scientific consciousness about the potential devastating effects of continued, uncontrolled use of chemical pesticides.

In 1966 a group of scientists and lawyers organized the Environmental Defense Fund in an effort to seek legal action against the use of DDT. After a prolonged struggle, they finally won a court ruling in 1972 ending nearly all uses of DDT in the United States. In that same year amendments to the Federal Insecticide, Fungicide, and Rodenticide Act gave the Environmental Protection Agency authority to develop a comprehensive program to regulate the use of pesticides. By 1978, agricultural uses in the United States of aldrin, dieldrin, chlordane, heptachlor, and mirex had all been banned as a result of animal studies linking those substances to cancer, mutations, and fetal deaths.

In response to the attack on chlorinated organics, the pesticide producers introduced a new series of products called organophosphates and carbamates. Most of them are more biodegradable than chlorinated hydrocarbons; however, they tend to be much more acutely toxic to humans. As a result, pesticide-related deaths and illnesses have become a serious problem among agricultural field workers. Insect pests have once again demonstrated their genetic versatility by rapidly developing immunities to these new insecticides.

The manufacture of pesticides has also caused serious environmental problems. The poisoning of workers and the contamination of the James River in Virginia by residues of the pesticide kepone is probably the worst example in U.S. history. The 1984 Bhopal disaster, which resulted in more than 3,500 deaths and 200,000 injuries, was due to the release of methyl isocyanate being used by Union Carbide to manufacture a pesticide.

Since 1975 insecticide sales have leveled off in the United States due to regulatory efforts. The pesticide industry has responded by increasing sales overseas—especially in developing countries—and by switching to the promotion of herbicides (weed killers) locally. The shift from mechanical to chemical weed control has resulted in a booming market for herbicides. These plant poisons are not without environmental problems. Undoubtedly the worst ecological effect of the use of agricultural chemicals was the decimation of five million acres of the Vietnamese countryside by Agent Orange, a mixture of two potent herbicides, 12 million gallons of which was used by the U.S. Air Force during the war. Dioxin, a contaminant in Agent Orange, is a highly potent teratogen and carcinogen. Claims that Vietnamese people and U.S. soldiers still suffer from its effects continue to be hotly disputed.

William R. Furtick is among those who contend that we must rely on pesticides despite the problems they may cause. In his view, these chemicals will continue to be the principal means of reducing crop destruction in a hungry world. Shirley A. Briggs argues that the harmful effects of pesticide use have fully vindicated Rachel Carson's dire predictions and that safer, more effective means of pest control are available.

YES

William R. Furtick

UNCONTROLLED PESTS OR ADEQUATE FOOD?

> Civilisation as it is known today could not have evolved, nor can it survive, without an adequate food supply. Yet food is something that is taken for granted by most world leaders despite the fact that more than half of the population of the world is hungry. Man seems to insist on ignoring the lessons available from history. The invention of agriculture, however, did not permanently emancipate man from the fear of food shortages, hunger, and famine. Even in prehistoric times, population growth often must have threatened or exceeded man's ability to produce enough food. Then, when droughts or outbreaks of diseases and insect pests ravaged crops, famine resulted.
>
> That such catastrophies occurred periodically in ancient times is amply clear from the numerous biblical references. Thus, the Lord said: 'I have smitten you with blasting and mildew' (Amos 4:9); 'The seed is rotten under the clods, the garners are laid desolate, the barns are broken down; for the corn is withered . . . The beasts of the fields cry also unto thee: for the rivers of waters are dried up and the fire hath devoured the pastures of the wilderness' (Joel 1:17, 20).
>
> Plant diseases, drought, desolation, despair were recurrent catastrophies during the ages.

Thus spoke Norman Borlaug in opening his lecture on the occasion of being awarded the Nobel Peace Prize for 1970. His words turned out to be of short-range significance, for poor harvests in the Soviet Union and other areas of the world in 1972 and 1975 quickly caused most world leaders and the public at large to stop taking the food supply for granted. The high food prices which resulted had an impact on the families of both the more affluent and poorer sectors of the world population. All the world's housewives are now suffering from the little-known fact that each time they prepare a meal they have fed extra invisible guests who ate before them. These are the myriad crop pests that consumed from a quarter to one-half of the food before it ever reached the table. Neither the hungry nor the affluent can continue to pay this price, which is to receive only part of their daily bread.

The losses caused to our food supply by insects, crop diseases, rodents, weeds, and similar pests have recently gained the attention of world leaders in the highest councils. Dr. Henry Kissinger, the [former] American Secretary of State, emphasized the magnitude of these losses and proposed urgent action to reduce them both in his statement which opened the United Nations World Food Conference in Rome in November 1974, and before a special session of the United Nations General Assembly in September 1975. This concern was echoed by large numbers of world leaders at the World Food Conference and led to a demand by many that an adequate supply of pesticides to combat these losses be assured. This was formalized by a special resolution on pesticides adopted by the World Food Conference. A special session of the United Nations General Assembly ten months later adopted a special resolution on reducing food losses after harvest. This resolution states:

> Developing countries should accord high priority to agricultural and fisheries development, increase investment accordingly and adopt policies which give adequate incentives to agricultural producers. It is a responsibility of each State concerned, in accordance with its sovereign judgment and development plans and policies, to promote interaction between expansion of food production and socioeconomic reforms, with a view to achieving an integrated rural development. The further reduction of post-harvest food losses in developing countries should be undertaken as a matter of priority, with a view to reaching at least a 50 per cent reduction by 1985. All countries and competent international organizations should operate financially and technically in the effort to achieve this objective. Particular attention should be given to improvement in the systems of distribution of foodstuffs.

The rapid change in the world food supply between the overabundance in the decades of the 1950s and 1960s to that of precariously low reserve stocks which followed the poor harvests of 1972 and 1975 resulted in a concerted international effort to increase world food production as an urgent priority. With new emphasis on production, there was a growing realization that protection not only of the present production but also of the future increased production from destruction by pests is a special imperative.

THE ROLE OF PLANT PROTECTION IN INCREASING FOOD PRODUCTION

Although the magnitude of losses from pests has not been adequately measured even in the most highly developed countries, these losses are recognized as being substantial. In developing countries, the pre-harvest and post-harvest crop losses are estimated by FAO to be in the region of 30 per cent or more of potential production. Even in the most highly developed agricultural countries they are still large. These are caused by diverse species of arthropod and vertebrate animals such as insects and rodents; weedy terrestrial and aquatic plants; plant diseases caused by bacteria, fungi, viruses, and microplasm; and plant nematodes. These losses do not include the impact on food production caused by the low efficiency of agricultural workers, who suffer from various vector-borne diseases and interal and external parasites. The same factors are important in livestock production and cause major losses in the production of meat; and in some areas vector-transmitted diseases make large geographic areas unfit for certain types of livestock production. To these direct losses must be added the increased food costs to consumers, particularly in the rapidly expanding urban centres, who must pay higher prices as a result of the wastage and loss of production caused by pests.

Leading plant protection experts called together by the FAO after the World Food Conference recognized that pesticides will during the foreseeable future remain a primary measure for combating losses from pests. With this in mind, it should be noted that in 1974 all developing countries combined used only about 10 per cent of the world pesticide production, and most of this was used in public health vector-control programmes against flies and mosquitoes, and on export cash crops. Both the importance of pesticides for pest control and the need for their greater use in a proper manner if developing countries are to meet their food production needs are readily apparent. All the same, several complex problems are involved.

The control of pests with both pesticides and non-pesticide control measures requires a more highly developed infrastructure and better manpower training than is generally true for most other agricultural inputs. Without adequate pest control a large measure of the benefit from other inputs may be lost. These components of agricultural modernization must therefore proceed in harmony. This can be readily observed in developing countries, where until recently nearly all pesticides used in agriculture have been on cash crops. As the high-yielding varieties of food crops have been introduced and used more widely, there has been a rapid increase in pesticide use on this small but increasing component of food crop production.

RECENT WORLD FOOD PRODUCTION TRENDS

It would be somewhat simplistic to suggest that in the 1970s the world food situation suddenly took a wrong turn under the impact of bad weather. An analysis of production trends during the past few decades indicates that more deep-seated problems were also accumulating, with the industrialized countries producing more food than they could consume or export and the developing countries facing food import bills that were growing larger every year. The industrial countries could not sell all the food they were producing, and the developing countries could not produce fast enough to supply their needs.

Contrary to popular belief, the difference did not lie in the dynamics of agricultural production. The developing countries, in spite of their difficulties, were expanding their agricultural output in the 1950s and 1960s just as fast as the industrial countries—a truly remarkable achievement. The difference lay in the rates of food growth demand, which increased by 2.5 per cent per annum in the industrial countries as against 3.5 per cent in the developing countries, mainly because of faster population growth in the latter.

During the period following the Second World War, the performance of agriculture has differed greatly from country to country. In more than 20 of the developing countries the rate of growth of food production exceeded 4 per cent per annum, while in about a dozen countries it was less than 2 per cent during the period 1953-71. In about 30 developing countries food output grew faster than food demand, and in about 50 it grew faster than population. Even in the food crisis year of 1972 the developing countries' food production was 20 per cent higher than in 1966, the latest previous year of widespread bad weather, so that progress even between the troughs in the long-term trend had kept ahead of population growth.

These considerable achievements reflected dramatic and effective application of technology in many countries. However, it was not vigorous or effective enough, and this shows how enormous is the strain put upon the agricultural sector in countries whose population and general economic activity are both expanding rapidly. Not only was the effort insufficient to meet the rise in domestic food demand, with the consequence of large and ever larger food imports, but it also meant a disappointing performance in agricultural exports, which are the chief source of foreign exchange for most developing countries.

In addition to the accumulating problems of food production, there is the equally vital issue of the nutritional adequacy of supplies within countries and the extent of undernutrition and malnutrition. Taking a conservative view, it would appear that out of nearly 100 developing countries, about 60 had a deficit on food energy supplies in 1970. In the Far East and Africa, 25 per cent and 30 per cent of the population is estimated to suffer from significant undernutrition. Altogether in the developing world (excluding the Asian centrally planned economies for which insufficient information is available) malnutrition affects at least 460 million people.

THE FOOD PROBLEM OF THE FUTURE

Although the long-term average increase in world food population has been greater than the growth of population ever since the Second World War, the margin was smaller in the 1960s than in the 1950s. The increase in food production slowed down in the 1960s in every major region except Africa and North America. In part this reflects the large element of post-war recovery in the early 1950s. In the industrial countries population growth has declined, and part of the slower increase in production in the 1960s was due to deliberate government policies. In the developing countries, on the other hand, it occurred in spite of accelerated population growth and government policies aimed generally at increasing the rate of food introduction.

Thus, although total food production has increased at about the same rate in the industrial and the developing countries, on a *per capita* basis the increase has been much smaller in the latter. This has meant that the already large difference in the actual level of production per head between these two groups of countries has widened still further: in the developing countries it was little more than one-quarter of that in the industrial countries in 1971-3, as compared with about one-third in 1961-3.

The fact that for so long a period food production in the developing countries as a whole kept ahead of a rate of population growth unprecedented in world history is a tremendous achievement. In many individual countries, however, developments have been much less favorable.

Domestic agricultural production is the main determinant of the level of available food supplies in most developing countries, but it has other important roles in their overall economic and social development, including the earning and saving of foreign exchange, and the provision of employment and of much of the capital needed for the development of the rest of the economy. However, diversion of this capital from agriculture may have contributed to reducing further the increase in food production. The importance given to agricultural production is reflected by the average targets in the various national development plans being set at 4 per cent for the developing countries as a whole. Even so, only about one-third of the developing countries have reached their targets.

Increasing production depends basically on expanding the inputs of the different factors of production: land and water, labour, material inputs, the various types of capital, and technological know-how, including pesticide use. For farmers in developing countries, and especially the numerous small farmers, the possibility of using—and the incentive to use—more inputs in turn depends to a great extent on the infrastructure and services provided by governments. Many government development budgets have tended in the past to neglect agriculture in favour of industry. More recently there has been a widespread tendency to increase the emphasis on agriculture.

To satisfy the increase in world demand generated by even a medium level of population and income growth, the world's agriculture must provide, during each decade remaining in this century, an additional annual output of about 200 million tonnes of cereals, 30-40 million tonnes of sugar, about 100 million tonnes of vegetables, and nearly the same amount of fruits per year; plus about 50 million tonnes of meat and 100 million tonnes of milk—together with the increased livestock feed to produce these products. It is therefore quite apparent that to meet food production needs, particularly in developing countries, will require great efforts both by individual countries and through increased international efforts to assist them to modernize their agriculture.

We know that it is technically possible to increase food production to accommodate the needs of a growing world population for some years ahead with current technology. The new technological packages that are being introduced as a result of increased agricultural research by national and international research centres have the potential for greatly increasing yield. So it is not technology that is the major limiting factor, but the social and economic constraints which retard the wide use of that technology. These problems must be solved in the developing countries, because if the industrial countries could increase their agricultural production substantially to meet the increasing needs of the developing countries it would probably be impossible to find a means of financing and transporting this volume of food.

AGRICULTURAL CHANGE IN THE DEVELOPING WORLD

Although it might be feasible to supply the increasing food needs of developing countries through stepped-up production in the highly developed agricultural countries, the idea has been generally discarded because of the almost insurmountable problems involved. The monetary requirement within a decade would become so enormous as to appear insoluble for both the exporters and the recipients. The general inadequacy of the developing countries' ports, storage facilities, internal transport, and general infrastructure to handle the volume of imports necessary means that a staggering capital investment and organized effort would be needed in a short time. A rapid acceleration of agricultural development in the food deficit countries therefore appears to be the best solution to the inevitable need for greatly increased food supplies.

This is not a sudden realization, nor has the world agricultural community been caught totally unprepared. There has been a steadily growing and increasingly coordinated effort by the various international organizations, the donor countries, and the private sector to help the developing countries in their efforts to improve and modernize agriculture. An example of this has been the coordinated development of a worldwide network of international agricultural research centres of the highest quality, financed through the Consultative Group on International Agricultural Research, which is sponsored by FAO, UNDP, and the World Bank. Since the World Food Conference, similar new coordinated efforts have been initiated to cover the non-research needs, such as investment and technical assistance.

It is not possible simply to transplant the technology of the highly developed agriculture of the industrial countries to fill the needs of the developing world. In many cases this has been tried and an unhappy lesson learned through expensive failures. There are various reasons why the requirements tend to be very different in developing countries. Some of these are due to cultural, social, economic, and political differences; others are the result of geography. Nearly all the industrial countries' agriculture is carried out in temperate climates. In contrast, the greatest concentrations of population and agricultural land in the developing countries are in tropical or sub-tropical areas. The agricultural practices—in fact to a large extent even the primary crops—tend to be totally different in the industrial and the developing countries.

However, there is one major similarity: the need for increased productivity per unit area through intensification of agriculture. Although there is still a substantial amount of potentially useful land that could be developed for agriculture, particularly in parts of Africa and South America, most of this could be developed only at a high capital cost. There is also a serious question about the suitability or advisability of diverting much of this land to agriculture; already there exists agricultural use of substantial areas of highly erodible land which should be returned to forest or similar conservation uses.

The year-round growing season in tropical areas facilitates a high degree of intensification of production through multiple cropping and other methods, if adequate water, fertility, and pest control can be provided. But under intensive agriculture, pest problems become more severe; and in the tropics, where no seasonal break is provided by winter to arrest the build-up of pest populations, a crop can be completely lost through attack by pests. The favourable climates for pests in the tropics have also led to a much greater diversity of pest problems in these areas.

PESTICIDES IN FOOD PRODUCTION

Although various methods are used to control pests in different pest management systems, pesticides are at present the most important factor in most control programmes, and this will be so for the foreseeable future. In 1974, the world pesticide market was about 5000 million U.S. dollars, of which about 40 per cent was in North America and only slightly less in Europe. The developing countries accounted for no more than 10 per cent of the total. The rate of increase in the use of pesticides in developing countries is considerably higher than in the intensive agricultural countries, and this trend is expected to continue.

Not only are pesticides the primary immediate weapon against pest losses, both on growing crops and during transport and storage after harvest; their use in chemical weed control is also a major factor in increased labour efficiency and reduced drudgery. The herbicides used for chemical weeding now represent more than half of all pesticides used. Adequate use of pesticides has a major stabilizing effect on agricultural production by preventing periodic massive losses caused by insect and disease outbreaks. Consistent production is very important for price stability.

Adequate and proper use of pesticides can be a major factor in achieving a better environment through improved public health resulting from control of fly, mosquito, and other vector-borne diseases; better health through adequate nutrition; supplies of reasonably-priced, wholesome food; and less pressure on the world's limited and precious land resources by making it possible to produce more food on less land. The proper use of pesticides would thus make it possible to confine agriculture to the most productive and agriculturally suitable land, and so eliminate the need for further destruction of forest to provide more arable land. The more erodible hill and other land could be kept out of agriculture, and marginal land of this type which is now cultivated could be returned to pasture, forest, or other conservation uses including recreation and reserves for wildlife.

The pesticide industry is one of very high technology and is research-intensive. This has to a large degree limited the primary pesticide manufacturing industry to the highly developed industrial countries. Public concern about the potential hazards that some people feel pesticides could exert on the environment has resulted in the pesticide industry being one of the most stringently regulated industries in the world.

Not only is there the need to increase world pesticide production to ensure an adequate food supply; there is also the need for continued high research investment to discover new, more efficient, and safer pesticides, and to find better and safer ways of using those which are currently available. The increasing array of divergent and more stringent world regulatory requirements threatens to take the time and investment required to develop new pesticides beyond the point that will justify the investment. Indeed, it is ironic that the research requirements placed on pesticides to ensure environmental safety have already often made it commercially unattractive to develop those highly specific pesticides which provide the greatest degree of environmental safety. Because of their very specificity, the market is limited and the cost-risk-benefit ratio does not justify investment.

The safety record in pesticides is, perhaps, nearly the best in any area of modern technology. Their benefit to mankind has been one of the greatest through their contribution to improved health and economical food production. They are urgently needed in the future to help ensure adequate food for survival.

Much of the anxiety about pesticides is derived from people who have unfounded fear about all synthetic chemicals who seek to consume or come into contact only with 'natural' substances. They should realize that nature is the most skilled organic synthesizing chemist known, and that man has so far been able to duplicate only some of the simpler natural organic compounds. Many of the natural compounds synthesized by plants and animals are highly toxic or persistent in the environment, or both; and many everyday foods contain natural toxins.

There needs to be an increased public awareness of the facts about pesticides: an awareness that will prevent fear and hysteria emanating from a few overzealous and misguided individuals from leading to a global tragedy that will entail mass suffering and starvation for the poorest people and serious harm to all.

The food we eat is limited to that derived from a small percentage of the world's animals and plants, because many of the latter are too toxic for man to consume. Even closely related plants in the same family as some of our common food crops are too poisonous to eat, and some of the most widely consumed tropical tubers must be soaked or cooked to destroy toxic substances before they are edible. Few people realize that potatoes and tomatoes, which can be eaten because they do not contain excessive levels of natural toxins, are close relatives of the nightshades, which are very poisonous. The same is true of the lettuce group, and also that group of closely related plants which include carrots, celery, parsley, and parsnips; these we can eat, but we cannot eat other plants in the group, such as the hemlocks: most schoolchildren know that Socrates committed suicide by consuming a tiny quantity of poisonous hemlock.

Many of the same factors are true of natural insecticides. A number of today's household sprays used to kill flies, mosquitoes, and similar insects contain a natural insecticide extracted from the pyrethrum plant. In the past, nicotine was widely used as an insecticide until it was replaced by safer and more efficient synthetic compounds. It is doubtful whether a natural compound like nicotine could qualify under the current stringent regulations for pesticides, for it is known to be toxic. With these thoughts in mind, it must be clear that a continuing supply of pesticides is a basic factor in our future welfare.

NO

Shirley A. Briggs

SILENT SPRING: THE VIEW FROM 1990

"We stand now where two roads diverge. But unlike the roads in Robert Frost's familiar poem, they are not equally fair. The road we have long been travelling is deceptively easy, a smooth superhighway on which we progress with great speed, but at its end lies disaster. The other fork in the road—the one 'less traveled by'—offers our last, our only chance to reach a destination that assures the preservation of our earth."

Rachel Carson
Silent Spring

Rachel Carson is credited with making 'ecology' a household word. In explaining how pesticides can affect the fabric of nature, *Silent Spring* gave to the general public an inspiring understanding of the natural order and its vulnerability to humanity's activities. Rachel Carson made people realize that industrial society is blind to this knowledge and that the makers and users of synthetic pesticides are extremely powerful and often ruthless.

Although there is now a growing acceptance of the message of *Silent Spring*, its opponents still maintain their assault against it. In the 28 years since its publication, essential backing for those scientists advocating sounder pesticide policies has come from ordinary people who have seen their own lives and the future of their children at stake. It is only through these people and the influential Government officials whom Rachel Carson reached directly through the book, that concrete changes have been made.

THE INITIAL IMPACT

Silent Spring's message was not to the liking of those institutions with close contacts with the chemical industry. Some reviewers argued that she exaggerated her case: others that the book was full of errors, although we at the Rachel Carson Council have yet to be shown a valid example. Others took a heavy-handed approach.

The Velsicol Chemical Company, for example, tried in the summer of 1962 to persuade Houghton Mifflin not to publish *Silent Spring*, objecting to its

From Shirley A. Briggs, "Silent Spring: The View from 1990," *The Ecologist*, vol. 20, no. 2 (March/April 1990). Reprinted by permission of *The Ecologist*, Worthyvale Manor, Camelford Cornwall, UK. Notes omitted. *The Ecologist* is distributed in North America by The MIT Press, 55 Hayward St., Cambridge, MA 02142. Subscriptions are $30 for individuals, $65 for institutions, and $25 for students.

material on chlordane. Louis A. McLean, Secretary and General Counsel, explained their stand:

> "Unfortunately, in addition to the sincere opinions by natural food faddists, Audubon groups and others, members of the chemical industry in this country and western Europe must deal with sinister influences, whose attacks on the chemical industry have a dual purpose: (1) to create the false impression that all industry is grasping and immoral, and (2) to reduce the use of agricultural chemicals in this country and in the countries of western Europe, so that our supply of food will be reduced to east-curtain parity. Many innocent groups are financed and led into attacks on the chemical industry by these sinister parties."

The same assertions are still made today by those who prefer to question their opponent's motives rather than face facts and issues squarely and objectively.

Claims from industry that we will be doomed without unlimited pesticides are hardly credible in view of estimates that losses to agricultural pests have doubled or tripled since the introduction of the new synthetic toxics. Formerly innocuous insects have become pests when their predators and parasites have been killed by pesticides, and many kinds of pests have become resistant to pesticides. Pimentel *et al* have calculated that even a complete stop in synthetic pesticide use would result in only relatively small increases in crop losses. With alternative farming methods, this crop loss could be reduced. The costs to producers would also be considerably lower. The many other costs measured in terms of damage to health and the environment, further question the claimed advantages of our current pesticide practices.

THE RISING TONNAGE OF POISONS

Silent Spring recorded a rise in the production of pesticide active ingredients in the United States from 124,259,000 pounds in 1947 to 637,666,000 pounds in 1960. By 1986, according to Environmental Protection Agency (EPA) figures, production had risen to 1.5 billion (thousand million) pounds for the range of products cited by Rachel Carson, and U.S. use was about 1.09 billion pounds, allowing for exports and the small amount imported. If wood preservatives (fungicides), disinfectants and sulphur, are taken into account, the figure for U.S. pesticide usage in 1988 is 2.7 billion pounds. U.S. pesticide production (not including the latter three categories) is about one quarter of the world total, so the annual burden on the earth must be about 6 billion pounds of these products. The US Environmental Protection Agency (EPA) does not offer production data on the expanded list, though the impact of the three additional classes of pesticides is also a major consideration.

Pesticides form only a part of the exploding volume of often toxic petrochemicals used in drugs, plastics, solvents and many other products. The overwhelming majority of these chemicals remain inadequately tested, if tested at all, for possible toxic effects to humans and other animals, and for their impact on the environment.

First in the chemical industry, and then in the general public, a belief has arisen that the use of toxic chemicals is required to permit our current way of life. Many in the industry seem to emphasize finding new ways to use toxic synthetic materials rather than envisioning new products for necessary functions and then devising non-toxic means of making them.

Our lives have become dominated by petrochemical products which have replaced those once made of traditional materials such as wood, glass, metal, and natural fibers. This has happened primarily because large-scale production has made products of synthetic materials cheaper in immediate purchase price than those made of traditional, non-toxic materials. However, when total costs are considered, including the adverse effects on human health and the environment, the synthetics may be far more costly.

ATTEMPTS AT CONTROL

The Environmental Protection Agency was established in 1970 to regulate environmental contaminants including pesticides. It was preceded in 1969 by the National Environmental Policy Act (NEPA) that requires the environmental cost of all Federal activities to be examined in an Environmental Impact Study (EIS). The Council on Environmental Quality was set up to evaluate these studies. Laws to enforce acceptable standards for clean air and water, reinforced more recently by the Superfund Law to deal with toxic waste dumping, have further defined EPA's purpose.

Some state and local governments followed with reinforcing legislation, setting in place the machinery to accomplish much of the reform suggested in *Silent Spring*. Where results have been disappointing, we find the usual impediments of inadequate determination and funding, and the formidable political and economic opposition from the industries and allied interests whose profits and reputations are at stake.

In 1972, the basic law authorizing Federal regulation of pesticides, the Federal Insecticide, Fungicide and Rodenticide Act (FIFRA) was drastically altered by Congress to require that all products then on the market be scrutinized and tested to fill all gaps in their data record. Pesticides had previously been checked only for immediate toxic effects on some non-target creatures. Since the U.S. Department of Agriculture, which had previously been in charge of pesticide registration, had never denied registration to any product submitted, whatever its record, the legislation called for an enormous testing and evaluation programme. Old products were to be brought up to standard, while all new pesticides submitted would be thoroughly tested for immediate and long-term effects on human health and the environment. Those chemicals found to pose "unreasonable risks" were either to be sharply restricted in their uses, or banned for all or most uses. All testing, which is the responsibility of the producers, was to be completed within a few years.

EPA'S PERFORMANCE

This legislation should have provided a good basis for regulating pesticide use and should have encouraged the increased use of biological controls, sounder mechanical procedures, and emphasis on pest-resistant crops.

But, this has not happened. A few exceptionally hazardous pesticides have been taken off the market or restricted in use, but very few have made their way through the whole battery of tests required, or if so, they have not had these results conclusively assessed by EPA.

There are about 1400 active pesticide ingredients involved, of which about 600 are used frequently enough to be of importance. About 150 of these are of critical importance because of the scale of use or hazard. By EPA definition, an 'active' ingredient is one that affects the target pest. So-called 'inert' ingredients, which include contaminants, solvents, surfactants or other material in the formulated commercial product, may be even more biologically potent and toxic to many forms of life. Producers were allowed to include these constituents under "trade secrets," and so were not required to reveal them (EPA was restricted from revealing them). After many complaints, in 1986 EPA finally issued a list of those active ingredients that they consider most dangerous and which will have to undergo testing.

In 1984, the National Academy of Sciences published a graph showing what percentage of various classes of chemicals in use in the United States had been tested sufficiently to permit a sound evaluation of human health hazards. Pesticides came out better than most of the categories, but even here only 10 per cent satisfied the basic requirements. There was some good data for 24 per cent, a scattering of information for another 28 per cent and *none* for 38 per cent.

In 1986, an assessment of current pesticide regulation from the General Accounting Office (GAO), stressed the limited nature of registration by EPA. Registration does not mean that a product is safe, non-toxic, harmless to people and pets, or is 'approved' by EPA. It is illegal for a producer to claim any of these attributes. EPA simply follows a prescribed formula for balancing risks and benefits from use of a pesticide. However great the risks, if the benefits appear greater, the pesticide may be registered.

The risks are often borne by those who are exposed to the pesticide at some distance from its application, from eating the crop on which it was used, drinking water into which it has seeped, touching contaminated soil, or breathing drifting spray. Benefits may be calculated in terms of the short-term profits to the crop grower who uses the product, but it often seems that the clearest advantage accrues to the manufacturer. In its report, GAO recommends that labels on pesticides be required to give more adequate precautions, and details as to whether or not the ingredients were thoroughly tested.

INADEQUATE TESTING

Progress in testing pesticides has been slowed by a number of factors. The EPA Office of Pesticide Programs took several years just to decide which tests should be made to fill specific data requirements. Guidelines for pesticides to be used on food crops were to be more stringent than for those on non-food crops, and products to be used on domestic animals were of more concern than those affecting wild animals. Laboratory standards were debated at length, and the rigour of testing was adjusted for different sites. Less concern was shown if products would be used in sparsely populated areas.

As the process dragged on, pressure increased to weaken the rules, and in 1980 the Reagan administration declared its intent to eliminate many environmental regulations. Many tests were made optional. When the final round of documents on testing procedures were sent out for public comment, they revealed that EPA did not plan to test non-herbicide pesticides for their possible adverse effects on plants. Yet the Rachel Carson Council found 83 non-herbicides known to harm plants, so we listed them in our comments. The lone person at EPA concerned with non-target plants phoned to ask for our references. Though most must have been listed in EPA files, he had no way to track them down since no one had thought to sort this information into the computer. It is not only lack of resources and concern that has stalled EPA efforts, but sometimes just a lack of imagination.

EPA convinced Congress that the registration process was intolerably slow because they claimed that full testing would have to be done on all of the approximately 45,000 commercial pesticide products. This, they claimed, at the rate they were going, might take well into the 21st century. In fact, testing is only done on the active ingredients, mainly among the 150–600 most used. Congress reacted by granting EPA's plea for "conditional registration." This allows a new product to be put on the market while the testing is being done; if the results are too alarming it can later be withdrawn, restricted, or banned. The purpose of the 1972 law, which directed EPA to proceed briskly with the testing of pesticides, was thus effectively compromised. Since 1978, when conditional registration was added to the law, it has often been used for new pesticides as well as for older ones. At the last count, about 40 per cent of new pesticides have been introduced onto the market while tests were still being completed. Some of these are extensively used.

Environmentalists have tried to strengthen FIFRA since 1972, but have effectively been thwarted by pressure from the producers and users of pesticides. In 1988, some of the main concerns of the environmental lobby were dealt with by amendments to the law. EPA is now required to complete the re-registration process by 1997. This new requirement had one immediate effect: many producers simply asked that their registrations be withdrawn. Either they knew that test results would be adverse, or they felt that profits from these products did not warrant the cost of the tests.

FRAUDULENT TESTING

A review of the often labyrinthine processes through which EPA operates raises some basic questions. Should one agency be given the combined functions of gathering information, assessing it, enforcing the resulting regulations and policing compliance? Is it reasonable to expect the producer who applies for registration of a product always to conduct the required testing honestly, or to hire a commercial laboratory to perform it without prejudice? Large amounts of money are at stake. Indeed, in 1978 it was found that some 200 major pesticides had been registered on the basis of fraudulent tests performed for various producers by the privately-owned Industrial Bio-Test Laboratory. As a result, some countries withdrew the suspect pesticides, but the United States and Canada divided the evidence between them and spent years reviewing the tests and having manufacturers repeat the many invalid tests. Even now, if we know that a product was registered on the basis of tests by the Industrial Bio-Test Laboratory, we should check its present credentials.

Silent Spring is credited with giving the essential impetus to the whole range of anti-pollution laws which came into force in the 1970s, so it is ironic that the law regulating pesticides, FIFRA, is the weakest of them all. Instead of focusing on the reduction of toxic contamination, as with the laws for air, water or dump sites, it has complex and even contradictory purposes that make it difficult to administer. The procedure for cancelling a pesticide registration allows the manufacturer numerous ways to delay or confound the process.

For many years, one clause of FIFRA made it almost impossible to take a product off the market even where it had been found to qualify for suspension. If EPA wanted a banned pesticide to be immediately removed from the market, the Agency had to compensate the producer and others holding stocks not only for the commercial value of the product, but also for the storage and disposal costs of the stocks. For only one chemical, this could amount to far more than the Agency's entire pesticide budget for one year. Usually therefore, EPA had to allow existing stocks to be sold until exhausted, no matter how dangerous. (Other manufacturers found to have a defective or dangerous product are not treated so indulgently. The Government can require the recall of defective automobiles, for example, without compensating the offending manufacturers.)

FIFRA AMENDMENTS

The 1988 amendments to FIFRA largely removed this problem. All holders of a pesticide at the time of suspension or cancellation are no longer to be indemnified, except for a few end-users, such as farmers, who may have bought the product unaware of its uncertain status. Such payments will come from the U.S. Treasury Judgement Fund, not the EPA operating budget. Any other payments will be made only by special authorization by Congress. These amendments also made other long-sought improvements, including tighter control for enforcing compliance, higher registration fees by which the industry will help pay more of the registration costs, and increased criminal penalties for knowingly violating any FIFRA requirement.

The 1972 revision of FIFRA made the pesticide user legally responsible for damage from his application, though this is still more honoured in principle than in practice. It also established the category of 'Restricted pesticides,' to be used only by certified operators. This was intended to protect the public from the most dangerous products, but it falls short of this goal in at least two respects: in the inadequate requirements for certification, and in the places where these products may still be used. Perhaps the least protected are the farm workers who must apply pesticides and work in treated fields, with little protective clothing and without a suitable delay allowed before re-entry into contaminated areas. Until a 1987 ruling by the Department of Labor, they were also usually denied sanitary and washing facilities and clean drinking water.

Finally, FIFRA is the only major environmental statute that does not provide for citizen suits against violations. This is an essential tool to assure Government compliance.

THE REGULATION OF DDT AND CHLORDANE

The history of the regulation of DDT and chlordane shows how the cumbersome system has operated. When EPA was established, a suit to ban DDT was already underway against the previous regulating agency, the U.S. Department of Agriculture, brought by the Environmental Defense Fund (EDF). With the switch of authority to EPA, the EDF lawyer found that he had allies in the competent young lawyers who first manned the EPA Office of General Counsel. Their opponents were still the Department of Agriculture and the manufacturers of DDT, both of which could muster formidable legal talent. Taking the very slides and test results with which the producers had claimed that DDT had only minor toxicity beyond insects, the expert pathologists enlisted by EDF and EPA demonstrated that the tests had in fact shown serious hazards, including cancer and the neurological damage common to the chlorinated hydrocarbons. Most alarming was evidence of irreversible and continuing harm to many forms of wildlife. Once in the environment, DDT could not be contained, and would continue to wreak its damage for decades as it persisted and moved about the earth.

Since DDT was removed from the United States market in 1972, some critically endangered species have begun to recover. Many of these are large, conspicuous birds at the top of food chains. They are especially susceptible to DDT but also easy for wildlife experts to count and test. Bald eagles and brown pelicans, though still threatened by other pesticides, carry far less DDT now than in the early 1970s.

After DDT was banned, the lawyers turned to the closely related pesticides for which they had enough evidence to ensure their cancellation. They repeated their method with success for aldrin/dieldrin (aldrin degrades to dieldrin; both are more toxic than DDT). Next came the closely related chlordane/heptachlor. As the comparable evidence was presented, the administrative law judge presiding was puzzled by evidence provided by the producer, Velsicol, who claimed to have found no positive data for cancer. As the impasse continued, the EDF and EPA lawyers feared that the judge's perplexity might lose them the case, with a domino effect as doubts cast on chlordane might reflect back on previous cases. They agreed to settle for a gradual phasing out of chlordane/heptachlor. This solution did not provide any legal finding of fact on carcinogenicity, protecting Velsicol from possible damage suits on that issue, and it did not have the finality of a legal decision through the regular cancellation process. Administrative decisions can be more easily reversed. This 1978 settlement gradually removed chlordane/heptachlor from the market by 1982 for agricultural and most other uses except termite control and dipping of non-food plants. Subsequent findings confirmed the carcinogenicity of chlordane/heptachlor, and the illegal withholding of evidence by Velsicol, but EPA did not re-open the case to put this information into the legal record.

THE WORLDWIDE IMPACT OF *SILENT SPRING*

The history of pesticide regulation and use in the United States, although inadequate, is on the whole more encouraging than the world picture. We have been helped by a responsible democratic form of government which has given us the Freedom of Information Act through which private citizens have managed to pry vital data on pesticides from Government agencies.

In other major industrialized countries, the situation is often less favourable. In a 1981 report, Ross Hume Hall pointed out some of the problems in Canada:

> "Pesticide companies have made it clear that for most of their products, the Canadian market does not warrant extensive testing in the Canadian environment. They state that if forced to pay for such testing, they may withdraw their products. The cost is borne by the public whether in price of pesticides or in taxes. Should it be left to the international chemical companies to set policy whether or not pesticides are tested under Canadian conditions?"

Our responsibility in the United States goes far beyond our borders. While we may not require testing in a range of conditions greater than our own, most other countries rely on us for precedents and data on pesticides even if they do not always follow our policies. The smaller and less industrialized countries are at special risk. They are under pressure to permit the sale of dangerous products and lack the facilities to appraise the hazard or enforce protective rules. In an effort to make the necessary information available, the United Nations Environment Programme (UNEP) has established a centre for information on toxics, and publishes the International Register of Potentially Toxic Chemicals, based on information from national governments and private sources.

Even UNEP finds that many national governments, no doubt at the insistence of the chemical companies, refuse to release much of their data, claiming they are 'trade secrets.' Though the United States Supreme Court has ruled that information on the toxicity and environmental fate of a pesticide is not a trade secret, and should be made available to the public, EPA insists that those seeking information should obtain it through the complicated process of the Freedom of Information Act.

We all pay the price of international laxity. When pesticides, the use of which the United States does not permit, are freely used in other countries, perhaps supplied by producers in the United States, much comes back in imported food. Air and water carry residues to the ends of the earth. It was not long into the era of synthetic pesticides that DDT was found in the Antarctic.

We continue to learn more about pesticides. We now know that the transformation products formed as pesticides break down in the environment and our own bodies are sometimes far more dangerous than their parent products. Resistance of pests to pesticides is now much more widespread than when *Silent Spring* was written.

DOCTORS AND LAWYERS

The response of most of the medical profession to the message of *Silent Spring* has been disappointing. Doctors are still taught little toxicology, and victims of chemical poisoning may have great difficulty finding a doctor who can recognize the symptoms, much less know what may be done about them. Whether from one severe exposure or a succession of small ones, human tolerance to pesticides may become so low that people cannot function in their usual home or work environment. General physical and mental health may deteriorate. For especially vulnerable people, finding a safe haven from the prevailing toxic exposure or the unthinking pesticide practices of neighbours may be almost impossible.

Victims of 'chemical trespass' need lawyers versed in the field, as well as doctors. The growth of environmental law has been impressive, but the need for lawyers to take individual cases still exceeds the supply. In 1973, a notable case was brought by the National Resources Defense Council, Environmental Defense Fund, National Audubon Society and Sierra Club against the U.S. Agency for International Development (AID) to make them stop using U.S. tax money to fund foreign governments' purchase of pesticides, the use of which is not permitted in the U.S. The environmentalists won the case, which stopped the practice and required AID to issue an Environmental Impact Statement (EIS) for their whole pesticide programme. From the many public comments on the EIS, AID has learned much and has adopted a more enlightened policy.

THE GROWTH OF CITIZENS' GROUPS

The Rachel Carson Council, established after Rachel Carson's death in 1964 in accordance with her wishes, for many years found itself almost alone in trying to keep abreast of pesticide research, in providing a wide range of information to individuals and groups, and in attending technical EPA hearings on proposed action. In recent years, however, the Council has been joined in much of this by new national and regional groups, notably the National Coalition Against Misuse of Pesticides, and the Northwest Coalition for Alternatives to Pesticides. They carry on Rachel Carson's tradition of combining a careful, scientific approach with public education. On the international scene, the Pesticide Action Network concentrates on problems in less industrialized countries.

We who deal directly with individuals and local groups have learned much from those who come to us for information. They have kept us in close touch with the growing number of people throughout the world who have understood the message of *Silent Spring*. We have been continually impressed by something that Rachel Carson had learned: even those with little academic or technical training have the capacity to grasp essential principles and to seek and comprehend very complex technical studies. The chemical industry defend their secrecy about their products by claiming that the public could not understand the data, and would just 'panic' if they were to be given it. In fact, what industry fears, is that once the issues are understood, and the data is available, citizens can organize themselves to persuade officials to take suitable action.

THE OTHER ROAD

Encouraging progress on the "road less travelled by" has come from private efforts to find sound pest control methods. Work toward non-chemical controls had languished since World War II when synthetic chemical pesticides were embraced as the solution to almost all agricultural problems. Only a few university and government departments continued research on the solutions outlined by Rachel Carson. The concept of Integrated Pest Management (IPM) was defined by scientists working with biological controls, notably the late Robert van den Bosch and his colleagues in California. Gradually their work gained credence and somewhat better funding, with the boost from *Silent Spring*.

IPM leaves toxic chemicals as a last resort to be handled as *precisely* as possible in *minimum* amounts. Like most ways of dealing cooperatively with nature,

IPM requires a detailed knowledge of pests, crops and localities, with continuous monitoring and experienced workers. The appeal of the quick-fix has to yield to concern for sustaining fertility and preventing the poisoning of the land, water and produce in the long-term. Attentive and industrious farmers, for example, can use low-till methods without resorting to the massive herbicide applications insisted upon by departments of agriculture. Some laws now specify IPM as the proper policy for state and national government, but the pursuit of this aim has not been vigorous.

In the years since *Silent Spring,* soil, water, air and the tissues of living creatures have become increasingly contaminated. We have set loose forces that we do not understand and cannot control. The dangers on the superhighway have increased, and the probability of diverting most traffic onto the other road has lessened. Whether we can do so depends on the growing numbers of people who understand Rachel Carson's message and are prepared to take action in line with their beliefs.

POSTSCRIPT

Is the Widespread Use of Pesticides Required to Feed the World's People?

Much of Furtick's concern is about food production in developing nations. In *Food First* (Houghton Mifflin, 1977), Frances Moore Lappé and Joseph Collins argue that pesticide use has actually exacerbated food shortages while poisoning the local population. They point out that the high cost of pesticides has contributed to a shift to exported cash crops at the expense of food crops consumed locally. People living in Guatemala, for example, where DDT and other pesticides banned in the United States are used intensively, have much higher average levels of these toxins in their bodies than was ever recorded for the U.S. population. For an exposé of the pesticide industry's practice of promoting the sale of chemicals in developing countries after their use was prohibited in the United States, read *Circle of Poisons*, D. Weir and M. Shapiro (Institute for Food and Development Policy, 1981).

A report frequently quoted by pesticide advocates is "Benefits and Costs of Pesticide Use in U.S. Food Production," by David Pimentel et al., *Bioscience*, December 1978. The conclusion reached by noted agricultural scientists Pimentel and his coworkers is that ending all U.S. pesticide use—a more extreme restriction than that advocated by integrated pest management (IPM) proponents—would cause an immediate increased annual crop loss worth $8.7 billion, whereas the cost of chemical controls is $2.2 billion. Those who cite these conclusions usually fail to mention that the authors admit they have not included the ecological and social costs of pesticide use and that much of the crop loss may be eliminated once natural pest predators reestablish a natural balance.

For further alternatives to intensive use of chemical pesticides, read "Getting Off the Pesticide Treadmill," by Michael Dover, *Technology Review* (November/December 1985) and "Revitalizing Biological Control," by L. E. Ehler, *Issues in Science and Technology* (Fall 1990). The success that can be achieved by using IPM in Third World countries is described in "A New Crop of Pest Controls," by Omar Sattaur, *New Scientist* (July 14, 1988).

The most recent heated dispute about an agricultural chemical resulted from the 1989 release of the Natural Resources Defense Council report, "Intolerable Risk: Pesticides in Our Children's Foods." Claims about health effects of Alar, used to promote the ripening of apples, caused the Uniroyal Chemical Company to cancel production of the product. Two conflicting assessments of this controversy are: "On Food Safety," by Peter Borrelli, *The Amicus Journal* (Spring 1989) and "Much Ado About Alar," by Joseph D. Rosen, *Issues in Science and Technology* (Fall 1990).

ISSUE 9

Is There a Cancer Epidemic Due to Industrial Chemicals in the Environment?

YES: Samuel S. Epstein, from "Losing the War Against Cancer," *The Ecologist* (March/June 1987)

NO: Elizabeth M. Whelan, from "The Charge of the Cancer Brigade," *National Review* (February 6, 1981)

ISSUE SUMMARY

YES: Physician Samuel Epstein claims that preventable environmental exposure is responsible for dramatically rising human cancer rates.
NO: Physician Elizabeth Whelan asserts that the claim that we are experiencing an environmentally caused cancer epidemic is a misconception.

Presently, about 100 natural and synthetic substances are known to increase the incidence of human cancer. Several hundred more have been shown to produce cancer in controlled tests on laboratory animals. Although it is known that a wide variety of chemicals present in the human environment are potential carcinogens, there is heated debate about the significance of this information. Environmentalists point to the proliferation of new synthetic industrial chemicals, most of which have never been tested for their carcinogenic potential, as cause for alarm. Industrialists emphasize the relatively small number of proven human carcinogens and contend that the only major environmental risk is from cigarette smoking. Public concern increased following reports of high cancer rates among people living near centers of industrial activity and among consumers of water laden with trace quantities of organic chemicals.

Several attempts have been made to determine what percentage of tumors result from inherited tendencies or internal biochemical malfunctions and what percentage are caused by external or environmental factors. The results of these efforts indicate that 60–90 percent are attributable to environmental factors. An undetermined fraction of the environmentally induced cancers result from exposure to naturally occurring substances such as sunlight or background nuclear radiation. The remainder are assumed to result from human contact with a small percentage of the millions of synthetic materials that have been added to the natural environment.

There are many obstacles preventing a precise assessment of the cancer risk due to synthetic chemicals. Carefully controlled exposure studies are one method of determining a causal relationship between a suspected substance and cancer. Although such experiments have been done on human prisoners and other powerless individuals in the past, they are now strictly forbidden.

Animal tests can be done but they are expensive and applying the results of such tests to human populations can be unreliable. Epidemiological studies can compare statistics of cancer incidence between a group of human beings which has been exposed to a particular substance and a control group matched for age and other variables. A few such studies of occupational exposure to highly potent carcinogens have actually produced unequivocal results—although usually it is impossible to adequately control for all the variables. Another particularly troublesome fact is that most human carcinogenicity appears to involve a latency period of 10 to 40 years between exposure and the development of the tumor.

There has definitely been an increase in the percentage of deaths attributed to cancer during the past several decades. Some investigators claim that when such factors as declining death rates due to other diseases and demographic changes are taken into account there remains no evidence for a significant increase in the incidence of cancer attributable to environmental factors other than smoking. Others examining the same data—working from somewhat different assumptions and studying particular types of cancer among specific population subgroups—do find such evidence. They warn that because there is a long latency period, the carcinogenic effects of the many industrial chemicals introduced during the past 30 years are just beginning to surface.

The regulation of environmental carcinogens presents many thorny issues. Most cancer experts believe that there is no threshold or minimum dose necessary for induction of cancer. Thus, standards can be set only to reduce estimated cancer incidence to some predetermined level. Whether this level should be one case per 100 exposed individuals, one per 1,000, or one per 1,000,000 is a value rather than a scientific question. Weighing the costs of regulation against the benefits of estimated reductions in cancer incidence is obviously a highly controversial business.

Samuel S. Epstein, professor of Occupational and Environmental Medicine at the University of Illinois Medical Center, is an active proponent of the stringent regulation of environmental carcinogens. His efforts contributed to the enactment of comprehensive legislation to control toxic substances. Dr. Elizabeth M. Whelan is executive director of the American Council on Science and Health, a private, industry-supported organization. She believes that most governmental cancer regulation efforts have been misguided.

YES — Samuel S. Epstein

LOSING THE WAR AGAINST CANCER

Cancer is now a major killing disease in the industrialised world and its rates are sharply rising. In contrast, there have been major reductions in deaths from cardiovascular disease, still the number one killer in the US, probably because of a recent decline in smoking and attention to diet and exercise.

With over 900,000 new cases and 450,000 US deaths last year, cancer has now reached epidemic proportions with an incidence of one in three and a mortality of one in four. Analysis of overall cancer rates, standardised for age, sex and ethnicity, has demonstrated a steady increase in cancer rates since the 1930s. In recent years, the incidence rate has risen sharply, by some 2 per cent a year, with mortality rates rising by 1 per cent a year.

Striking confirmation of these recent increases comes from estimates of the lifetime probability of getting cancer for people born at different times. For white males born in 1975 to 1985 for instance, the probability of developing cancer has risen from 30 to 36 per cent, whilst the probability of dying from cancer has risen from 19 to 23 per cent. Such increases in overall cancer rates are also reflected in the increasing incidence of cancers of the lung, breast, colon, prostate, testis, urinary bladder, kidney, and skin, and of malignant melanoma and lymphatic/hematopoietic malignancies, including non-Hodgkin's lymphoma. Lung cancer is responsible for about one-third of the overall recent increase in incidence rates. It should be stressed that some 75 per cent of all cancer deaths occur in people over 55 years, and that recent increases are largely restricted to these ages.

STATIC CURE RATES

The overall cancer "cure rate", as measured by survival for over five years following diagnosis, is currently 50 per cent for whites but only 38 per cent for blacks. There is no evidence of substantial improvements in treatment over the last few decades, during which the five-year survival and age-

From Samuel S. Epstein, "Losing the War Against Cancer," *The Ecologist*, vol. 17, no. 2 (1987). Reprinted by permission of *The Ecologist*, Worthyvale Manor, Camelford Cornwall, UK. *The Ecologist* is distributed in North America by The MIT Press, 55 Hayward St., Cambridge, MA 02142. Subscriptions are $30 for individuals, $65 for institutions, and $25 for students.

adjusted mortality rates for the major cancer killers (lung, breast and colon) and for most other organs, have remained essentially unchanged. The only improvements have been for cancer of the cervix, and for relatively rare cancers, such as testicular seminomas, Hodgkin's disease and childhood leukemias treated with radiation and/or chemotherapy. Apart from immediate toxicity, such treatment, while effective, can increase the subsequent risk of developing a second cancer by up to 100 times.

INCREASING CARCINOGENIC EXPOSURES

Cancer is an age-old and ubiquitous group of diseases. Its recognised causes and influences are multifactorial and include natural environmental carcinogens (such as aflatoxins and sunlight), lifestyle factors, genetic susceptibility, and more recently industrial chemicals. Apart from modern lifestyle factors, particularly smoking, increasing cancer rates reflect exposure to industrial chemicals and run-away modern technologies whose explosive growth has clearly outpaced the ability of society to control them. In addition to pervasive changes in patterns of living and diet, these poorly controlled technologies have induced profound and poorly reversible environmental degradation, and have resulted in progressive contamination of air, water, food and workplaces with toxic and carcinogenic chemicals, with resulting involuntary exposures.

With the dawn of the petrochemical era in the early 1940s, by when technologies including fractional distillation of petroleum, catalyic and thermal cracking and molecular splicing became commercially established, the annual US production of synthetic organic chemicals was about one billion pounds. By the 1950s, this had reached 30 billion pounds, and by the 1980s over 400 billion pounds annually. The overwhelming majority of these industrial chemicals has never been adequately tested—if tested at all—for chronic toxic, carcinogenic, mutagenic and teratogenic effects, let alone for ecological effects, and much of the limited available industrial data is at best suspect.

Occupational exposure to industrial carcinogens has clearly emerged as a major risk factor for cancer. The National Institute for Occupational Safety and Health (NIOSH) estimates that some 10 million workers are now exposed to 11 high volume carcinogens. Five to 10-fold increases in cancer rates have been demonstrated in some occupations. Also persuasive are British data on cancer mortality by socio-economic class, largely defined by occupation, which show that the lowest class, particularly among males, has appoximately twice the cancer mortality rate of the highest class.

Living near petrochemical and certain other industries in highly urbanised communities increases cancer risks, as demonstrated by the clustering of excess cancer rates. High levels of toxic and carcinogenic chemicals are deliberately discharged by a wide range of industries into the air of surrounding communities. Fall-out from such toxic air pollutants is also an important source of contamination of surface waters, particularly the Great Lakes. While there still are no regulatory requirements in the US for reporting and monitoring these emissions, unpublished government estimates indicate that they are in excess of 3 billion pounds annually.

Another example of the effects of runaway technology is the hazardous waste crisis. The volume of hazardous wastes disposed of every year in the USA has risen from under 1 million tons in 1940 to well over 300 million tons in the 1980s—more than 1 ton per US citizen per year. The industries involved—fossil fuel, metal mining and processing, nuclear, and petrochemical—have littered the entire land mass of the US with some 50,000 toxic waste landfills—20,000 of which are recognised as potentially hazardous—170,000 industrial impoundments (ponds, pits and lagoons), 7,000 underground injection wells, not to mention some 2.5 million underground gasoline tanks, many of which are leaking. Not surprisingly, an increasing number of rural and urban communities have found themselves located on or near hazardous waste sites, or downstream, down-gradient or down-wind from such sites. Particularly alarming is growing evidence of contamination of ground water from hazardous waste sites, contamination which poses grave hazards for centuries to come. Once contaminated, ground waters are difficult, and sometimes impossible, to clean up.

Environmental contamination with highly potent carcinogenic pesticides has reached alarming and pervasive proportions. Apart from high level exposure of workers in manufacturing, formulating and applicating industries, the contamination of ground and surface waters has become commonplace. Residues of ethylene dibromide in excess of 1,000 ppb in raw grains, cereals and citrus fruits have been well known to industry and the Environmental Protection Agency (EPA) for as long as ten years after its very high carcinogenicity was first demonstrated; not until 1984 however, did EPA develop a 30 ppb tolerance, which was rejected by the Commonwealth of Massachusetts and the States of New York and Florida, and replaced by much lower and less hazardous levels. While the exact numbers are uncertain, it is probable that tens of millions of homes nationwide are contaminated with varying levels of chlordane/heptachlor, pesticides are still registered by EPA for termite treatment. It should be noted that, on the basis of extensive hearings some 14 years earlier, the Agency concluded that exposure to chlordane/heptachlor posed an "imminent hazard" due to cancer (besides other chronic toxic effects) leading to a subsequent ban on their agricultural uses.

Much cancer today reflects events and exposures in the 1950s and 60s. The production, use and disposal of synthetic organic and other industrial carcinogens was then miniscule in terms of volume when compared to current levels, which will determine future cancer rates for younger populations now exposed. There is every reason to anticipate that even today's high cancer rates will be exceeded in coming decades. . . .

HOW INDUSTRY FIGHTS REGULATION

Twentieth century industry has aggressively pursued short-term economic goals, uncaring or unmindful of harm to workers, local communities and the environment. So far, industry has shifted responsibility for the damage it has caused onto society-at-large. Belated government efforts to control polluting industries have generally been neutralised by well-organised and well-financed opposition. With the exception of special purpose legislation for drugs, food additives and pesticides, there were no regulatory requirements for pre-testing industrial chemicals until the 1976 Toxic Substance Control Act, legislation which the industry had stalled for years, and which is now honoured more in the breach than in the observance.

Apart from the failure to pre-test most chemicals, a key characteristic of industry's anti-regulatory strategy has been the generation of self-serving and misleading data on toxicology and epidemiology, and on regulatory costs and cost-benefit analyses. The record of such unreliable and often fraudulent data is so extensive and well-documented as to justify the presumption that much industry data must be treated as suspect until proven otherwise.

Attempts by the Carter Administration to develop comprehensive, 'generic' regulation of occupational carcinogens, later reversed by the Reagan Administration, were attacked by the Manufacturing Chemists Association, which created the American Industrial Health Council to organise opposition. Such reactions generally reflected a short-sighted preoccupation with perceived self-interest rather than with efficiency and economy. The virtual uniformity of industry opposition to regulation is in marked contrast to the heterogeneity of the industries involved, both in terms of size and the diversity of their interests. Regulation has, in fact, generally resulted in substantial improvements in industrial efficiency and economy, particularly in large industries, by forcing development of technologies for recovery and recycling of valuable resources. A deplorable result of regulation, however, has been, and continues to be, the export of the restricted product or process to the so-called lesser developed countries.

Apart from well-documented evidence on control and manipulation of health and environmental information, industry has used various strategies to con the public into complacency and divert attention from their own recklessness and responsibility for the cancer epidemic. Key among these is the 'blame-the-victim' theory of cancer causation, developed by industry scientists and consultants and a group of pro-industry academics, and tacitly supported by the 'cancer establishment'. This theory emphasises faulty lifestyle, smoking, and fatty diet, sun bathing or genetic susceptibility, as the major causes of preventable cancer, while trivialising the role of involuntary exposures to occupational and environmental carcinogens. Another misleading diversion is the claim that there is no evidence of recently increasing cancer rates other than lung cancer, for which smoking is given the exclusive credit. While the role of lifestyle is obviously important and cannot be ignored, the scientific basis of this theory is as unsound as it is self-serving. Certainly, smoking is a major, but not the only, cause of lung cancer. But a wealth of evidence clearly incriminates the additional role of exposure to occupational carcinogens and carcinogenic community air pollutants. Hence, some 20 per cent of lung cancers occur in non-smokers; there have been major recent increases in lung cancer rates in non-smokers; an increasing percentage of lung cancer is of a histological type (adenocarcinoma) not usually associated with smoking; high lung cancer rates are found with certain occupational exposures independent of smoking; and excess lung cancer rates are found in communities where certain major industries are located. The chemical industry thus clearly uses tobacco as a smoke screen to divert attention from the role of carcinogenic chemicals in inducing lung cancer, besides other cancers.

When it comes to diet, the much touted role of high fat consumption, while clearly linked to heart disease, is based on tenuous and contradictory evidence with regard to breast and colon cancers. The evidence certainly does not justify the wild claims by lifestyle theorists that some 30 to 40 per cent of all cancers are due to faulty diet. For instance, a 1982 National Academy of Sciences report concluded that " . . . in the only human studies in which the total fibre consumption was quantified, no association was found between total fibre consumption and colon cancer." Similarily, a large scale 1987 study, based on the eating habits of nearly 90,000 nurses, concluded that "—there is no association between dietary fat and breast cancer."

Another illustration of grossly misleading strategies relates to the identification of chemical carcinogens. When a particular chemical or product is threatened with regulation on the basis of animal carcinogenicity tests, the industry invariably challenges the significance of these tests, while routinely using negative test results as proof of safety. At the same time industry insists on the need for long-term epidemiological investigations to obtain definitive human evidence of carcinogenicity. To test this apparent reliance on direct human evidence, researchers at Mt Sinai Hospital in New York compiled a list of some 100 chemicals accepted as carcinogenic on the basis of animal tests, but for which no epidemiological information is available, and sent this list to some 80 major chemical industries. Respondents were asked whether any of the listed carcinogens were in use and, if so, whether epidemiological studies had been conducted, whether they are being conducted, or whether it was intended to conduct them in the future, and if not, why not. The responses were revealing. The great majority of those industries using particular carcinogens replied that they had done no epidemiological studies, were not doing any, and didn't intend to do any for various reasons, including alleged difficulty, impractibility, expense, or because of their belief that these chemicals could not possibly be carcinogenic to humans. A perfect catch-22; knock the animal tests and insist on human studies, but make sure that the human studies are never done.

Industry positions are vigorously advocated by trade associations, such as the Chemical Manufacturers Association, public relations firms, such as Hill and Knowlton, from organisations, such as the American Council on Science and Health (the contributions of whose director, Whelan, have been aptly characterised as "voodoo science"), and lay writers such as Efron (who charges that the American scientific community has been terrorised into submission by environmental "apocalyptics"). Disturbingly, another major source of support for anti-regulatory strategies is a stable of academic consultants who advance the industry position in arenas including the scientific literature, federal advisory committees, and regulatory and congressional hearings.

GOVERNMENT AND THE CANCER EPIDEMIC

Presidents play a powerful role in setting national public health priorities, not unnaturally reflecting their own political agendas. Reagan, however, is unique in having run for office on an ideological anti-regulatory platform, and in having then systematically used his office to implement this ideology, often in contravention to the spirit and letter of the law. Reagan has thus neutralised legislative mandates on controls of toxic and carcinogenic exposure by direct frontal assaults on regulatory agencies. Strategies employed include: staffing senior positions with unqualified, ideologically selected staff hostile to their agency mandates; budget cutting; insisting on formal cost-benefit analyses which focus on industry costs with little (or biased) consideration of the costs of failing to regulate and which effectively stall the regulatory process; illegal, behind closed doors meetings with industry; and making regulation dependent on the Office of Management and Budget with its subservience to the White House.

An example is the White House decision to block the $1.3 million 1984 request by the National Institute for Occupational Safety and Health (NIOSH) to notify some 200,000 workers of risks from previously undisclosed exposure to workplace carcinogens, as identified in some 60 government studies, in order to enable medical follow-up and early diagnosis of cancer. The reason for this refusal of modest funding seems to have been a desire to shield corporations from possible legal claims. . . .

The US Congress has become sensitive to public health and environmental concerns, as exemplified in a plethora of legislation in recent decades. Such legislation has evolved fragmentarily, reflecting particular interests and priorities. New laws have focused on individual environmental media—air, water, food or the workplace—or on individual classes of products or contaminants, such as pesticides or air pollutants, with little or no consideration of needs for more comprehensive and integrated approaches. Furthermore, the legislative language has traditionally been ambiguous, thus allowing maximal regulatory discretion to bureaucracies which in some instances, have subsequently become closely associated with or even "captured" by the regulated industries. A noteworthy exception is the 1958 Delaney Amendment to the Federal Food Drug and Cosmetic Act, with its absolute prohibition against the deliberate introduction of any level of carcinogen into the food supply. Even so, the Reagan FDA is redefining the Delaney Amendment to allow carcinogenic food additives at levels alleged to be devoid of significant risk.

Congress has also tended to abdicate decision-making to scientific authority (or perceived authority), rather than questioning its basis in the open political arena. Of particular importance was passage of the 1971 Cancer Act in response to orchestrated pressures from the "cancer establishment", the National Cancer Institute (NCI), American Cancer Society (ACS), and clinicians aggressively pushing chemotherapy as a primary cancer treatment. The cancer establishment misled Congress into the unfounded and simplistic view that the cure for cancer was just around the corner, *provided* that Congress made available massive funding for cancer treatment research. The Act did just this, while failing to emphasise needs for cancer prevention, and also gave the NCI virtual autonomy from the parent National Institutes of Health, while establishing a direct chain of command between the NCI and the White

House. Some 16 years and billions of dollars later, Congress still has not yet appreciated that the poorly-informed special interests of the cancer establishment have minimised the importance of critically needed cancer prevention efforts, and have singularly failed to support such efforts.

Until recently, state governments have largely deferred to federal authority, exercising relatively minor roles in cancer prevention. Reagan's federal deregulatory efforts have begun to reverse this relationship. Regulatory actions against carcinogens are now emerging at the state level, such as the banning of chlordane/heptachlor and aldrin/dieldrin for termite treatment by Massachusetts and New York, the banning of daminozide (Alar) for apple ripening and tough restrictions on ethlylene dibromide food tolerances by Massachusetts, and the introduction of informative occupational labeling laws by various states, such as the "right-to-know" workplace legislation of New Jersey.

In some cases, such state initiatives have evoked federal preemption by restricted regulations—such as the 1983 Hazard Communication Standard of the Occupational Safety and Health Administration—despite Reagan's avowed commitment to getting big government off the backs of the people. In February 1987, a coalition of labour and citizen organisations asked the US Court of Appeals to enforce its 18 month old order directing OSHA to expand coverage of its communication standard from manufacturing to all workers. In an apparent about face turn, the Chemical Manufacturers Association is supporting the expansion in conformity with regulations developed for various states.

THE CANCER ESTABLISHMENT: STANDING IN THE WAY OF PREVENTION

The cancer establishment still continues to mislead the public and Congress into believing that "we are winning the war against cancer", with "victory" possible only given more time and money. The NCI and ACS also insist that there have been major advances in treatment and cure of cancer, and that there has been no increase in cancer rates (with the exception of lung cancer which is exclusively attributed to smoking). Yet, the facts show just the contrary.

The cancer establishment periodically beats the drum to announce the latest "cancer cure" and dramatic "break-through". These announcements reflect optimism and wishful thinking, rather than reality. The extravagant and counterproductive claims for Interferon as the magic cancer bullet of the late 1970s have been followed by the unpublicised recognition of its limited role in cancer treatment. The latest NCI "break-through" claims for interleukin-2 as a cancer cure are grossly inflated and rest on questionable data. These claims fail to reflect the devastating toxicity and lethality of this drug, and gloss over the high treatment costs, which can run into six figures.

Equally questionable are claims by the NCI and ACS that overall cancer survival rates have improved dramatically over recent years. These claims, based on "rubber numbers" according to one prominent critic, ignore factors such as "lead-time bias", earlier diagnosis of cancer resulting in apparently prolonged survival even in the absence of any treatment, and the "over-diagnosis" of essentially benign tumours, particularly those of the prostate, breast and thyroid, as malignant. Recently the director of the NCI, DeVita, resorted to blaming community doctors for using inadequate doses of chemotherapy drugs as the "real" reason why cancer cure rates are no better than they are.

The NCI misrepresentations are well reflected in budgetary priorities which are largely and disproportionately directed to cancer treatment research—to the neglect of cancer prevention. Even the very modest funding on cancer prevention is largely directed to endorsing industry's "blame-the-victim" concept of cancer causation. Thus, the NCI exaggerates the role of tobacco for a wide range of other cancers besides lung cancer, and treats as fact the slim and contradictory evidence relating diet to colon, breast, and other cancers.

Apparently still oblivious to mounting criticisms, the NCI continues vigorously to propagate these misrepresentations. A 1986 NCI document on cancer control objectives, the executive summary of which fails to even mention environmental and occupational exposures to carcinogens and focuses on diet and tobacco as the major causes of cancer, rashly promises that annual cancer mortality rates could be reduced by 50 per cent by the year 2000.

More disturbing than indifference to cancer prevention is evidence uncovered in September 1982 by Congressman Dave Obey that the NCI has pressured the International Agency for Research on Cancer (IARC) funded in part by the NCI, to downplay the carcinogenicity of benzene and formaldehyde in IARC monographs which review and rank the carcinogenicity data on industrial and other chemicals. Such evidence is noteworthy since, contrary to the scientific literature and its own explicit guidelines, IARC has also down-graded the carcinogenicity of other carcinogenic industrial chemicals, such as the pesticides aldrin/dieldrin and chlordane/heptachlor, and the solvents trichloroethylene and perchloroethylene.

Over the last decade, consistent with its low priorities for cancer prevention, the NCI has played little or no role in providing the data base in support of critical federal or state legislation and regulation on cancer prevention. Examples where NCI scientific input could have been reasonably expected include attempts to prevent the exposure of much of the US population to carcinogenic pesticides, such as ethylene dibromide residues in food or chlordane/heptachlor in the air of a high percentage of the many million homes treated for termites, and also exposure to industrial discharges of carcinogenic air pollutants or drinking water contaminants. . . .

THE LIFESTYLE ACADEMICS

The lifestyle academics are a group of conservative scientists including Sir Richard Doll, the Warden and Director of the industry-financed Green College, Oxford, his protege R. Peto, a statistician also from Oxford, and more recently Bruce Ames, a California geneticist. The purist pretensions of "the lifestylers" for critical objectivity are only exceeded by their apparent indifference to or rejection of a steadily accumulating body of information on the permeation of the environment and workplace with industrial carcinogens, and the impact of such involuntary exposures on human cancer.

Consciously or subconsciously, these academics have become the mouthpiece for industry interests, urging regulatory inaction and public complacency. Among the more noteworthy contributions of these academics is a series of publications claiming that smoking and fatty diet are each responsible for 30-40 per cent of all cancers; that sunlight, drugs and personal susceptibility account for another 10 per cent, leaving only a few per cent unaccounted for which, just for want of any other better reason, was then ascribed to occupation. According to the lifestylers, this then proves that occupation is an unimportant cause of cancer, which really does not warrant much regulatory concern.

Apart from circulatory referencing each other as authority for these wild guesses, the lifestylers have never attempted to develop any estimates of how many workers are exposed to defined levels of specific carcinogens. Without such estimates there is no way of attempting to determine just how much cancer is due to occupational exposure. . . .

Bruce Ames is a geneticist who, in the 1970s, developed bacterial assays for mutagenicity which he advocated as short-term tests for carcinogens. He then published a series of articles warning of increasing cancer rates and of the essential need for tough regulation of industrial carcinogens, such as the fire retardant Tris and the fumigant ethylene dibromide. By the 1980s, however, Ames did an unexplained 180 degree turn, now claiming just the opposite, that overall cancer rates are not increasing, that industrial carcinogens are unimportant causes of cancer which do not need regulating, and that the real causes of cancer are natural dietary carcinogens, largely because mutagens can be found in a variety of foods.

WHAT TO DO ABOUT CANCER

The cancer epidemic poses the nation with a grave and growing crisis of enormous cost to health, life and the economy. In my 1979 book, *The Politics of Cancer* I concluded with the following specific recommendations designed to reduce the toll of preventable cancer:

- Cancer must be regarded as an essentially preventable disease;
- The hidden political and economic factors which have blocked and continue to block attempts to prevent cancer must be recognised;
- The ineffective past track record of government in cancer prevention must be recognised.
- The critical roles in cancer prevention that public interest groups and informed labour leadership have exercised must be recognised and their further efforts fully encouraged and supported.
- Congress must resolve the major inconsistencies in a wide range of laws on environmental and occupational carcinogens;
- Substantially higher federal priorities for the prevention of cancer must be developed;
- Policies of the various federal agencies with responsibilities in cancer prevention must be effectively integrated and coordinated.
- Top business management must recognise the essential similarities between their long-term interest and goals and those of society. Prevention of occupational cancer and cancer in the community-at-large is of primary importance to both;
- The American Cancer Society must be influenced to balance its preoccupation with treatment with activist programmes designed to prevent cancer;

• The medical and scientific community must accept a higher degree of responsibility and involvement in the prevention of cancer by actions on both the professional and political levels;

• Medical schools and schools of public health must be persuaded to reorient their educational and training programmes from the diagnosis and treatment of disease and cancer to prevention;

• Chemicals in consumer products and in the work-place must be clearly and simply identified and labeled;

• New approaches must be developed for obtaining and for retaining honest and scientifically reliable data on the carcinogenicity and toxicity of new chemicals, in addition to those untested or poorly tested chemicals already in commerce; such data must be made accessible to public scrutiny. Maximum legal penalties should be directed against all those responsible, directly and indirectly, for distortion or manipulation of toxicological and epidemiological data on the basis of which decisions on human safety and risk are based.

• Apart from actions on a political level, we all have limited personal options. To some extent, it may be possible to reduce our own chances of developing cancer by making informed changes in lifestyle, in our use of consumer products, and in our work;

• The major determinants of preventable cancer are political and economic, rather than scientific, and as such must be addressed in the open political arena. Cancer prevention must become a major election issue, on a par with inflation.

A decade later, these goals still stand as valid, but none have been achieved while cancer rates have steadily risen. To prevent similar conclusions a decade from now, the cancer prevention rhetoric must be translated into reality. . . .

Cancer is essentially a preventable disease. Given high national priority, this goal will be achieved.

NO

Elizabeth M. Whelan

THE CHARGE OF THE CANCER BRIGADE

The topic of cancer weighs heavily on the regulatory mind. Indeed, federal agencies are now waging an all-out war on the nation's "cancer epidemic." Interagency committees are battling to determine a national policy on chemical carcinogens. Stringent and expensive regulations are being aimed at chemicals in our air, food, and workplaces. War has been declared, but, unfortunately, it's a misdirected war against a misidentified enemy.

The Food and Drug Administration (FDA) and the Environmental Protection Agency (EPA) have a long history of waging war against allegedly carcinogenic chemicals. The FDA's banning of red dye #2, cyclamates, and the cattle growth-stimulant DES, and the proposed banning of saccharin, were all in the name of "protecting us from cancer." The EPA's banning of the pesticides aldrin/dieldrin, heptachlor, and chlordane, and its frequently reformulated air carcinogen standards, have the same intent. Now, however, we have more agencies and committees in on the act.

In July 1979, the Interagency Regulatory Liaison Group (IRLG) issued a report entitled "Scientific Bases for Identification of Potential Carcinogens and Estimation of Risks." In August, the Toxic Substances Strategy Committee (TSSC) of the Council on Environmental Quality (CEQ) reported to the President on "Cancers and Carcinogens: Prevention Policy." In October, the Regulatory Council issued a Statement on Regulation of Chemical Carcinogens." In late June 1980, the TSSC issued a second report to the President, including a section on cancer.

The aim of these reports was to establish a uniform federal policy on cancer and specifically on the regulation of man-made carcinogens in the environment. For instance, in the hope that all the agencies which regulate carcinogens would act consistently, the Regulatory Council report attempted to define policies in four major areas of activity: 1) determining whether a chemical may cause cancer, 2) assessing the risk of cancer to humans, 3) establishing regulatory priorities, and 4) undertaking regulatory activities.

The reports also called for stronger federal efforts to protect the public from hazardous chemicals. In particular, the second TSSC report recommended new legislation and other, additional steps to deal with the "growing problem" of toxic chemicals, with special emphasis on carcinogens, in the environment, in the work-place, and in the home. . . .

In January 1980, the Occupational Safety and Health Administration (OSHA) issued a proposal for banning or severely restricting "carcinogens" in the workplace. This policy, the strictest of the new measures, would require industries to invest large sums of money to protect their workers from allegedly harmful chemicals.

The most notable feature of the OSHA policy is its emphasis on urgency. In the interest of accelerating the regulation of specific suspect chemicals, OSHA would anticipate major scientific judgments about carcinogens, to reduce the areas for debate. Suspect chemicals would be classified automatically in Category I (requiring that worker exposure be reduced to zero if "suitable substitutes" exist; otherwise, to the "lowest feasible level") if there is any evidence that they cause cancer in humans, or if there are positive results from a single long-term animal test, supported by any evidence from short-term tests.

If both positive and negative animal tests exist, in different species, the negative results would be disregarded. Differences between benign and malignant tumors would be ignored. No distinction would be made between substances that initiate cancer and those that promote existing cancers. Positive animal tests would be considered valid even if the doses of chemicals used were so high that most of the animals died of acute poisoning. Clearly OSHA considers the need for *immediate* regulation of workplace carcinogens to be so great that it would ignore major scientific disagreement.

The proliferation of these government proposals and their desperate sense of urgency imply that the government is building a major regulatory policy on two basic assumptions: first, that there is a cancer epidemic in the U.S.; and second, that this alarming development is related to the presence of man-made carcinogens—food additives, pesticides, air pollutants, workplace hazards—in our environment. An examination of these assumptions is in order.

The news media frequently refer to our nation's "cancer epidemic." Fortunately, the idea that we are experiencing an epidemic of cancer is a misconception. While it is true that cancer deaths were less common in the early years of this century, life expectancy was limited then, and cancer is primarily a disease of older adults. In the good old days, most people died of communicable diseases before cancer had a chance to strike. When statistics are age-adjusted (to take into account the growing proportion of older Americans) the combined incidence rate for all cancer sites in both sexes actually *decreased* 3.9 per cent between 1947 and 1971.

More recent, unpublished data indicate that there was a 1.3 per cent average annual increase of all types of cancer between 1969 and 1976. However, the largest portion of this increase is explained by a rise in lung cancer in women, primarily attributable to cigarette smoking. (Smoking became popular among women long after men took up the habit, and the effects of cigarettes on women are only now showing up in the incidence rates of lung cancer.) If lung-cancer deaths are eliminated from the calculation, then the combined death rate from cancer at all other sites combined remains relatively stable or is slightly declining. Moreover, even if one does include the figures for lung cancer, the age-adjusted death rate from cancer in the U.S. increased from 125 per 100,000 in 1950 to 132 per 100,000 in 1976. It strains credibility to describe this as an epidemic.

The current regulatory proposals imply (and sometimes explicitly state) that the overwhelming majority of cancers in the U.S. are environmentally caused. The EPA stated that "although the uncertainties are great, estimates . . . have suggested that 60 to 90 per cent of all cancers may be due to environmental factors." EPA administrator and Regulatory Council chairman Douglas M. Costle, when releasing the latter group's report, said: "We must significantly reduce our exposure to environmental causes of cancer that may account for 60 to 90 per cent of the incidence of this disease."

The problem here is with the word "environment." To many non-scientists, and apparently to some government agencies, environment is a synonym for synthetic chemicals. But to scientists, environment is a much broader term referring to everything except heredity and the body's own biochemistry. The 60 per cent figure quoted above was derived from the work of Dr. John Higginson, founding director of the World Health Organization's International Agency for Research on Cancer. During the 1950's, Dr. Higginson compared the incidence of certain tumors in black Africans and black Americans and concluded that about two-thirds of all cancers had an environmental cause. However, as Dr. Higginson has been forced to clarify repeatedly, he was using the term environment in the broad scientific sense, not in the more limited way in which it is used by the layman. Such factors as cigarette smoking, diet, exposure to sunlight, drugs, viruses, radioactivity, sexual habits, and reproductive patterns are all part of the "environment."

But when references to environmental cancer are included in government proposals, they frequently do not acknowledge that environment includes many voluntary factors. Instead, they imply that environment means a smorgasbord of involuntarily imposed carcinogenic chemicals in our food, air, and workplaces. Such a link between chemicals and cancer suits the needs of the agencies because it indicates a fertile field for the type of regulations they are prepared to impose. But the link is not fully consistent with the facts.

While frequently the target of regulation, food additives, when consumed in the amounts normally added to food, have never been shown to cause cancer or any other illness in human beings. Indeed, there is suggestive laboratory and epidemiologic evidence that the use of some additives, specifically the antioxidants BHA and BHT, has contributed to the dramatic decline in stomach cancer in this country over the past four decades. Likewise, although pesticides in concentrated form are certainly toxic (after all, their purpose is to kill the insects and other pest with whom we compete for our food supply), there isn't a single documented case of cancer of any other human illness caused by the minute pesticide residues present on foods.

Air pollution is certainly undesirable, and it may play a role in diseases such as bronchitis and asthma. But again, to our knowledge, general air pollution does not cause lung cancer or any other form of cancer. A recent report on a twenty-year study conducted by the American Cancer Society concluded that "Air pollution was not found to be a great culprit in causing lung cancer. General air pollution had little effect in a comparison [of cancer rates] between urban and rural people. Smoking is the key factor [in lung cancer]."

It has been known for centuries that long-term high-dose exposure to some occupational substances can lead to human cancer. In the eighteenth century Percivall Pott noted, in what is considered a classic report, that chimney sweeps had an unusually high rate of scrotal cancer. We now know that these boys were in regular contact with benzo(a)pyrene and related carcinogens, and that poor hygiene, which was the norm contributed to the intensity of the exposure.

In modern times, cancers have been caused by high-dose occupational exposures to asbestos, vinyl chloride, and several other substances. Thought the 1960's and 1970's, both industry and government have become increasingly aware of this type of hazard, and steps have been taken to reduce the exposure of workers to known carcinogens. Techniques for assessing the potential hazards presented by new chemicals introduced into commerce have also been developed, and with a continuing awareness of the potential dangers of high-dose exposures in the workplace, the hazards presented by new chemicals will also be controlled.

Fortunately, the total number of occupationally induced cancers is generally agreed to be relatively small. The estimate accepted by many authorities, including WHO's International Agency for Research on Cancer, is that occupational exposures account for 1 to 5 per cent of all cancer deaths. In many cases, the current victims were exposed to workplace carcinogens several decades ago, when worker-protection practices were often sloppy, and when many hazards were unrecognized. With the improved safety standards of the last two decades, and with our current ability to identify hazardous new chemicals before workers are exposed to them, there is every reason to hope that the rate of occupationally induced cancer will decline.

However, the general agreement that the incidence of occupational cancer is relatively low and probably decreasing was shaken by statistical bombshell from Washington. In a 1978 speech to the AFL-CIO National Conference on Occupational Safety and Health, then-Secretary of Health, Education, and Welfare Joseph Califano announced: "If the full consequences of occupational exposure in the present and recent past are taken into account, estimates of the [total contribution of occupational exposures many be] at least 20 percent of all cancer in the U.S.—and perhaps more—may be work-related." Subsequent press coverage focused on the Secretary's phrase "and perhaps more" and implied that the figure might be more like 40 per cent.

Where did Secretary Califano get his numbers? His figures, which have since shown up in almost every federal document dealing with cancer and the environment, come from an unreviewed, unpublished paper entitled " Estimates of the Fraction of Cancer in the United States Related to Occupational Factors." Nine scientists, with substantial credentials, evidently contributed in some fashion to the "estimates paper," but no one has publicly admitted authorship.

This paper has been the focus of widespread professional criticism and scorn. It was heavily criticized at a conference on "Cancer Prevention—Quantitative Aspects" shortly after Secretary Califano's announcement. There, Dr. John Cairns, Director of the Mill Hill Laboratory of Cancer Research in London, said, "There are several parts of it which seem to be manifestly silly, and anyone who could sit down and do calculations could see how stupid it is.' The distinguished epidemiologist Sir Richard Doll flatly stated: "I regard it as nonsense." In an editorial in *Nature,* Dr. Richard Peto of Oxford University said that "The document was based on the flimsiest of arguments, and reached conclusions which are grossly at variance with easily accessible data . . . a group of reasonable men can unfortunately collectively generate an unreasonable report." An editorial in the prestigious medical journal *Lancet* said that "its framework is insubstantial, remarkably so," and concluded, with admirable British courtesy, that "it is sad to see such a fragile report under such distinguished names."

The "estimates paper" is used as a major justification for the new, remarkably inflexible OSHA proposal for the regulation of workplace carcinogens. Although the hazard of occupational cancer is and will continue to be of real concern, this and other OSHA actions have tended toward excess.

OSHA's recent attempt to lower the permissible concentration of benzene in factory air from ten parts per million to one part per million is another example of this agency's overzealous regulation. In this case, the Supreme Court overruled OSHA, because the agency did not justify the new standard well enough. The Justices wanted OSHA to show that the existing standard involved significant risks, and that the new standard would reduce these risks. Without such evidence, OSHA might be forcing enormous expenditures to eliminate insignificant hazards.

No one will win the regulatory war on cancer. Strenuous new regulation of chemicals in our food, air, and general environment will not conquer cancer, because it will not eliminate any major risk factors. Judicious regulation of workplace chemicals is necessary, but regulatory overkill of occupational cancer hazards will accomplish little. The known cancer risk factors are primarily voluntary practices, such as cigarette smoking, which cannot be controlled by regulation; and the causes of many cancers are unknown. Funds intended for "cancer prevention" might be better spent on education of the public about voluntary risks than on excessive regulation. They would certainly be better spent on research. Our understanding of this disease and its causes is far from complete. Many brilliant, promising avenues of research are being pursued and are in need of funds. The current and exciting work on interferon as a possible means of treating some forms of cancer is merely one example.

But if the regulators continue their misguided war on cancer, there will be many casualties. Businessmen will face higher costs, fewer jobs and limited production all as results of the new regulatory policies. The consumer will face higher prices, tax increases, and fewer products—a diminished standard of living and no new health benefits. Scientists will also suffer if funds and priorities are diverted away from the research lab and into regulation. Perhaps the most important result will be the perpetuation of a major misunderstanding if the outpouring of regulations gives the false impression that cancer is being conquered through prevention. We should not develop a sense of security when we are merely bombing the wrong targets. Cancer is not an epidemic, but it is an important health problem in the United States. If we're going to have a war against cancer, we should use tactics that will enable us to win.

POSTSCRIPT

Is There a Cancer Epidemic Due to Industrial Chemicals in the Environment?

Epstein refers to the American Council on Science and Health, of which Whelan is director, as a "front organization" because it is a private, industry-related group rather than an official government organization, as its name suggests. By referring to red dye #2, DES, aldrin, dieldrin, heptachlor, and chlordane as "allegedly carcinogenic chemicals," Dr. Whelan questions the scientific judgment of panels of cancer experts who carefully reviewed the evidence before deciding that each of these substances was capable of inducing cancer in laboratory animals. Former Health, Education and Welfare (HEW) secretary Califano may have exaggerated the percentage of cancer cases due to occupational exposure, but even if the correct estimate is one to five percent—as Whelan contends—is it appropriate to refer to this as a relatively small effect?

Epstein is the author of several books on the subject of environmental and occupational health, including *The Politics of Cancer* (Anchor Press, 1979). Whelan has also written or been the coauthor of several popular books—one of which, *Preventing Cancer* (Norton, 1977), includes a more detailed presentation of her views on the significance of environmental factors.

Bruce Ames, a noted geneticist, is referred to by Epstein as a "life-styler" because he has recently attempted to demonstrate that natural substances in foods we eat may pose more of a carcinogenic risk than industrial chemicals. A controversial proposal for ranking carcinogens is described in an article that Ames wrote with R. Magaw and L. Gold for the April 17, 1987, issue of *Science*. For one of several critical responses to that article, see the letter by E. Silbergeld in the September 18, 1987, issue of the same journal.

Recently published epidemiological studies present evidence showing an increase in brain tumor incidence and a general rise in cancer deaths in people over age 55. Dr. Devra L. Davis is the coauthor of two such reports published in *The Lancet* (August 25, 1990) and *The American Journal of Industrial Medicine* (December 1990). Her suggestion that diet and environmental factors may be responsible for these changes is disputed by other medical scientists.

In December 1990 the Environmental Protection Agency (EPA) released its preliminary findings concerning several research studies that appear to demonstrate a statistical link between certain types of human cancer and exposure to low-level electromagnetic radiation from electric utility power lines and transformers or even from home appliances. The conclusion is that there is sufficient evidence of a correlation to warrant further research.

For a brief report illustrating the public confusion resulting from conflicting findings on carcinogenesis, see "Diet, Workplace Called Cancer Factors," *Chemical and Engineering News* (June 21, 1982). An interesting report on the political and scientific confusion over the saccharin issue is "Latest Saccharin Tests Kill FDA Proposal," *Science* (April 11, 1980).

Two rather different proposals on how cancer-causing agents should be regulated are described in "Carcinogens in the Environment," by B. Commoner, *Chemtech* (February 1977) and "The Regulation of Carcinogenic Hazards," by G. B. Gori, *Science* (April 1980). For an interesting report on the carcinogen regulation policy of the EPA under the Reagan administration, see "Formaldehyde," by Bette Hileman, in the October 1982 issue of *Environmental Science and Technology*.

ISSUE 10

Is Immediate Legislative Action Needed to Combat the Effects of Acid Rain?

YES: Jon R. Luoma, from "Acid Murder No Longer a Mystery," *Audubon* (November 1988)

NO: A. Denny Ellerman, from Testimony before the Subcommittee on Environmental Protection, U.S. Senate (June 17, 1987)

ISSUE SUMMARY

YES: Science writer Jon R. Luoma argues that research has produced convincing evidence that legislation requiring large cutbacks in acid gas emissions is urgently needed to prevent continued destruction of forest and lake ecosystems.

NO: Coal association executive A. Denny Ellerman claims that sulfur oxide emissions are presently declining and further restrictions would produce marginal benefits at great cost to industry and the public.

Efforts to control the effects of local air pollution may, ironically, have exacerbated another serious, but less tractable, problem. To comply with the provisions of the Clean Air Act, industry built tall smokestacks. This had the intended effect of reducing local pollutant concentrations. But in the late 1960s, scientists began to uncover evidence that the sulfur and nitrogen oxides in the plumes from smokestacks in heavily industrialized areas were returning to earth hundreds of miles away in the form of sulfuric and nitric acid-contaminated precipitation.

Acid rain, as this phenomenon is generally labeled, has become the subject of intense political and scientific debate. There is general agreement that in regions such as Minnesota, northeastern United States, southeastern Canada, and Scandinavia, rain and snow tend to be considerably more acidic than in areas that are not downwind from coal-burning power plants and nonferrous metal smelting industries. Despite continued denials from a few scientists employed by these industries, there is a growing consensus that sulfur and nitrogen oxide emissions are the principal cause of the acidity. Despite more than 20 years of study, much uncertainty remains about the complex chemistry and the many atmospheric and meteorological variables involved. There is a dispute about whether or not acidity of precipitation in the regions already affected is increasing. The phenomenon does, however, appear to be spreading.

The effects of acid rain are also a subject of controversy. Best documented are the dramatic ecological consequences in the growing number of lakes and streams that are becoming acidic. Reports of presently dead lakes, which teemed with aquatic life only a few decades ago, continue to mount. Less well documented are the sensitivities of forest systems and agricultural crops to acid precipitation. Recent reports from Canada, Vermont, and New Hampshire appear to link the rapid decline of such important forest species as red spruce and balsam fir to increasing surface water acidity.

The acid rain problem presents a classic question of what to do about a serious environmental problem which transcends state and national boundaries. It is also sufficiently complex so that a complete understanding will require prolonged scientific study, and it appears to require very costly remedial action. Former president Reagan's first budget director, David Stockman, angered many environmentalists by presenting a very crass economic analysis of the issue. He asked if it was worth spending billions of dollars to control emissions from Ohio utilities in view of the limited economic value of the fish that no longer exist in 170 lakes in New York State. The Canadian government sees the matter very differently and has demanded that the United States join Canada, Germany, Sweden, and Norway in committing to reduce sulfur oxide emissions by 50%. The amendments to the Clean Air Act, which were finally passed after nearly a decade of acrimonious debate, contain provisions designed to meet this goal.

Jon R. Luoma, a science writer who has followed the acid rain debate for *Audubon* magazine, reports some of the evidence that he believes supports the need to regulate sources of acidic emissions. A. Denny Ellerman, executive vice president of the National Coal Association, articulates the position that the energy industry took in the 1987 congressional debate on environmental protection. He claims that existing regulations are already reducing sulfur oxide release and that further restriction is unwarranted.

YES

Jon R. Luoma

ACID MURDER NO LONGER A MYSTERY

It's been nearly a decade since acid rain surged into news media and public consciousness with reports from scientists that the rains and snows—once symbols of purity—could quite literally poison entire freshwater ecosystems. Captured in huge weather systems, acid-forming pollutants could travel hundreds, or even thousands, of miles from their sources to pollute waters across state or national boundaries.

Now, a near-decade and hundreds of millions of research dollars later, concerns about acid rain have broadened to include threats to wider regions of freshwater lakes and streams, threats to forests, and threats to public health. A clear scientific consensus that acid rain is at least a long-term threat to some aquatic ecosystems has solidified.

In many other ways, very little has changed. The key sources of acid rain remain sulfur dioxide from poorly controlled fossil-fuel-burning power plants and nitrogen oxides from a range of combustion sources, including industrial furnaces and cars. The responsible industries, citing high costs, steadfastly oppose tighter controls, instead calling for more research.

Attempts at federally legislating new controls may have faltered because the pathways of acidification are initially so subtle that widespread damage is not evident. Chemical compounds naturally present in even some of the most sensitive lakes, streams, and watersheds can neutralize acids, often for many years. Only when those neutralizers are used up will a lake quite suddenly begin to turn acid. And although lakes with little remaining neutralizing capacity number in the tens of thousands in North America, lakes actually acidified to date number only in the hundreds. Similarly, research scientists have learned that visible symptoms of forest destruction become obvious only after damage is well under way.

What follows is a compendium of new developments on the acid rain front.

From Jon R. Luoma, "Acid Murder No Longer a Mystery," *Audubon* (November 1988).

NEWS FROM MOUNT MITCHELL

When Audubon last visited Mount Mitchell in North Carolina ("Forests Are Dying, But is Acid Rain to Blame?" March 1987), plant pathologist Robert I. Bruck was speculating that it might be many years before he had enough data to make clear projections about whether acid rain—or any kind of air pollution—was causing the forest devastation there, so complex was the issue. He said that he'd let us know when he felt scientifically confident to make a bold statement—maybe a decade hence.

"Time's up," Bruck fairly growled in a telephone interview from his prefab headquarters on the 6,684-foot mountain this summer. "You would not believe how this place has changed even in the time since you were up here."

As it turns out, Bruck and a team of scientists, with their networks of towers, collectors, monitors, tubes, and wires laced through the skeletal trees, have discovered persistent high levels of pollutants—notably ozone and atmospheric acids—that appear to correlate strongly to a forest die-off that has increased by some 30 percent since we visited the mountain.

"It's plain," he says, "that no one has proved, or ever will, that air pollution is killing the trees up here. But far more quickly than we ever expected, we've ended up with a highly correlated bunch of data—high levels of air pollution correlated to a decline we're watching in progress."

A trip up Mount Mitchell, eastern North America's highest peak, is a thumbnail experience of the kind of biotic transition one might see on a 1,500-mile automobile journey northward from the sultry southeastern United States to frigid Labrador. Although Mount Mitchell sits at about the same latitude as California's Mojave Desert, the trees on and around the blustery mountaintop are hardy near-Arctic species, red spruce and Fraser fir.

And like many of the ridgetop trees along the entire stretch of the Appalachians—like trees in many of the forests of Europe—they are in evident decline. During my 1986 visit, the fir forest at the summit was eerily dominated by death—stark, brown hulks and deadfalls creaking in the ever-present wind.

According to Bruck, it has gotten worse. The damage extends even further down the mountain into more of the pure red spruce stands and, in some cases, all the way down to the line where hardwood forests begin. According to other reports, Appalachian spruce-fir forests from the White Mountains in New Hampshire to the Great Smokies south of Mount Mitchell are showing signs of reduced growth and general decline.

The information gathered during the federally funded mega-study that Bruck has been coordinating on Mount Mitchell has him saying that he is "ninety percent certain" that air pollutants are killing the southern Appalachian ridgetop spruce and fir forests.

Two years ago Bruck was wondering if some other factor or complex of factors—insects, fungi, climate, or even forest management practices—might be responsible.

But his data now shows that more than half of the time ozone levels on the mountain exceed those at which tree damage has been proven to occur in contolled laboratory studies. Frequently, levels increase to more than double the minimum damage limit. Acidity in the clouds that bathe the summit of the mountain eight out of every ten days has also been extraordinary: ranging from a worst case of pH 2.12 to a best case of 2.9. In other words, on the *best* days of cloud cover, acidity has been somewhat more than that of vinegar.

"Let me tell you about an experiment thirty yards from where I'm sitting on top of the mountain," Bruck said. He described a set of large outdoor chambers, sealed in clear plastic to create a controlled greenhouse, in which young trees have been placed. The usual, and apparently polluted, mountain air is pumped into one chamber without alteration. A second chamber receives mountain air filtered though activated carbon, which removes ozone and some other pollutants. "We set up that experiment only six weeks ago. We're already getting fifty percent growth suppression in the chamber receiving ambient [unfiltered] air."

He also reports that the research team noted widespread burning of new needletips on conifers after particularly acid air masses passed through. Analysis of the burnt needles in Environmental Protection Agency laboratories revealed extremely high levels of sulfate, a compound associated with acid rain.

Mountaintops are subject to greater pollutant deposits because they are frequently bathed in polluted cloud water. Extensive forest destruction in Europe began in the 1970s on the mountaintops but has extended to stands at lower altitudes. Many scientists studying the problem feel that air pollution does not kill trees directly but rather weakens them to the point where, like punch-drunk fighters, they are no longer able to withstand normal episodes of moderate drought or insects or diseases that they could otherwise easily resist.

KILLER MOSS, ACID SOILS

Now come reports from field reseachers of problems stemming from acid mosses and acid soils.

Lee Klinger, working with the National Center for Atmospheric Research in Boulder, Colorado, claims to have correlated forest diebacks with the presence of three kinds of mosses: sphagnum, polystrichum, and aulocomnium. He says that the mosses alone are not killing trees but were virtually always present in more than a hundred dying forests that he examined, and that they appear to be a key part of a forest death syndrome.

The mosses produce organic acids which appear to gang up with inorganic acids in polluted rain to mobilize aluminum naturally present but harmlessly bound up in most soils. Additional aluminum appears to be falling out of the atmosphere bound to dust particles. The mobilized aluminum is toxic to the fine feeder roots of most trees. In fact, says Klinger, mats of killer mosses inevitably overlay networks of dead feeder roots. Furthermore, sphagnum acts as a sponge, saturating the soil just beneath the moss and creating an anaerobic, or oxygen-starved, soil environment, which also helps kill roots. The mosses, he says, occur naturally in forests and may even be part of an extremely slow plant succession process that, over centuries or millennia, turns old forests to bogs. But the present moss invasion appears to be promoted and greatly speeded up by acidic rainfall.

"In general," Klinger says, "mosses require acid conditions for their establishment—so it appears that as the soils become acidified, there are more places for the mosses to get established." He also suggests that nitrates that form from nitrogen oxide pollutants in the rain fertilize the mosses, which have no roots but are entirely nourished by atmospheric chemicals.

Meanwhile, some scientists have begun to change their minds about acid soils. Until recently, most researchers looking into acid rain's terrestrial effects have assumed that trees and other plants growing in naturally acid soils—including trees found in many northern coniferous forests—had more resistance to pollutants.

But now Daniel Richter of Duke University reports that his laboratory experiments show that naturally acid soils, such as those often found at high elevations in parts of the Appalachians, are highly sensitive to chemical imbalances caused by the addition of more acids from precipitation. According to Richter, highly acid forest soils that become further acidified can become virtually "infertile" through a complex series of chemical reactions that deprives them of nutrients. At the same time, the soils are assaulted by an overload of mobilized and root-toxic aluminum.

IT KEEPS GETTING WORSE

New York was one of the first states to become deeply concerned about the effects that acid rain could have on its surface waters, and particularly on the pristine but poorly buffered lakes of the huge and beautiful Adirondack Park. In the 1970s the New York Department of Environmental Conservation jolted the conservation community by reporting that more than two hundred lakes, most of them in the western Adirondacks, had already become too acidic for fish to survive, and that many more appeared to be threatened.

In a new study, three years in the making, the New York DEC reports that fully 25 percent of the lakes and ponds in the Adirondack Mountains are now so acidic that they cannot support fish life. Another 20 percent have lost most of their acid-buffering capacity and therefore appear doomed if acid input continues.

Massachusetts has reported that almost 20 percent—or about eight hundred—of the state's ponds, lakes, and rivers are vulnerable to acid deposition and could become acidified within the next forty years. Already, according to Environmental Affairs Secretary James Hoyte, surveys have located 217 acidified bodies of water that cannot support natural communities. Particularly alarming to state officials is data showing that more than 50 percent of the state's thirty-four drinking-water reservoirs have lost much of their acid-buffering capacity since 1940. The largest of these, Quabbin Reservoir, has lost about three-fourths of its buffering capacity, and the Massachusetts Executive Office of Environmental Affairs now estimates that twenty years remain before the reservoir loses all of its capacity to handle acids. Acidification tends to occur within a few years after buffering capacity is lost.

Meanwhile, Environmental Protection Agency researchers, in a recent report, have identified new and surprising acidification sites in the mid-Atlantic states of Virginia, Delaware, Pennsylvania, Maryland, and West Virginia. The report shows that 2.7 percent of sampled stream miles in the mid-Atlantic region are already acidic, with the afflicted number as high as 10 percent at higher elevations. Despite the fact that rain acidity levels in the region are among the highest in the nation, it has long been assumed that soils in much of the mid-Atlantic could buffer the acidity effectively. The study notes that the stream damage is "probably associated with atmospheric acid deposition."

The Pennsylvania Fish Commission estimated in 1987 that half the state's streams will not be able to support fish life by the year 2000 unless acid deposits decline. Now seventy-eight of the two hundred members of the Pennsylvania House of Representatives are cosponsoring a bill to slash emissions of sulfur dioxide by more than half, in a program that would be phased in over a period of thirteen years.

Pennsylvania has long been in an acid rain quandary. Its emissions of sulfur dioxide are the second worst in the nation, and both its industrial emitters and its powerful coal industry have strongly opposed regulation. But unlike many other high-emission areas, which contain few easily acidified waters, the state has long recognized that many of its woodland streams are acid-sensitive. Pennsylvania government insiders expect vigorous opposition from the pollution lobby on the bill: The state coal association issued a statement immediately after the bill was introduced pointing to government "facts" that watershed acidification was not going to get any worse.

FERTILIZING CHESAPEAKE BAY

Two-thirds of the excess acidity in precipitation falling on the eastern United States comes from sulfur dioxide. So SO_2 has long occupied the attention of most of those concerned with the problem.

But a new study from the Environmental Defense Fund points out that there are plenty of reasons to be concerned about oxides of nitrogen, which not only produce the other one-third of the acid rain problem but are key to the formation of ground-level ozone. (While ozone in the stratosphere is, indeed, necessary to shield the Earth from excess ultraviolet radiation, low-level ozone is not only a health hazard but a proven multibillion-dollar destroyer of agricultural crops and, possibly, forests.)

The EDF study looked at neither of those aspects of nitrogen oxides, but rather at their ability also to function as fertilizers—in this case, unwanted fertilizers of the already pollution-ravaged Chesapeake Bay. The resulting report, based on data from government studies, calculated that one-fourth of the total nitrogen entering Chesapeake Bay comes from excess atmospheric nitrogen oxides, which are produced by combustion in automobile engines, in power plant furnaces, and in virtually all high-temperature burners.

The deposited nitrogen, in turn, is one of the key nutrient pollutants feeding algae in the waters of Chesapeake Bay—to such an extent that biological oxygen demand is up and water quality and fish and shellfish survival are down. The entire process is accelerating eutrophication, a rapid and premature aging process in the bay.

According to the EDF's analysis, the nitrogen input to the bay from air pollution exceeds the contribution from the sewage-treatment-plant effluents and, in fact, exceeds the contribution from all sources but agricultural fertilizer runoff. Although the study was limited to Chesapeake Bay, it suggests that other bays and estuaries may also be suffering from the fertilizing effects of nitrogen raining from the skies.

AN HONEST FEDERAL SUMMARY?

In the history of acid rain research, surely the strangest few days came in September 1987, when the National Acid Precipitation Assessment Program finally released a long-awaited "interim report."

NAPAP, set up late in the Carter Administration to coordinate federal acid rain research, had already been criticized by the General Accounting Office for foot-dragging. When the interim assessment was finally released—two years late—it ignited a veritable firestorm of scientific protest.

There are few complaints about the veracity of the three main volumes of the study. But the slim fourth volume, an "executive summary" was at the heart of the heat. J. Lawrence Kulp, then the NAPAP director and a former Weyerhaeuser executive appointed by Ronald Reagan, had written much of the summary himself.

Critics were especially disturbed that the summary's tally of "acid lakes" counted only those so acid that adult fish could not survive, with far less emphasis on the more numerous acidified waters where amphibian and insect life are harmed, fish reproduction is destroyed, and ecosystem food webs are disrupted.

Further, some critics were outraged that the summary promoted as fact an unproven chemical "steady state" theory, favored by Kulp, that lakes in the Northeast would not become more acidic.

Some prominent researchers spoke out, including several vociferously in the pages of *Science*, which had obtained a pre-release copy. Just days after the summary's release, J. Lawrence Kulp resigned.

The new NAPAP director is one James R. Mahoney. Last April, Mahoney offered a pleasant surprise to many conservationist observers. Testifying before a congressional subcommittee, he offered to prepare a new summary. "I believe the executive summary can and should be expanded to be more representative of all the data available," he said, adding that he "would not subscribe . . . at this time" to Kulp's assertion that acid rain would not harm more northeastern lakes.

Mahoney and Representative James Scheuer, chairman of the subcommittee, later agreed that a shorter report responding to the scientific criticisms would make more practical sense than a wholly republished "interim" summary. That because Mahoney has agreed to gear up his staff to accelerate the massive final assessment to meet the original 1990 deadline, despite the previous delays.

NAPAP plans, this time around, to be more active in soliciting criticisms and comments, according to staff ecologist Patricia Irving. "We want this to be a [scientific] consensus document in every sense," she says.

Scheuer, the congressman, appears to agree that the summary, at least, needs some work. He called that 1987 version "intellectually dishonest."

WAIT FOR CLEAN COAL?

A broad and bipartisan coalition of representatives and senators has introduced compromise legislation in an attempt to break through the political blockade that has stalled all attempts to control acid rain at its source. The new legislation scales back on earlier calls for an annual reduction of 12 million tons of sulfur dioxide. The new target would be 10 million tons per year, or about a 35 percent decrease from current levels.

Connie Mahan of National Audubon Society's government relations office says the society is supporting the bill "even though we are not happy with scaling back by two tons. We went into the 100th Congress believing that the political will for solving this problem was finally taking shape. We continue to hope that by the 101st Congress we'll have acid rain legislation."

The coal and electric power lobbies are continuing to fight legislation which would require them to reduce sulfur dioxide emissions at costs that could exceed $100 million for each poorly controlled fossil-fuel-burning power plant. Instead, they are promoting "clean coal technologies" now in the research and testing stage that could control pollution in the combustion process at much lower costs. However, they have not suggested that large-scale clean coal technology will be available in the foreseeable future.

"We're very suspicious that the promise of future technological improvements is being used as an excuse for not introducing technology that's already known," says Jan Beyea, National Audubon Society senior staff scientist. Beyea points out that the dirtiest power plants, in terms of acid gas emissions, are pre-1980s facilities that were grandfathered at high emission rates by the Clean Air Act. "The real need is for control technology on these older plants. A technology that's useful by the year 2010 isn't going to be of much use. And we don't want to see North America's forests go the way of the European forests."

From his mountaintop headquarters in western North Carolina, Bob Bruck would seem to agree. "People are going to have to start understanding that this is not like some kind of disease where we're going to give all the trees a pill and cure it. We are going to have to decide as a society how to come up with the most logical and reasonable way of implementing what looks like the best solution."

NO

A. Denny Ellerman

TESTIMONY ON BEHALF OF THE NATIONAL COAL ASSOCIATION

My name is A. Denny Ellerman. I am Executive Vice President of the National Coal Association, recently merged with the Mining and Reclamation Council of America. The National Coal Association represents producers of coal, both large and small, both east and west, both low and high sulfur. Our membership produces about 60% of the nation's coal.

I'm sure it will come as no surprise to you that the National Coal Association is opposed to the enactment of S. 321, S. 300 and S. 316. This is the same position we have taken with regard to similar legislation introduced in previous years and we see nothing in the provisions of these or other bills, or in the continuing scientific research concerning this issue to warrant a change in our position.

The legislation before this subcommittee continues to be, as was its predecessor, a proposal to impose a nationwide limit on emissions of SO_2 [*sulfur dioxide—Ed.*] from sources not now subject to new source performance standards without regard to whether such limit is required to achieve the National Ambient Air Quality Standards for SO_2 to protect public health and welfare. The proposed legislation will force a reduction in aggregate SO_2 emissions that will be costly and will have little effect on the localized and limited instances of ecological distress which are the supposed beneficiaries of this and other legislation proposed to "control" acid rain.

I might note also, not with respect to S. 321 but to S. 1351, the "Clean Air Standards Attainment Act of 1987," recently introduced by Senator Mitchell, that this proposal contains requirements for EPA to (1) implement a one-hour SO_2 standard, (2) establish a secondary standard for fine particles, and (3) promulgate a standard for acid aerosols in addition to the NAAQS for

From U.S. Senate. Committee on Environment and Public Works. Subcommittee on Environmental Protection. Hearing, June 17, 1987. Washington, DC: Government Printing Office, 1987.

particulate matter. These provisions are not related to ozone non-attainment. They constitute an "acid rain" control bill as severe in magnitude of forced reductions as other proposals introduced in this Congress. Although proposed as a health standard, and currently being evaluated as such by EPA, the one-hour standard is often spoken of as an alternative approach to so-called "acid rain control." This proposal is being properly considered by EPA, who should be allowed, after appropriate peer and public review and comment, to come to a decision on this issue. The National Coal Association would oppose any provisions such as contained in Title III of S. 1351, to force prematurely the issuing of this standard.

With respect to the continuing scientific debate, we continue to look forward to the publication of the interim report of the National Acid Precipitation Assessment Program which will provide the best summary of the state of our scientific knowledge following what has been the very significant research on acid rain conducted since 1981. Rather than debate the uncertainties of cause and effect concerning acidity in the environment, the costs of the proposed reduction of SO_2 emissions, or particular provisions of S. 321, all of which will be amply covered by other witnesses before you today, I would like to direct my remarks to the specific findings of the National Coal Association with regard to the reduction of SO_2 emissions in the United States. S. 321 appears to be motivated, at least in part, by a perception that the Clean Air Act is not working; and that absent additional controls such as those proposed, emissions of SO_2 will rise. This is an erroneous view of the efficacy of the Clean Air Act and of the trend in SO_2 emissions.

In fact, the efficacy of the Clean Air Act has been consistently underestimated. We recently came across a quote from the House Committee Report on the 1977 Clean Air Act Amendments in which it was stated that absent this particular provision (in this case, the new source performance standard and relying solely on the 1970 act) SO_2 emissions from coal-fired power plants would double between 1970 and 1980 and triple by 1990. In fact, three years later, in 1980, far from doubling, these emissions were 2% *lower* than they were in 1970 despite a 78% increase in the use of coal. By 1985, when the revised NSPS called for in the 1977 Amendments had just begun to have effect, emissions were still lower, and in 1990 they will be lower still, not triple the 1970 level.

Similar forecasts of rising SO_2 emissions in conjunction with calls for additional legislative controls on SO_2 emissions have characterized the acid rain debate. By 1980, we knew that SO_2 emissions from coal-fired power plants were not going to be double the 1970 level and we knew that total SO_2 emissions had fallen substantially despite rising coal use. But the forecasts, although adjusted downward, continued to show rising SO_2 emissions. And the argument was advanced that they would rise in the 1980s because of the large anticipated increase in coal use and because of the expectation that with most plants in compliance with respect to sulfur dioxide by 1980, no further reductions of SO_2 emissions could be expected to result from the Clean Air Act, absent additional acid rain controls.

Last year in similar testimony before this Committee, NCA provided a detailed analysis and study of SO_2 emissions from coal-fired power plants from 1980 to 1985. Contrary to the forecasts of rising emissions, that study found that SO_2 emissions from coal-fired power plants continued to decline. EPA later verified those findings. Specifically, between 1980 and 1985, while the nation experienced a significant increase in coal use—in 1985, electric utilities consumed 23% (125 million tons) more coal than they did in 1980—emissions of SO_2 from coal-fired power plants actually declined by nearly 3%. The actual level of SO_2 emissions from coal-fired power plants in 1985 was 10% below the EPA projection for 1985 and, more significantly, below the 1980 level of emissions.

NCA has just completed the same analysis for 1986 data and, with your permission, Mr. Chairman, I would submit this study for inclusion in the record.

SO_2 emissions from coal-fired utilities continued their decline in 1986, and this last year's [1987] decline was greater than the SO_2 emissions decline experienced from 1980 through 1985. The magnitude of the 1986 decline reflects the combined effect at a small, 1.2%, reduction in utility coal consumption coupled with the more important, however, a continuing decline in SO_2 emission rates.

It should be noted that nearly 65% of the decline in SO_2 power plant emissions since 1980 has occurred in existing plants. There are the plants not subject to new source performance standards, that are the target of acid rain control legislation. They are the same plants of which it was assumed once in compliance that no further reductions could be anticipated. The causes of the unexpected decline in SO_2 emissions from coal-fired power plants is illustrated by the graph attached as Exhibit A.* Had SO_2 emission rates not declined between 1980 and 1986, total SO_2 emissions would have risen to nearly 20 million tons based on increased coal use. Instead, SO_2 emissions fell nearly a million tons. As explained in the attached NCA study, "Reduction in Sulfur Dioxide Emissions at Coal-fired Power Plants—The Trend Continues," the cause of the decline can be attributed to four distinct causes:

1. *The NSPS effect:* the result of the emissions limit imposed on new plants that is more stringent than that obtaining for existing plants;
2. *The Demand shift:* the result chiefly of the movement of economic activity and electricity demand to regions which use lower sulfur coals;
3. *The Sulfur shift:* the result of substituting lower sulfur coals in power plants throughout the country; and
4. *The Scrubber effect:* the result of retrofitting new technology, in this case, scrubbers onto plants not subject to the new source performance standard.

When we first attempted to analyze trends in SO_2 emissions from coal-fired power plants, based on 1984 data, we hazarded a guess at 1990 emissions based on a continuation of the trends in evidence between 1980 and 1984. Based on two more years' data, we are more confident in asserting that SO_2 emissions from coal-fired power plants will be lower in 1990 than they were in 1980. This is a view not shared yet by EPA; however, I am compelled to note that our record to date on this matter has been better than theirs. . . .

EPA's latest interim base case forecast has recognized what is now history with respect to 1985, and they have lowered

*[*Ellerman displays graph, which has been omitted here, to further support position.—*ED.]

the forecast for 1990 considerably, but they persist in anticipating a rather sharp increase in SO_2 emissions in the late 1980s when growth in coal demand by electric utilities will be decidedly sluggish.

That increase did not occur in 1986. Instead, as noted, SO_2 emissions fell by another half-million tons, so that to reach the EPA forecast for 1990, we will have to experience an increase in SO_2 emissions of a million and a half tons in the next four years. This is very unlikely to occur given the sluggish outlook for coal demand by electric utilities and the continuation of the now 13-year unbroken trend of declining SO_2 emissions *rates* from coal-fired power plants.

The argument with respect to SO_2 emissions from coal-fired power plants is now most properly discussed with respect to the post-1990 period. We have not extended our projections into the 1990s, not so much out of fear of the results, but out of respect for the uncertainties involved in any such prediction. Based on the experience to date, we believe that the constantly revised projections of rising SO_2 emissions must be greatly discounted. Although not stated as such, they are, at best, worst-case projections, which are consistent only in ignoring the potential for things not to turn out as badly as we feared they might. With respect to sulfur dioxide emissions from coal-fired power plants in the 1990's, we would do well to remember the following:

1. *The Clean Air Act* will continue in effect, and particularly those provisions of the 1977 Amendments, such as the PSD requirements, which will keep emissions from rising to levels otherwise permitted by National Ambient Air Quality Standards;

2. *Emerging New Technologies* of coal combustion and sulfur control will permit, in conjunction with the Clean Air Act, even lower average emission rates; and

3. *Coal is not guaranteed the next increment* of generating capacity if oil and gas prices continue to be lower than what we have come to expect before 1986.

In summary, gentlemen, there is no crisis of rising SO_2 emissions, and consequently, no reason for enacting acid rain controls based on such reasoning. There is some possibility that the SO_2 emissions could begin to increase again in the 1990s, but it is no more than a possibility. It is equally possible, and indeed more probable, that those emissions will continue to decline.

Acid rain controls, such as proposed in S. 321 or similar legislation, will cause a more precipitate decline in SO_2 emissions from electric power plants than what would occur otherwise. These additional controls, essentially a more stringent emissions limit on non-NSPS plants, or their forced retirement, will cost a great deal and yield marginal benefits, if any at all, with respect to acidic waters, forest decline, or other instances of ecological distress all too easily ascribed to "acid rain." On this basis, the National Coal Association recommends that S. 321 and similar legislation designed to "control acid rain" not be considered favorably by this committee.

POSTSCRIPT

Is Immediate Legislative Action Needed to Combat the Effects of Acid Rain?

In the May 1986 issue of *Environment*, soil scientist Arthur H. Johnson reported on the results of a study that had recently been released by the National Research Council confirming the impacts of acid rain on surface water ecosystems. Johnson, who has served on several national panels studying the acid rain problem, sums up the controversy over regulation as follows:

> In balancing the pros and cons of instituting controls, it seems to me that the issue boils down to one of values more than to information that science can provide. Those who believe that pride in practicing good stewardship of the environment is worth the monetary cost can demonstrate the reality of adverse environmental effects. Those who insist that more needs to be known before controls are instituted do so because they are willing to take risks with the future environment in order to avoid near-term economic costs.

A factor that helped to persuade legislators is the accumulating evidence that acid rain has human health implications. This aspect of the issue is explored in another article by Jon Luoma in the July 1988 issue of *Audubon*.

Research findings on the pollutant sources responsible for acid precipitation in New England are described in an article by Kenneth A. Rahn and Douglas H. Lowenthal in *Natural History* (July 1986).

For an article that focuses on the technology available to solve the acid rain problem, see "The Challenge of Acid Rain," by Volker A. Mohnen, *Scientific American* (August 1988).

Many scientists believe that forest and agricultural damage is not caused by acidic gases alone, but by a combination of air pollutants. Evidence for this view is presented by James MacKenzie and Mohamed El-Ashry in "Ill Winds—Air Pollution's Toll on Trees and Crops," *Technology Review* (April 1989).

Recent media attention to the global warming issue has decreased reporting on acid rain. In "Whatever Happened to Acid?" *New Scientist* (September 15, 1990), science writer Fred Pierce reports that in Europe the debate about acid precipitation is very much alive.

ISSUE 11

Should Women Be Excluded from Jobs That Could Be Hazardous to a Fetus?

YES: Hugh M. Finneran, from "Title VII and Restrictions on Employment of Fertile Women," *Labor Law Journal* (April 1980)

NO: Carolyn Marshall, from "Fetal Protection Policies: An Excuse for Workplace Hazard," *The Nation* (April 25, 1987)

ISSUE SUMMARY

YES: Corporate counsel Hugh M. Finneran believes women should be excluded from occupations that threaten the unborn.
NO: Health writer Carolyn Marshall counters that reproductive toxins are hazardous to men as well as women and that cleaning up the workplace is the only acceptable response to conditions that threaten the fetus.

The workplace continues to be the bitterest battleground for struggles over environmental safety and health issues. Individual workers, their unions, and independent occupational safety and health committees are pitted against corporate executives, their insurers, company doctors, and industry-controlled health and safety organizations.

Until 1970 the federal government resisted efforts by labor lobbyists promoting comprehensive legislation for worker protection. The degree to which the states attempted to prevent work-related illness and injury through regulation varied. In general, the pattern was to respond after the fact to only the most glaring examples of workplace hazards. On December 29, 1970, the Occupational Safety and Health Act was signed into law, finally establishing a leading role for the federal government in this important environmental realm. The regulations were to be exercised through the newly created Occupational Health and Safety Administration (OSHA). As might be expected, there is a clear dichotomy between workers, who are the potential victims of unhealthy industrial environments, and owners, who must pay the cost of health and safety measures, and this has resulted in ongoing controversy. OSHA has been singled out by the opponents of federal regulation as the number-one target. Pro-industry groups claim that the law has proven ineffective and is a drag on the economy. Labor leaders assert that much progress has been made in reducing hazards, despite inadequate funding of the enforcement effort and serious defects in the law.

Over the years the principal targets of occupational health advocates have changed. Initial efforts to deal with the multiple hazards in sweatshops gave way to programs aimed at reducing the toll of such specific problems as black lung disease, produced by coal dust exposure, and phosphorus poisoning, found among workers in the match industry. Recent regulatory efforts have included reducing worker exposure to vinyl chloride (used to make plastics) and asbestos (a widely used industrial material), because both of these substances were unequivocally linked to human cancer and a variety of other serious ailments. Since World War II the rapid increase in the yearly introduction of potentially hazardous synthetic industrial chemicals has greatly complicated the task of health and safety regulation.

A new controversy has developed as a result of recent industrial initiatives aimed at reducing work-related disease by screening out individuals or groups suspected of being more susceptible to the hazard in question. One such program was an unsuccessful attempt by the Johns-Manville corporation to impose a mandatory ban on cigarette smoking among the workers at one of its asbestos plants because smokers have been shown to be much more susceptible than nonsmokers to asbestos-related diseases Present industrial efforts to detect genetic traits that may make some workers more sensitive than others to certain workplace chemicals may result in more sophisticated screening programs. All of these strategies raise the question of whether it is ethical to combat occupational health hazards by discriminating against specific groups of workers.

Among the special problems that women workers face is possible exposure to substances that pose a reproductive risk. Several corporate employers have responded by excluding fertile women from certain jobs. At least 24 toxic substances have been identified as workplace hazards to pregnant women or unborn children. The Equal Employment Opportunities Commission (EEOC), in conjunction with OSHA, responded to the charge that denying jobs to all fertile women is a violation of the Civil Rights Act by issuing a set of proposed guidelines defining very specific conditions under which only pregnant women could be involuntarily transferred to safer jobs. Employers have reacted by contending that fertility, not pregnancy, should be the criterion for exclusion. In the following articles Hugh M. Finneran, labor counsel for PPG Industries Inc., presents legal arguments in support of the employers, and journalist Carolyn Marshall, who is writing a book on reproductive hazards, demands that workplaces be made safe for *all* employees.

YES

Hugh M. Finneran

TITLE VII AND RESTRICTIONS ON EMPLOYMENT OF FERTILE WOMEN

During the decade of the 1970s, there was a rapid expansion of the female work force accompanied by a simultaneous expansion of scientific knowledge concerning hazards of exposure to toxic substances in the workplace. Health hazards in industry present serious legal, medical, and sociological issues.

Recently, a dramatic awareness of the hazards to the employee's reproductive capacity, i.e., miscarriage, stillbirth, and birth defects, has materialized. The hazard to the reproductive capacity and fetal damage is not a unique problem for female workers. Rather, it is a problem which may impact upon all workers. This article, however, will restrict its analysis to factual situations where the employer considers the problems of exposure to chemicals as uniquely, or primarily, arising out of the female physiology and either restricts or refuses to hire females with childbearing ability. Physical conditions other than chemical substances may also be harmful to the fetus, i.e., radiation, heat stress, vibration, and noise, but will not be treated in this article.

When an employer elects to restrict or refuse to hire a female because of concerns about reproductive hazards, there are serious issues of discrimination which must be considered and analyzed by the employer's legal counsel. At this time, there has been no reported court decision in which the judiciary has considered the legal consequences of such restrictions on female workers. In the event that the employer's legal judgment is wrong, the financial consequences will be tremendous, since the alleged discriminatees in a class action will be numerous.

Title VII of the Civil Rights Act of 1964 incorporates two theories of discrimination which must be considered in a legal analysis of restrictions (the term "restriction" includes a refusal to hire) placed on females because of health hazards. These are: disparate treatment and policies, practices, or procedures with disparate impact not justified by business necessity.

From Hugh M. Finneran, "Title VII and Restrictions on Employment of Fertile Women," *Labor Law Journal*, vol. 31, no. 4 (April 1980).

Two types of substances will be considered in this article: teratogens and mutagens. Teratogens are substances that can harm the fetus after conception by entering the placenta. Mutagens are substances that can cause a change in the genetic material in living cells.

DISPARATE TREATMENT

The Supreme Court in *International Brotherhood of Teamsters v. United States* stated: "Disparate treatment . . . is the most easily understood type of discrimination. The employer simply treats some people less favorably than others because of their race, color, religion, sex, or national origin. Proof of discriminatory motive is critical, although it can in some situations be inferred from the mere fact of differences in treatment. . . ."

The Equal Employment Opportunity Commission and the United States Department of Labor on February 1, 1980, issued, for comment, Interpretive Guidelines on Employment Discrimination and Reproductive Hazards. "An employer/contractor whose work environment involves employee exposure to reproductive hazards shall not discriminate on the basis of sex (including pregnancy or childbearing capacity) in hiring, work assignment, or other conditions of employment."

An employer's policy of protecting female employees from reproductive hazards by depriving them of employment opportunities without any scientific data is a per se violation of Title VII. The Guidelines' position, however, is that the exclusion of women with childbearing ability from the workplace is a per se violation. To arrive at such a conclusion without an analysis of the precise scientific and medical evidence is an erroneous and indefensible legal standard. Thus, an employer's exclusion of females on the basis of their susceptibility to the mutagenic effects of a toxic substance should not be a per se violation but should be analyzed under the rubric of disparate treatment or adverse impact.

One line of inquiry under the disparate treatment analysis would be whether the mutagenic substance has reproductive hazards for male and female employees. If the particular chemical substance has a mutagenic effect on male and female employees, the obvious question is why female workers are treated differently. The answer may be scientifically explained, but it raises the issue of disparate treatment. Indeed, the employer should consider whether there are any other substances in the workplace, other than the substance relied on to exclude the female, which have mutagenic effects on males.

In essence, if the basis for the exclusion is the mutagenic characteristics of a substance, the employer would have to treat all employees, male and female, who are exposed to mutagenic effects in the same manner. The employer may face a serious possibility of a Title VII violation for disparate treatment unless the scientific justification for the differential treatment is very persuasive.

In establishing a prima facie case of sex discrimination, under the principles of *McDonnell Douglas Corp. v. Green* a female must show that: she belongs to a protected class; she applied or was qualified for a job for which the employer was seeking applicants; and despite her qualifications, she was rejected. She also must prove that, after her rejection, the employer continued to seek applicants with her qualifications.

Applying the *McDonnell Douglas* principles to a restriction on female employment, the female could establish a prima facie case of sex discrimination if a chemical substance has a mutagenic effect on the males but only females are excluded from exposure to the hazard by the employer's restrictive policy. In this assumed factual situation, the very basis for the restriction would be applicable to either of sex discrimination, the employer has the burden of proving the existence of a business necessity or a bona fide occupational qualification. Of course, proof of compelling scientific data that the degree or severity of risk was substantially greater might alter the existence of a prima facie case, but the court more likely would consider such evidence as an affirmative defense.

GENDER-BASED CLASSIFICATION

Varying the factual assumptions, let us consider the existence of a work environment in which the chemical substance is a teratogen and an employer restricts the employment of females with childbearing ability. In these circumstances, the employer could argue that the exclusion is based on a neutral health factor rather than sex-based criteria. Since teratogens by definition harm a fetus after conception, the safety hazard is present only for females with childbearing ability and cannot affect males or females without childbearing ability. Thus, a strong argument could be presented that the exclusion of females based upon the teratogenic effect of a chemical substance is a health classification and is not gender based.

In *Geduldig v. Aiello*, the Supreme Court ruled that the exclusion of pregnancy-related disabilities from a state disability system was not sex discrimination but was a distinction based on physical condition "by dividing potential recipients into two groups—pregnant women and non-pregnant persons." Likewise, *General Electric Co. v. Gilbert* viewed pregnancy classifications as not being gender based.

At least one commentator has criticized the relevance of *Gilbert* and *Aiello* to the restriction of female employment in toxic workplaces, because the classification suffers from overinclusiveness since "many women in the excluded class delay or plan to avoid childbearing and thus face no additional risk at all." This contention is small comfort to an employer, however, since women have been known to change their plans and birth-control techniques are not universally effective.

Furthermore, some teratogens are cumulative and remain in the body long after the exposure has ceased. The legal issue is more complex where there is a restriction on the employment of a woman with childbearing ability where teratogens are present but mutagens with adverse reproductive effects present in the workplace affect males on whom no restrictions are placed.

The Pregnancy Disability Amendment to Title VII may have a bearing on the issue of whether the classification is gender based. "The terms 'because of sex' or 'on the basis of sex' include, but are not limited to, because of or on the basis of pregnancy, childbirth, or related medical conditions. . . ."

The Pregnancy Amendment to Title VII does not state expressly that the terms "because of sex" or "on the basis of sex" includes a woman's childbearing ability or potential. The Guidelines, however, interpret "childbearing capacity" as prohibited by the Amendment. Such an interpretation is not without some doubt as to its validity. Nevertheless, if the Guidelines' construction is correct, a distinction based on childbearing ability would be considered gender-based disparate treatment. The practical consequences may be minimal since exclusions or restrictions on the employment of females with childbearing ability has a disparate impact and is best analyzed in this context.

DISPARATE IMPACT

The Supreme Court in *Griggs v. Duke Power Co.* held: "Under the Act, practices, procedures, or tests neutral on their face, and even neutral in terms of intent, cannot be maintained if they operate to 'freeze' the status quo of prior discriminatory employment practices." Thus, *Griggs* ruled that the employer's requirement of a high school diploma or passage of a test as a condition of employment was a prima facie race violation of Title VII, unless these requirements are a "business necessity." "The Act proscribes not only overt discrimination but also practices that are fair in form but discriminatory in operation. The touchstone is business necessity."

In *Dothard v. Rawlinson,* the Supreme Court held that the employer violated Title VII by requiring a minimum height of five feet two inches and a weight of 120 pounds for prison guards since the policy had a disparate impact on women. Likewise, *Nashville Gas Co. v. Satty* is relevant to the issue. In *Satty,* the employer denied accumulated seniority to female employees returning from pregnancy leaves of absence. The Court held that an employer may not "burden female employees in such a way as to deprive them of employment opportunities because of their different role." The conclusion appears inescapable that an employer's restriction on the employment of women with childbearing ability, and this includes restrictions limited to specific jobs, is a prima facie violation of Title VII's proscriptions against sex discrimination under *Griggs, Dothard,* and *Satty.*

BONA FIDE OCCUPATIONAL QUALIFICATION

Two affirmative defenses must be considered: bona fide occupational qualification and business necessity. Title VII provides an affirmative defense to a charge of sex discrimination where sex "is a bona fide occupational qualification reasonably necessary to the normal operation of that particular business or enterprise. . . ."

The Guidelines state: "narrow exception [for BFOQ] pertains only to situations where all or substantially all of the protected class is unable to perform the duties of the job in question. Such cannot be the case in the reproductive hazards setting, where exclusions are based on the premise of danger to the employee or fetus and not on the ability to perform." Under *Weeks v. Southern Bell Telephone & Telegraph Co.,* an employer relying on the bona fide occupational qualification exception "has the burden of proving that he had reasonable cause to believe, that is, a factual basis for believing, that all or substantially all women would be unable to perform safely and efficiently the duties of the job involved."

In the absence of medical evidence to the contrary, an employer's assumption is that all, or substantially all, females have the capacity of bearing children. Thus, the area of controversy will probably center on the issue of whether the safety of the fetus or future generations is reasonably necessary to the normal operation of the employer's business. However, plaintiffs may argue that all or substantially all females are not at risk since not all females plan to have a family.

Courts have sustained decisions by bus companies not to hire drivers over specified ages as being a BFOQ justified by increased safety hazards for third persons. In *Hodgson v. Greyhound Lines, Inc.*, the company refused to consider applications for intercity bus drivers from individuals thirty-five years of age or older. The Seventh Circuit held that the company was not guilty of age discrimination, since its hiring policy was a BFOQ justified by the increased hazards to third persons caused by hiring older drivers. "Greyhound must demonstrate that it has a rational basis in fact to believe that elimination of its maximum hiring age will increase the likelihood of risk of harm to its passengers. Greyhound need only demonstrate however a minimal increase in risk of harm for it is enough to show that elimination of the hiring policy might jeopardize the life of one more person than might otherwise occur under the present hiring practice."

The Fifth Circuit in *Usery v. Tamiami Trail Tours, Inc.*, in upholding the company's refusal to hire bus drivers over forty years of age, found that the policy was a BFOQ. The company had demonstrated "that the passenger-endangering characteristics of over-forty job applicants cannot practically be ascertained by some hiring test other than automatic exclusion on the basis of age."

The language of the BFOQ exception under the Age Discrimination Act is essentially the same as the language of the BFOQ exception under Title VII of the Civil Rights Act. Cases in the airline industry also have considered third-party safety as a sufficient BFOQ in situations involving involuntary pregnancy leaves of absence for flight attendants.

The concept of concern for third parties is sufficiently elastic to include the unborn. It is submitted that society, including employers, has an obligation to avoid action which will have an adverse effect on the health and well-being of future generations. With all the present concerns about the protection of our environment and endangered species, an enlightened judiciary should not callously turn its back on generations unborn. Indeed, on the more mundane and pragmatic basis, it is of the essence of a business venture to operate safely in a manner which avoids costly tort liability.

BUSINESS NECESSITY

The business necessity defense may also justify the exclusionary or restrictive practice. In order to prove this defense, the employer has the burden of establishing that: the practice is necessary to the safe and efficient operation of the business; the purpose must be sufficiently compelling to override the adverse impact; and the practice must carry out the business purpose. The employer also must establish that there are not acceptable alternative policies or practices which would better accomplish the business purpose or accomplish it with lesser adverse impact on the protected class.

PRENATAL INJURY

Since the safe and efficient operation is premised on the need to protect the fetus, tort law relating to prenatal injuries is pertinent. The potential tort liability bears on the necessity for the exclusion. The law of Texas will be reviewed in regard to prenatal injuries. Texas was selected because of its large petrochemical industry.

The parents of a child suffering prenatal injuries resulting in its death have cause of action under the Texas wrongful death statute, provided the child was born alive and was viable at the time the injury was inflicted. In so ruling, the court stated that the statutory requirement of the Texas wrongful death statute, that the deceased has suffered an injury for which he could have recovered damages had he survived, was met. This holding of necessity implied that the Texas Supreme Court recognized a cause of action for a surviving child who is born alive with a birth defect caused by prenatal injuries. For a child born with birth defects, the cause of action exists for prenatal injuries at any time during pregnancy.

The Texas courts apparently have not yet decided whether parents have a cause of action under the wrongful death statute in cases where a child is stillborn due to prenatal injuries. The inquiry in such a case would revolve around the issue of whether a fetus is a person within the meaning of the wrongful death statute. Other state courts interpreting their wrongful death statutes have split on the issue.

Assuming that liability is established, Texas courts allow surviving parents to recover damages under the wrongful death act to compensate them for the pecuniary value of the child's service that would have been rendered during minority, less the cost and expense of the child's support, education, and maintenance, as well as economic benefits reasonably expected to have been contributed after reaching majority.

While it is generally held that some evidence of pecuniary loss is necessary to support a wrongful death judgment, the Texas courts have recognized that such proof cannot be supplied with any certainty or accuracy in cases involving young infants. Therefore, they leave the damages question largely to the discretion of the jury. Of course, a prenatally injured infant who manages to survive would be able to sue for his own personal injuries, including pain and suffering, loss of earning capacity, and any other damages, if applicable. Recognizing the "deep pocket syndrome," employers have a reasonable basis for being concerned about large tort recoveries.

The female employee's willing and informed consent to the assumption of the risk is not binding to the unborn child. Hence, obtaining a waiver from the female employee is an act with no legal significance other than documenting the employer's awareness of the unavoidably unsafe condition of the workplace for the fetus for use against the employer in tort litigation.

The employer should not be required to assume the risk of significant tort liability which could threaten the very existence of the enterprise, depending on the financial assets of the employer and the severity of injuries. Courts have required employers in discrimination cases to assume additional expense to achieve compliance with Title VII (costs of validation studies, loss of customer patronage, and training costs), but it is submitted that the magnitude of the risks of exposure to prenatal injuries and reproduction hazards should result in a different decision. The financial impact on the employer is important but certainly not the most important factor. A lifetime of suffering by future generations is worthy of societal concern. The Civil Rights Act does not exist in a vacuum.

Whether the purpose of the restriction is sufficiently compelling to override the adverse impact on women and is necessary to accomplish the employer's business purpose of ensuring a safe workplace without reproductive hazards will be decided by the scientific and medical data relating to the severity of the health hazard of the particular substance.

LESS RESTRICTIVE ALTERNATIVES

Under the business necessity principles of *Robinson v. Lorillard,* the employer must demonstrate the absence of "less restrictive alternatives" before relying on the affirmative defense. The Guidelines indicate that four factors should be considered. These are: whether the employer is complying with applicable occupational federal, state, and local safety and health laws; respirators or other protective devices are used to minimize or eliminate the hazard; product substitution is used; and affected employees are transferred without loss of pay or other benefits to areas of the plant where the reproductive hazard is minimal or nonexistent.

The employer's obligation to comply with its safety obligations under the Occupational Safety and Health Act is eminently reasonable, provided that it is recognized that the employer's obligation under OSHA only requires the use of technologically and economically feasible engineering and administrative controls. If engineering and administrative controls are not feasible, the employer must protect his employees by the use of personal protective devices. It is fair and reasonable to require an employer to satisfy his legal obligations under safety and health laws before excluding females from the workplace.

To suggest, however, that the employers change their products or provide rate retention for employees restricted from hazardous exposure is extreme and without legislative support. If Congress had intended to require substitution of products and rate retention for employees under Title VII, it would have done so explicitly. When, as here, these matters are at best tangentially related to nondiscrimination, Title VII is silent on the subject, and wages and rates of pay and seniority of workers transferred to jobs other than their usual jobs are mandatory subjects of collective bargaining, then a reasonable interpretation of the legislation is that Title VII does not impose this obligation of management.

If an employer intends to sustain his business necessity defense, there must be evidence that the employer has explored the feasible alternatives to imposing restrictions on the employment of fertile females. One alternative which must be considered is a system for individual screening and evaluation with restrictions imposed on the female only if she becomes pregnant. Serious medical questions are posed by this alternative. Indeed, for some teratogenic substances the first weeks of pregnancy are the most critical. During this period, a woman may not know that she is pregnant, and sophisticated tests may not reveal the pregnancy. The administration of such a program might raise serious personnel problems since female employees might object to continuous monitoring to determine whether they are pregnant.

CONCLUSION

The decade of the 1970s was the era of the testing cases under Title VII. The decade of the 1980s will be the era of large class actions involving the exclusion of fertile females from exposure to reproductive hazards.

On the extreme of one side will be those arguing that Title VII rejects these protections as Victorian, romantic paternalism which deprives the individual woman of the power to decide whether the economic benefits justify the risks. On the other extreme, some employers will argue that any possible risk of harm to the female's offspring require her exclusion.

An informed judiciary should consider not only the economic interests of the female employee and the employer but the societal concern for the quality and happiness of future generations as well. The Supreme Court in *Roe v. Wade* recognized that a state may properly assert important interests in protecting potential life. After evaluating the level, duration, and manner of exposure in the specific employer's workplace, if there is reputable scientific evidence of a recognized reproductive hazard, either from a mutagen with significantly greater risk for female workers or a teratogen, the employer should be allowed to exclude females from that workplace if the business necessity criteria are satisfied. The employer should have the right and, indeed, the duty and obligation to operate his facility with due concern for the safety and health of future generations.

NO

Carolyn Marshall

AN EXCUSE FOR WORKPLACE HAZARD

Last January the American Telephone and Telegraph Company, one of the world's largest semiconductor manufacturers, barred pregnant women from production lines at many of its computer chip plants around the country. Company officials said they were responding to a growing concern over the exposure of workers to chemicals that might lead to miscarriages, birth defects or other harm to the fetus.

A.T.&T.'s action was prompted by a controversial health study conducted in Massachusetts last year. That study, which examined employee health at a plant of the Digital Equipment Corporation in Hudson, Massachusetts, showed an increased incidence of miscarriages among production workers who use a variety of highly toxic liquids and gases to make computer chips. The study failed to pinpoint the specific cause of the problems, but the results were still alarming. In one group of sixty-seven women the rate of miscarriage was 39 percent, nearly twice the national average.

The Digital study shook the semiconductor industry. In California's Silicon Valley, where chip firms employ more than 52,000 workers, most of them women, many companies offered job transfers to pregnant production workers. Like A.T.&T., many companies announced or reaffirmed a commitment to in-house policies that bar pregnant women, and in some cases all fertile women, from potentially hazardous jobs.

Although these events raise serious questions about chemical hazards in the computer industry, they also dramatize the larger issue of the link between workplace hazards and human reproductive failure. At the same time, the practice of excluding women from jobs raises troubling questions. Are dangerous substances in the workplace in fact more threatening to pregnant women than to other workers? Is the fetus at risk only through the mother, or is a father's exposure of equal concern? What is industry doing to mitigate the hazards?

Actions like A.T.&T.'s are "just an excuse not to clean up the workplace." says Marvin Legator, a geneticist at the University of Texas at Galveston and a leading researcher on reproductive hazards. "The driving force behind such policies is the fear of lawsuits."

The government estimates that 15 million to 20 million jobs in the United States expose workers to chemicals that might cause reproductive injury. According to the National Institute of Occupational Safety and Health (NIOSH), 9 million workers are exposed to radiofrequency/microwave radiation, which causes embryonic death and impaired fertility in animals; at least 500,000 workers are exposed to glycol ethers, known to cause testicular atrophy and birth defects in animals; and some 200,000 hospital and industrial employees work with anesthetic gases and ethylene oxide, both linked to miscarriage in humans.

No one knows how often workers suffer miscarriage or infertility due to chemicals in the workplace, or how many of their children are born with defects, but in 1985 the Federal Centers for Disease Control called human reproductive failure a "widespread and serious" problem, and one of the ten most prevalent work-related diseases.

At the same time, government statistics and university studies show a marked, and as yet unexplained, increase in reproductive disorders. Of the sixteen major birth defects monitored by the Centers since 1970—including heart, limb and brain deformities—seven have increased at rates ranging from 20 percent to 300 percent. Infertility is widespread, with one in every six American couples involuntarily infertile, unable to conceive after one year of trying. And numerous studies show a drop in both the quality and quantity of sperm, with some research suggesting a decline of 80 percent in just fifty years. No one seems to be able to explain the meaning of these statistical anomalies. Some experts believe variations are due to better and more conscientious reporting. Others say record-keeping techniques haven't improved enough to account for changes of the magnitude now being discovered.

Moreover, there is a growing list of "clusters"—the increased incidence of a particular reproductive disorder discovered in small communities and workplaces. At one semiconductor plant in New Mexico, 200 miscarriages were reported in a one-year period. At least twelve miscarriage clusters have been observed in the United States and Canada among workers who use video display terminals, which emit low-level radiation. Farmworkers exposed to pesticides have been the subject of investigations into birth defect clusters; and miscarriage and birth defect clusters have been reported among laboratory and hospital workers who use toxic chemicals and gases. "Workers frequently report they have recognized a cluster or an unusual number of reproductive problems," says Mike Silverstein, assistant director of health and safety for the United Automobile Workers. "It's less frequent that scientists undertake a study to find out what's going on."

A major source of funding for scientific research comes from the government, but studies examining the effects on reproduction of pollution or workplace hazards aren't a priority for the Reagan Administration. Researchers at the Centers for Disease Control have for years complained that the White House Office of Management and Budget had obstructed environmental and occupational health research through its powers under the Paperwork Reduction Act.

Last year Congress finally looked into the matter, asking researchers at the Harvard School of Public Health and the Mount Sinai School of Medicine to study the O.M.B.'s pattern of rejecting research proposed between 1984 and 1986. The final report, completed in September for the House Committee on Oversight and Investigations, found that the "OMB has delayed, impeded and thwarted governmental research" on environmental and occupational issues, and exhibited a pattern of "demonstrable bias" in its review of the Centers' research. The O.M.B. repeatedly refused to fund studies on reproductive hazards, the researchers said.

Funding inadequacies aside, there are other obstacles to understanding the link between toxic substances and health—among them, ignorance about the effects of industrial chemicals. According to Dr. Philip J. Landrigan, one of the authors of the O.M.B. survey and director of the division of environmental and occupational medicine at Mount Sinai, "too few chemicals are adequately tested for toxicity before they get on the market," making it difficult to substantiate cause and effect at cluster sites. Landrigan points to a 1984 study by the National Research Council that found that in a random sample of 55,000 or so pesticides, chemicals, drugs and cosmetics in commercial use, less that 20 percent had enough toxicity data to assess fully the effects on humans. Based on that sample the council said that only 6 percent of all chemicals in commercial use had been tested for reproductive effects and only 9 percent had been tested for genetic effects.

There is some data on chemicals that cause reproductive injury. NIOSH reports that at least fifty widely used industrial chemicals, including heavy metals, solvents, pesticides and chemical intermediates like vinyl chloride, cause reproductive failure in animals. Scores of other chemicals are suspected of interfering with reproductive processes. And scientists believe that approximately 90 percent of all substances known to cause cancer also cause genetic mutations.

Federal law does not require chemicals to be tested for reproductive risks before workers use them. And without comprehensive toxicity tests to link exposure and deleterious health effects, there is little basis for the government to regulate. Of the more than 59,000 chemicals on the Registry of Toxic Effects of Chemical Substances compiled by NIOSH, the Occupational Safety and Health Administration has developed safety standards for only twenty-three substances, four of which are regulated for hazards to human reproduction.

Meanwhile, industry has come up with its own solution. Many large corporations have adopted exclusionary policies, often referred to as "fetal protection policies" or F.P.P.s, designed to protect unborn children. Like A.T.&T., a number of firms bar pregnant or fertile women from areas where there is a potential reproductive hazard. Other companies offer job transfers to pregnant women who feel they are at risk, although often without guarantee of equal pay. In a few extreme cases, firms have restricted employment to women who can present medical evidence of sterility. And some firms hire only men for certain jobs.

No one disputes that the fetus is extremely sensitive to toxins and that certain chemicals damage the developing child. Those chemicals, known as teratogens (derived from the Greek word *terato,* meaning "monster"), act only after the sperm has fertilized the egg, crossing the placental barrier to damage the fetus, causing birth defects. But research shows that toxins can affect human reproduction in other ways. Gametoxins decrease or damage sperm or ova, reducing fertility. Mutagens have no visible effect on parents, but damage the genes that are passed along to children and future generations, resulting in birth defects, cancer or other disorders. Other substances alter menstrual cycles or destroy sex drive.

In 1985 an Environmental Protection Agency study of data for nineteen pesticides and five industrial chemicals found no cases in which damage to the fetus was the only documented effect, implying, the study said, that attempts to protect only women from teratogens were misguided. The study, prompted by criticism of the E.P.A.'s gender-specific approach to regulation of reproductive risk, concluded that "if a pesticide is extensively tested for health effects, it is likely to show additional positive effects other than teratogenicity." Thus, teratogenic effects almost always occur in tandem with reproductive, genetic and carcinogenic effects.

No one knows precisely how many companies have F.P.P.s. Although the computer industry is now in the spotlight, the government believes exclusionary policies are most prevalent in the lead, chemical, petrochemical and oil refining industries. Firms that reportedly have such policies are Dow, Du Pont, Allied Chemical, B.F. Goodrich, Union Carbide and Monsanto. A 1985 report by the Congressional Office of Technology Assessment found that at least fifteen major U.S. corporations and numerous hospitals have formal F.P.P.s. Critics suspect many more companies have unwritten policies and that their use will increase as scientists reveal how reproductive health is threatened by a chemically complex workplace.

"Every company in every industry has some sort of a pregnancy policy," and Sheila Sandow, a spokeswoman for the Semiconductor Industry Association, a trade group that represents U.S.-based semiconductor manufacturers. "Pregnancy is a condition that warrants special attention as a protective measure. If you drink and you're pregnant, you should stop drinking; if you smoke, you should stop smoking. And in some cases, if you work, you should stop working."

That attitude angers critics because it assumes fetal harm occurs through material exposure alone. But a growing body of evidence suggests otherwise. Men who manufactured the pesticides DBCP and Kepone, for example, became sterile or fathered deformed children. The wives of men exposed to vinyl chloride suffered a high incidence of stillbirths and miscarriages. Many male veterans of the Vietnam War believe the defoliant Agent Orange caused their children's birth defects.

There is also evidence that male reproduction failure is connected with carbon disulfide, methylene chloride, EDB, arsine gas and the category of chemicals known as glycol ethers, which are widely used in the semiconductor industry. An OSHA study found that of the twenty-six chemicals from which industry reports it bars women of childbearing age, twenty-one also cause male infertility or genetic damage.

Because so many substances also affect men, critics see industry's exclusion of women as both sexist and expedient. They point out that F.P.P.s are most common in industries in which women are new to the work force. The same chemicals are used in the garment industry, in hospitals and in dental offices, where the work force is mostly female, yet no one considers protecting the fetus by excluding women from those areas.

Joan Bertin, an attorney for the American Civil Liberties Union, and other labor experts argue that fetal policies violate two key federal laws. Because F.P.P.s suggest the workplace is unsafe for the fetus, they counter the Occupational Safety and Health Act, which entitled "every working man and woman" to a "safe and healthful" workplace. The policies also violate federal discrimination laws, critics say, because they treat men and women differently "without convincing justification." In legal tests to date, the courts have ruled that exclusionary policies are discriminatory unless companies can demonstrate that the chemicals in question are uniquely hazardous to the fetus—proof that is almost impossible to provide.

Federal rules designed to guide employers in developing nondiscriminatory policies that protect both men and women from reproductive hazards were proposed by OSHA, the Labor Department and the Equal Opportunity Commission during the Carter Administration. But the controversial guidelines, strongly opposed by industry, were permitted to die after Reagan took office. In the absence of federal regulation, unions have taken the issue to court.

The U.A.W., for example, is suing Johnson Controls, because the company's nine battery-making plants exclude all women of "childbearing capacity" from jobs that expose them to lead unless the women can demonstrate they cannot become pregnant. Although the company says its policy protects the fetus, the class-action suit, filed in a Federal District Court in Wisconsin, claims the policy is discriminatory. U.A.W. attorneys point to medical evidence that lead harms both the male and female reproductive systems. They are requesting reduced exposure for all employees.

Another lawsuit, filed in Federal District Court in Texas, charges that the Rio Grande Cancer Treatment Center discriminated against a female radiation therapist. The hospital fired the highly qualified nurse when she became pregnant, saying that exposure to X-rays could cause miscarriage, birth defects or genetic damage. But the lawsuit alleges that there is ample evidence that the risk of such damage from radiation is often higher for men. Attorneys for the nurse say the hospital could have protected her with a temporary job transfer or by carefully monitoring her radiation exposure.

Currently environmentalists are investigating whether to sue the government over its revised regulation of the herbicide dinoceb. The E.P.A. issued an emergency ban of the chemical last October, saying it posed a "very serious risk" of birth defects and male sterility for farmworkers. But on March 30 the agency reversed its decision, saying that dinoceb could be used on certain crops as long as women of childbearing capacity under the age of 45 were excluded from mixing, loading or applying it.

In the end, labor experts say the only acceptable solution is to reduce toxic exposure to levels that are safe for all workers. Ideally, that would be accomplished through both increased funding for research and workplace-safety laws. But for now, labor and environmental groups say they will continue to sue. Ironically, should rulings favor industry, female workers will be forced to fight for the right to share with men equal access to reproductive toxins.

POSTSCRIPT

Should Women Be Excluded from Jobs That Could Be Hazardous to a Fetus?

Both Marshall and Finneran agree that on-the-job exposure to mutagens and teratogens is a serious problem for employees. They disagree on how employers should respond to this problem.

Finneran rejects as extreme the requirement that employers demonstrate that they can't make the workplace safer by switching to other products or transfer the employees in question to jobs at equal salaries. He also emphasizes that OSHA requires only "economically feasible" control technologies to reduce exposure to hazardous substances. The rub, of course, is that employers and employees rarely agree on what is economically feasible.

The charge that the proposed job exclusions are a new manifestation of industry's historical practice of sex discrimination is supported by the fact, pointed out by Marshall, that the issue has been raised with respect to traditionally male jobs, whereas no such action is contemplated in the lower-salaried, female-dominated allied health field, where similar hazards have been documented. Since workplace toxins can also affect the male reproductive system, Marshall concludes that elimination of the hazards is the only acceptable way of protecting both today's workers and future generations.

A California court recently ruled that a lead battery maker's "fetal protection" policy that denies jobs to fertile women is illegal sex discrimination. The Supreme Court has agreed to rule on an appeal of a federal court decision that upheld the identical policy of Johnson Controls, a Milwaukee manufacturer that owns the California plant.

A comprehensive look at health problems in the workplace is *Work Is Dangerous to Your Health* (Vintage, 1973), by Jeanne Stellman and Susan Daum. The history of occupational hazard regulation and the political struggle that resulted in the Occupational Safety and Health Act is documented in *Bitter Wages* (Grossman, 1973), by Joseph Page and Mary-Win O'Brien.

For a discussion of all sides of the worker screening issue, see "Industrial Screening Programs for Workers," by Mary Lavine, *Environment* (June 1982); "Industrial Genetic Screening," by Jon Beckwith, *Science for the People* (March/April 1990); and "Genetic Screening; Medical Promise Amid Legal and Ethical Questions," by Rudy M. Baum, *Chemical and Engineering News* (August 7, 1989).

UNITED NATIONS/Photo by M. TZOVARAS (NJ)

PART 3

Disposing of Wastes

Modern industrial societies generate many types of waste. Manufacturing and construction activities yield hazardous liquid and solid residues; treatment of raw sewage produces sludge; mining operations generate mountains of tailings; radioactive waste results from the use of nuclear isotopes in medicine and in the nuclear power and nuclear weapons industries. Each of these forms of waste, in addition to ordinary household garbage, contains toxins and pathogens that are potentially serious sources of air and water pollution if they are not disposed of properly. We must now deal with the legacy of waste contamination problems that have resulted from years of neglect and inappropriate waste disposal methods. This section exposes some of the major controversies concerning proposed solutions to three important waste categories.

Hazardous Waste: Are Cleanup Efforts Succeeding?

Municipal Waste: Should Incineration Be a Part of Waste Disposal Methods?

Nuclear Waste: Is Yucca Mountain an Appropriate Site for Nuclear Waste Disposal?

ISSUE 12

Hazardous Waste: Are Cleanup Efforts Succeeding?

YES: Robert G. Wright, from "Waste Management: A Cooperative Cleanup for Superfund Sites," *Environment* (June 1990)

NO: Robert J. Mentzinger, from "GEMS Landfill: A Superfund Failure," *Public Citizen* (May/June 1990)

ISSUE SUMMARY

YES: Labor union hazardous waste director Robert G. Wright describes a partnership among government, labor, and management to produce a training program for waste workers that has proven effective in past cleanup efforts.

NO: Researcher Robert J. Mentzinger chronicles the problems at one toxic waste site as an introduction to a highly critical assessment of current government cleanup programs.

The potentially disastrous consequences of improper hazardous waste disposal burst upon the consciousness of the American public in the late 1970s. The problem was dramatized by the evacuation of dozens of residents of Niagara Falls, New York, whose health was being threatened by chemicals leaking from the abandoned Love Canal, used for many years as an industrial waste dump. Awakened to the dangers posed by chemical dumping, numerous communities bordering on industrial manufacturing areas across the country began to discover and report local sites where chemicals had been disposed of in open lagoons or were leaking from disintegrating steel drums. Such esoteric chemical names as dioxins and PCBs have become part of the common lexicon, and numerous local citizens' groups have been mobilized over efforts to prevent human exposure to these and other toxins.

The expansion of the industrial use of synthetic chemicals following World War II resulted in the need to dispose of vast quantities of wastes laden with organic and inorganic chemical toxins. For the most part, industry adopted a casual attitude toward this problem and, in the absence of regulatory restraint, chose the least expensive means available. Little attention was paid to the ultimate fate of chemicals that could seep into surface water or groundwater. It has been estimated that less than 10 percent of the waste was disposed of in an environmentally sound manner.

The magnitude of the problem is truly mind-boggling. Over 275 million tons of hazardous waste is produced in the United States each year. As many as 10,000 dump sites may pose a serious threat to public health, according to the federal Office of Technology Assessment. Other government estimates indicate that over 350,000 waste sites may ultimately require corrective action at a cost that could easily exceed $100 billion.

The congressional response to the hazardous waste threat is embodied in two complex legislative initiatives. The Resource Conservation and Recovery Act (RCRA) of 1976 mandated action by the Environmental Protection Agency (EPA) to create "cradle to grave" oversight of newly generated waste, and the Comprehensive Environmental Response, Compensation, and Liability Act, commonly called Superfund, gave the EPA broad authority to clean up existing hazardous waste sites. The implementation of this legislation has been severely criticized by environmental organizations, citizens' groups, and members of Congress who have accused the EPA of foot-dragging and a variety of politically motivated improprieties. Less than 20 percent of the original $1.6 billion Superfund allocation was actually spent on waste cleanup.

Tougher amendments designed to close loopholes in the RCRA were passed in 1984, but it is widely acknowledged that we are a long way from providing assurance of safe disposal of industrial waste and of cleaning up the enormous legacy of years of careless dumping. After allowing the Superfund to expire in 1985, Congress finally passed a more stringent set of reauthorization measures in late 1986, raising the appropriation to a compromise figure of $8.5 billion for the next five years. It wasn't long, however, before environmentalists complained that the underfunded EPA was not strictly adhering to the terms of the new legislation. The question "How clean is clean enough?" has become a heated issue of debate.

On the optimistic side, union environmental official Robert G. Wright describes a training program for cleanup workers jointly developed by labor and industry under government guidance that appears to be a success. In contrast, Robert J. Mentzinger, a researcher for a public interest group, provides some grim examples to support his contention that the revised Superfund legislation continues to be a dismal failure.

YES

Robert G. Wright

WASTE MANAGEMENT: A COOPERATIVE CLEANUP FOR SUPERFUND SITES

The Superfund legislation of 1980 helped the Environmental Protection Agency (EPA) to identify the country's worst environmental injuries, but cleaning them up has proved difficult, even dangerous. A solid step toward safely and effectively rehabilitating these most abused sites came in 1986, when amendments to the Superfund legislation provided $50 million, to be spent over five years, to develop and implement a training program for workers engaged in toxic and hazardous waste cleanup. These funds, administered by EPA through the National Institute for Environmental Health Sciences (NIEHS), were to be used to unite government, labor, and management interests in a cooperative partnership to address national problems of hazardous waste sites.

Under government guidance, the Laborers International Union of North America (LIUNA) and the Associated General Contractors (AGC) joined forces to design an education and training program for safely handling, controlling, removing, and disposing of toxic and hazardous wastes. Workers would clean up Superfund contamination areas and respond to emergencies such as overturned trucks or derailed trains carrying hazardous materials.

Although EPA did establish certain minimum training standards for prospective environmental workers, LIUNA/AGC exceeded those mandated requirements in developing their program. Primarily, LIUNA/AGC wanted to emphasize that the workers targeted for this training would face the greatest potential exposure to toxic contaminants and hazardous waste. This concern reflects LIUNA's traditional commitment to the personal well-being of its members and was reinforced by AGC members who would bear the primary responsibility for ensuring that such work was performed in strict accordance with procedures regulated by EPA and the Occupational Safety and Health Administration (OSHA).

Additionally, LIUNA/AGC hoped to anticipate the federal government's increasingly stringent training requirements for toxic and hazardous waste workers. They also wanted to accommodate an expected range of local regulations and requirements for licensing or certifying such workers. Training was subsequently focused on five general kinds of work:

- waste handling and processing at active and inactive hazardous substance treatment, storage, and disposal facilities;
- cleanup, removal, and other remedial actions at waste sites;
- emergency response involving hazardous substances;
- risk assessment at hazardous substance disposal sites, including possible remedial actions or cleanup by state and local personnel; and
- transportation of hazardous wastes.

Although national in scope, the program was designed to accommodate differences among various regions of the country. A system of regional facilities was decided on, which provided a mechanism to address generic problems posed by different kinds of sites while also being able to focus on specific problems raised by particular sites in each region. Six existing LIUNA-affiliated training facilities were designated for trainees within each region: Pomfret Center, Connecticut (Northeast region); Belton, Missouri (Midwest); Kingston, Washington (Northwest); Anza, California (Far West); Livonia, Louisiana (Southeast); and Jamesburg, New Jersey (Mid-Atlantic).

Facilities were selected on the basis of several considerations. Planners looked for on-site residential, classroom, and hands-on training capabilities. More important, however, was the availability of experienced, dedicated instructors who would be able to support the unique prerequisites of this program.

An 80-hour curriculum—twice the length of similar occupational training programs—was developed to provide workers with a thorough understanding of the inherent dangers of working in toxic environments. Educational techniques, tailored to the trainees' needs, were included to enhance learning and limit potential risk to the cleanup workers and other site personnel. The final curriculum addressed hazard recognition; principles of toxicology, biological monitoring, and risk assessment; safe work practices and general site safety; engineering controls and hazardous waste operations; site safety plans and standard operating procedures; decontamination practices and procedures; emergency procedures, first aid, and self-rescue; safe use of field equipment; handling, storage, and transportation of hazardous wastes; use, care, and limitations of personal protective clothing and equipment; techniques to comply with safety procedures; and rights and responsibilities of workers, under OSHA, as regulated by EPA.

This general curriculum was supplemented by related materials on medical monitoring and the role of community relations. In particular, concerns for the welfare of local residents and the general public were emphasized throughout.

To ensure that the training duplicated actual toxic and hazardous waste working environments, simulated Superfund sites were constructed at each of the regional training facilities. (The mock sites were built by workers receiving basic or upgrade instruction in construction-related skills in other LIUNA-affiliated training programs, thus controlling costs through shared resources.) Taking advantage of the realistic but simulated hazardous waste sites, hands-on training demonstrations and techniques were developed that included use and maintenance of personal protective equipment, basic instrumentation, and site safety plans, as well as decontamination procedures. In addition, site procedures and behavior were recorded on video equipment for evaluation by instructors and trainees.

An environmental worker refresher course also is being developed, to be made available to participants in the original program. The refresher course will not only reinforce the skills initially developed, but will also provide a way to track the original participants' retention of skills.

TODAY, ONLY TWO YEARS INTO THE FIVE-year initiation program, the training already has begun to prove itself. At Superfund sites like the Rocky Mountain Arsenal in Colorado and the site at Weldon Springs, Missouri, the training has demonstrated its validity in practical applications. The remaining three years of national training will see the program introduced to additional, strategically located training facilities in the LIUNA network.

After completing the five-year effort funded by the government, LIUNA/AGC is committed to continuing the environmental worker training permanently. This will be accomplished without additional federal funds, using resources provided through LIUNA's network of collective bargaining relationships. Negotiated cents-per-hour contributions support nearly 70 existing LIUNA-affiliated training programs throughout the United States and Canada.

The overall success of the present partnership between government, labor, and management can best be measured from the perspective of each of the participants.

The federal government has a long-term national mechanism to address important environmental concerns. Under EPA, this has been accomplished in a responsible and practical way, making the most of private resources that otherwise would have been unavailable to such a program. Furthermore, the commitment of labor and management to institutionalize the program within the private sector ensures taxpayers of a continuing return on their initial investment.

LIUNA's objectives as an organized labor group are furthered with the addition of a new generation of members, environmental workers, who are productively employed throughout America in the emerging and expanding environmental industry.

Benefits for private sector management include qualified, licensed or certified environmental workers for toxic and hazardous waste projects in all areas of the country.

Beyond these immediate horizons, however, is the knowledge that current and future generations will benefit from this unique, innovative, and cooperative effort to clean up and rehabilitate the environment shared by all. The program serves as a model for putting our limited resources to work in a coordinated approach to the important problems of our larger society.

NO

Robert J. Mentzinger

GEMS LANDFILL: A SUPERFUND FAILURE

Loretta Fortuna wants out. Sickened by odors wafting from the red toxic pools near her home, angered by the political battleground she's had to maneuver, grief-stricken by two miscarriages within a year and perpetually worried for the health of her two small sons, she's finally had enough.

Fortuna has spent the past year waging a campaign on behalf of her family and neighbors to get a leaking waste dump near her home cleaned up. But now she has decided to move elsewhere, a disheartened casualty of a frustrating battle.

The Gloucester Environmental Management Services landfill in Gloucester Township, New Jersey—known as GEMS—is but one site among 1,175 toxic nightmares nationwide waiting to be cleaned up under the auspices of "Superfund" legislation passed by Congress in 1980.

But to citizens like Fortuna who live near these sites, it's clear that the nation's long legacy of careless waste disposal habits is much too large a problem to be entrusted to Superfund's extremely complex bureaucracy.

When she started her hunt for a new home, Fortuna says: "I got out that map of New Jersey which lists the locations of all the state Superfund sites, and I drew circles out to three miles around each one. I found a house in the only town in South Jersey which didn't fall into one of those circles."

Fortuna is bitter over being forced from her "dream home." Her husband still spends evenings alone there—separated from Loretta who has moved into the new house with their sons. Loretta's husband must stay in their old, dangerous home to guard against vandalism. He is waiting patiently for financial compensation promised by Superfund to lessen the monetary blow caused by the uprooting of the family.

"We built this place to raise our kids," Loretta said of their house. "We added a $6,000 kitchen and wanted to spend our life here."

GEMS: ONE OF 30,000 TOXIC HOT SPOTS

The landfill, located not 1,500 ft. from Fortuna's home, symbolizes America's growing waste disposal problem. GEMS is an unlined municipal landfill—

one of thousands that have operated throughout the United States for decades. But now these sites are becoming potential health hazards to millions of Americans. The Environmental Protection Agency says more than half of all Superfund sites are landfills. At GEMS, decomposing garbage has begun to leach from the unlined pit into streams and groundwater. A malodorous melange of industrial chemicals was found on-site, dumped there sometimes illegally by a succession of private contractors who ran GEMS throughout the 1970s with minimal state regulation.

A list of chemicals measured in the groundwater below the site includes high levels of toxins and carcinogens like toluene, benzene, and lead—byproducts of decomposing trash.

Superfund's National Priorities List ranks the nation's waste sites by the severity of their contamination and puts GEMS near the top, in 12th place.

But though the waste dump holds high priority and is located in the backyard of a powerful politician, cleanup at GEMS has languished.

James J. Florio, the principal author of the 1980 Superfund bill and governor of New Jersey, lives only a half-mile from GEMS, in Blackwood, N.J.

Florio campaigned successfully as a champion of toxic waste cleanup in a state which has spent $1.1 billion in federal and state money cleaning its 109 toxic "hot spots"—more sites than any other state. Yet GEMS still poses a threat to Florio's 38,000 neighbors who live within three miles of its closely guarded gates.

In New Jersey, as in 47 other states containing Superfund sites, the promise of Superfund has diminished. Indeed, the verdict on the GEMS cleanup is similar to that at Superfund sites around the country: relief has been long in coming and, even when it comes, may never address long-term effects to Gloucester Township's residents or its environment.

Loretta Fortuna cannot understand why she and her neighbors have had to endure this living horror: the risk of an insidious poisoning of their bodies as wastes from one of America's worst landfills seeps into the wind, water and soil around their suburban homes.

SUPERFUND'S INADEQUATE SOLUTION

There are numerous problems with Superfund.

Cleanups are not taking place at enough sites; tangled, overlapping bureaucracies are delaying cleanup at the sites chosen; EPA's choices of cleanup methods are under scrutiny for their inconsistency and proven ineffectiveness and the role of contractors has been criticized as a "hidden" process—outside of governmental accountability. Also, cleanup costs have spiraled.

Of 30,000 contaminated sites across the nation identified as possible Superfund targets, 1,193 sites were singled out by EPA investigators for their severity of contamination and proximity to population centers.

Of this narrow group of sites, 1,175 were actually placed on the list for cleanup, according to the Rand Corporation, a policy analysis firm.

Cleanups have been initiated at only 177 of these 1,175 sites, and only 18 cleanups have been completed to date, a full decade after the Superfund program was initiated, according to Rand.

And some of the "clean" sites have had to be reopened for additional work because it was later discovered the contamination had not been effectively controlled, the Rand report said.

In all, "it takes over eight years from the time these sites enter the federal process to the time definitive cleanup work begins," and another 25 months for the work to reach actual completion, said the Rand report.

There is also concern over bureaucratic irregularities in Superfund: in some cases the states have the lead role in developing remediation plans, while in others it is the EPA that is the "lead agency." This creates what Fortuna calls "layers of bureaucracy," which complicate the cleanup process and frustrate efforts by citizens to gather information on the site.

"I'd call DEP (New Jersey's Department of Environmental Protection) and ask them for the results of their air sampling, and they'd say EPA was responsible for that. I'd call EPA, and they wouldn't know anything. All they'd tell me was that DEP was the lead agency," Fortuna says, adding that she has encountered other problems from DEP: their slow reaction to requests for a health study and the refusal of their managers to allow her on-site.

In situations where EPA has had the "lead," critics charge that the agency has chosen several improper cleanup methods.

A report released by several environmental groups, which analyzed all 75 "Records of Decision" made by the EPA in 1987, found that 68 percent of the remedies selected "failed to use any treatment whatsoever on the sources of contamination." Only in six cases did the EPA recommend a technology which would clean up the site "to the maximum extent practicable," as required by the original Superfund law of 1980.

In most instances, EPA chose methods based on cost-effectiveness without taking many local conditions into account.

In Bayou Sorrel, La., for example, rather than excavate wastes and remove them from the site, EPA chose to "cap and contain"—place an "impermeable" ceramic cap over the waste—despite the fact that the site is located on a floodplain which submerges the dangerous waste under water several times a year.

Perhaps the most stinging criticism of Superfund has been based on its reliance on the private sector—often the polluters themselves—to carry out much of the cleanup work.

In the Superfund process, polluters identified by the EPA as "principally-responsible parties" can hire and supervise their own contractors to clean up the site, as long as EPA specifications on the cleanup method are followed.

The Office of Technology Assessment said in a report released in January 1988 that contractors receive between "80 and 90 percent" of Superfund's budget each year, transactions which are largely tucked away from government process and "hidden from public scrutiny and accountability."

Superfund contractors, the report said, "have enormous influence over Superfund, perhaps more than Congress, the public, environmental groups, the news media, and other institutions."

In addition, costs for cleanups have routinely exceeded EPA estimates. At GEMS, the guesstimate made by the EPA in their Record of Decision—$27.9 million—may double by the time of completion to about $50 million.

EPA Administrator William K. Reilly, in a speech last November, called Superfund "a program that unfortunately has been on a rollercoaster," but "nonetheless, a program which, increasingly, has a good story to tell."

Reilly defended Superfund, saying that corrective work is now in progress at 80 percent of the sites on a National Priorities List, and citing "a new comprehensive Superfund strategy: streamline the bureaucracy, strengthen enforcement, open the agency to community involvement, and involve the states."

Reilly pledged to implement several specific goals recommended in a report by Sens. Frank Lautenberg, D-N.J., and Dave Durenberger, R-Minn., including hiring 500 new Superfund staff people and recovering $300 million per year from polluters by the end of 1993.

"A TRULY REGRETTABLE SITUATION"

From 1957 until 1970, the GEMS Landfill—essentially a large pit with no protective liner—was used mostly for the disposal of household garbage. In 1970, however, Gloucester Township allowed toxic wastes to be shipped to GEMS.

When a chemical fire erupted on the site that same year, New Jersey's DEP stepped in and stopped chemical wastes from entering the site. From 1970 until at least 1974, however, liquid industrial wastes from as far away as Philadelphia were still being disposed of at GEMS. Rohm-Haas, Owens-Corning and duPont were three companies fingered by the EPA as polluters.

By 1980, it was apparent that liquid waste seeping from the dumpsite was flowing into streams and into the backyards of newly-constructed homes near the site's boundary.

To make matters worse, DEP officials then allowed the city of Philadelphia to dump solid waste sludge at GEMS that was later found to contain DDD—a carcinogenic pesticide similar to DDT. Finally, later in the year, DEP made the decision to close the dump.

Simultaneously, Gloucester Township officials continued to approve the building of homes near the dump, many of which went on the market well after 1980.

But these new homeowners encountered an unexpected problem. Alarmed by a smell like rotten eggs and experiencing dizziness and shortness of breath, the Gloucester Township citizens began demanding health studies.

A study conducted by Rutgers University at the behest of people living near GEMS illustrated the dangers of living near a toxic waste dump. It was, the study said, "the most serious in terms of stress and psychological impairment among all the environmental hazards that have been studied," due to the "great uncertainty regarding the extent of the damage" and the "prolonged duration of victimization."

Nosebleeds, headaches, and breathing abnormalities were commonplace in Gloucester Township. Whether they were brought on by the dump's chemicals or the residents' anxiety over them is uncertain.

John MacDonald, a DEP attorney, called GEMS a "truly regrettable situation."

"But this isn't a Love Canal or a Times Beach," MacDonald said, referring to two 1980s toxic waste disasters in New York and Missouri which caused the widespread contamination and eventual evacuation of whole neighborhoods.

"The perceived consequences of these situations are often more serious than any real physical harm that may result," MacDonald said.

Yet in 1983, Fortuna, as well as five of her neighbors, had miscarriages.

Even more dramatic evidence of the growing turmoil in Gloucester Township was provided when basements in the Fox Chase II neighborhood began filling with methane and exploding.

"I'd be driving out of the development, and there'd be a fire truck blocking the road," Fortuna says. "This was happening rather frequently."

WAITING FOR A MIRACLE

Relief for the citizens of Gloucester Township may be too little and come too late under Superfund.

The EPA issued its Record of Decision on GEMS in September of 1985. While a methane-collection system has been in place for two years, additional controls—such as placement of the "impermeable" cap, excavation of existing liquid wastes, and construction of a groundwater pumping and treatment system to rid the site of the leached wastes—still have not been implemented.

By the time cleanup reaches the "advanced stage," and water treatment systems are in place, it will be well into the 1990s, about in line with the 10-year average it takes Superfund sites to go from start to finish, according to the Rand report.

EPA regulations require that a health study be conducted to help determine which cleanup methods should be used.

But Fortuna and some neighbors claim that EPA never did a health study, and the agency can't properly determine which health-based standard could justify their choice of cleanup methods.

The EPA counters that it did conduct a health study, but records show a draft was not completed until December, 1988—fully three years after the Record of Decision.

EMOTIONAL WASTELAND

Two of Gloucester Township's younger citizens—Steven Ash, 9, and Bridget Wing, 10—prop their bikes against a sand pile which spills from a corner of the fence surrounding GEMS.

Ash, who lives about 50 ft. from the fence, answers quickly when asked if he ever goes adventuring to the other side.

"No way."

"Why not?"

"Cause it's poison."

Several feet down the fence from them, medium-height stalks of withered corn sprout from the ground, twenty ft. from the landfill boundary.

Not far from the corn, a dirt road is marked by a sign which encourages entry and reads "Briar Lake: Houses Starting at $109,000."

Down that road, a single contemporary split-level home sits barren among the landfill skyline: windows tightly boarded. Several other abandoned homes dot the area. Many are being refurbished and placed back on the market.

At the landfill gate, two policemen prohibit entry. Men in "zoot suits"—lead-lined outfits often worn by astronauts to protect them from a harsh environment—are seen on top of a huge mound of dirt from afar.

These scenes indicate that, despite all the knowledge, legislation, and technical know-how, the area around GEMS may never be as safe as it once was for human inhabitance.

But the evidence of emotional devastation is far clearer. The GEMS landfill—and the inadequate bureaucratic solution of Superfund—have made life for residents like Loretta Fortuna one long toxic nightmare.

"Who would want to live here?" asks Fortuna, displaying her neighborhood and standing next to the "For Sale" sign on her front lawn. "I'm happy just to get out with my family."

POSTSCRIPT

Hazardous Waste: Are Cleanup Efforts Succeeding?

Cleaning up existing waste sites and properly disposing of newly generated waste require the construction of facilities such as hazardous waste landfills or incinerators to receive the wastes. The proposed siting of such facilities has invariably resulted in organized opposition from local residents. This problem is discussed in "Not In My Backyard," by Carl La Vo, *National Wildlife* (April/May 1988).

Mentzinger's report of serious deficiencies in current cleanup efforts is reinforced in "Superfund, Superflop," *U.S. News & World Report* (Feb. 6, 1989).

Another serious dimension of the problem is the growing use of developing nations as the dumping grounds for waste from the United States and other wealthier nations. For an exposé of this practice, see *Global Dumping Ground,* by Bill Moyers (Center for Investigative Reporting, 1990).

Some critics have argued that the waste cleanup effort has been hampered by an unwillingness to employ new technological methods of waste treatment and disposal and to dispense with land dumping. Bruce Piasecki and Jerry Gravander make this argument in "The Missing Links: Restructuring Hazardous Waste Controls in America," *Technology Review* (October 1985). Piasecki's book *Beyond Dumping: New Strategies for Controlling Toxic Contamination* (Greenwood Press, 1984) explores this question in greater depth.

The major impact of liabilities of new hazardous waste legislation on American business is detailed in "The Hidden Liability of Hazardous Waste Cleanup," by Gordon F. Bloom, in *Technology Review* (February/March 1986). Carl Pope suggests that the way to reduce such liabilities is to provide economic incentives for industry to switch to safer chemicals in "An Immodest Proposal," *Sierra* (September/October 1985).

The reduction of hazardous waste generation is explored by Joel S. Hirschhorn in an article in the April 1988 issue of *Technology Review.* In "Source Reduction Research Partnership: A Unique Joint Venture," *Environment* (November 1989), Azita Yazdani describes a cooperative study by an environmental organization and a water agency that has developed a comprehensive hazardous waste reduction plan for industry.

Both the Superfund and RCRA are again due for revision and reauthorization. Environmentalists are hopeful that the glaring inadequacies will be corrected this time around.

ISSUE 13

Municipal Waste: Should Incineration Be a Part of Waste Disposal Methods?

YES: John Shortsleeve and Robert Roche, from "Analyzing the Integrated Approach," *Waste Age* (March 1990)

NO: Neil Seldman, from "Waste Management: Mass Burn Is Dying," *Environment* (September 1989)

ISSUE SUMMARY

YES: Incineration industry executives John Shortsleeve and Robert Roche argue that an integrated system that incinerates the residue from a municipal waste recycling and composting operation is the best disposal option.
NO: Waste disposal consultant Neil Seldman claims that intensive recycling and composting can do the job without the use of costly and hazardous mass incineration technology.

Since prehistoric times, the predominant method of dealing with refuse has been to simply dump it in some out-of-the-way spot. Worldwide, land disposal still accounts for the overwhelming majority of domestic waste. In the United States roughly 90 percent of residential and commercial waste is disposed of in some type of landfill, ranging from a simple open pit to so-called sanitary landfills where the waste is compacted and covered with a layer of clean soil. In a very small percentage of cases, landfills may have clay or plastic liners to reduce leaching of toxins into groundwater.

By the last quarter of the nineteenth century, odoriferous, vermin-invested garbage dumps, in increasingly congested urban areas, were identified as a public health threat. Large-scale incineration of municipal waste was introduced at that time in both Europe and the United States as an alternative disposal method. By 1970, over 300 such central garbage incinerators existed in American cities, in addition to the thousands of waste incinerators that had been built into large apartment buildings.

Virtually all of these early garbage furnaces were built without devices to control air pollution. During the period of heightened consciousness about urban air quality following World War II, restrictions began to be imposed on garbage burning. By 1980, the new national and local air pollution regulations had reduced the number of large United States municipal waste incinerators to fewer than 80. Better designed and more efficiently operated landfills took up the slack.

During the past decade, an increasing number of U.S. cities have been unable to locate suitable accessible locations to build new landfills. This has coincided with growing concern about the threat to both groundwater and surface water from toxic chemicals in leachate and runoff from dumpsites. Legislative restrictions in many parts of the country now mandate costly design and testing criteria for landfills. In many cases, communities have been forced to shut down their local landfill (some of which had grown into small mountains) and to ship their wastes tens or even hundreds of miles to disposal sites.

Municipal waste disposal costs are skyrocketing. The lack of long-range planning, typical of America's urban development, has resulted in a crisis situation. Energetic entrepreneurs have seized upon this situation, and with the help of local and state politicians are promoting European-developed incineration technology as the panacea for the garbage problem.

Ironically, the proliferation of heat-recovery waste incinerators in this country coincides with increasing concern in Europe about the air pollution produced by burning modern waste, which contains increasing amounts of plastics and other exotic chemical components. Airborne emissions from garbage incinerators include toxic metals such as cadmium, lead, and mercury, as well as highly toxic dioxins and other organic carcinogens. The addition of sophisticated pollution control devices has reduced the air pollution potential at the expense of trapping these toxins in the incinerator ash, which presents a very troublesome disposal problem.

Recycling, which has until recently been dismissed as a minor waste disposal alternative, has recently been shown to be a viable major option. The Environmental Protection Agency (EPA) and several states have established hierarchies of waste disposal options with the goal of using waste reduction and recycling for as much as 50 percent of the material in the waste stream. Several environmental groups are urging even greater reliance on recycling, citing studies that show that more than 90 percent of municipal waste can theoretically be put to productive use if large-scale composting is included among the options considered.

John Shortsleeve and Robert Roche are among the incineration industry executives who argue that integrating recycling and composting with heat-recovery incineration results in the most efficient and economical waste disposal system. Waste utilization consultant Neil Seldman disagrees. He argues that by using the full potential of recycling and composting, communities can eliminate the need for costly, polluting incinerators.

YES

John Shortsleeve and Robert Roche

ANALYZING THE INTEGRATED APPROACH

As solid waste managers increasingly include recycling and composting in their plans, waste-to-energy companies become more aware of the commercial possibilities of these approaches *and* of the synergy when combined with burning and landfilling in an integrated solid waste program.

Unfortunately, on a national level, the U.S. has not yet adopted an integrated program. While most Western European nations recycle, compost, and burn 80% of their waste (landfilling only 20%), the U.S. figures are reversed. We recover only 20% of our annual 165 million tons of refuse, and landfill 80%.

Legislators and policy makers in Washington, D.C. are currently debating whether or not to mandate a specified level of recycling as a matter of federal law. The current markup of the clean air act requires states to certify that they have a program in place to achieve 25% recycling by 1994 (Senate Bill 196). We think this type of legislation makes sense.

Our reliance on landfilling can't last much longer. Since 1979, the number of operating landfills has plummeted by 72%. EPA projects that in five years, the number of landfills will be slashed again by more than half. By the year 1994, says the agency, only 2,275 landfills will remain of the 19,500 that were operating in 1979. Even now, in most regions, it is a near political impossibility to site a new landfill or expand an old one.

A more promising approach to the problem of MSW [municipal solid waste] disposal is "integrated resource recovery,"—a coordinated mix of recycling, composting, and waste-to-energy. Of the 20% of our waste stream not going to landfills, half (or 10% of the total) is being recycled already.

Thus far, however, recycling, composting, and waste-to-energy have evolved in the U.S. as separate technologies.

From John Shortsleeve and Robert Roche, "Analyzing the Integrated Approach," *Waste Age* (March 1990). Reprinted by permission of *Waste Age* magazine.

COMPARING INTEGRATED vs. MASS-BURN

Foster Wheeler has developed an approach that lets us look at the economic interaction of these three technologies when integrated under unified management. This approach also lets us compare integrated resource recovery with conventional mass-burning, relative to construction and operating costs, and environmental impact.

For this article, we modeled two facilities, each designed to take in 1,000 tpd of refuse. One would burn all 1,000 tons. The other, using the integrated approach, would burn just 650 tons a day, recycling and composting the rest.

In the waste stream going to the integrated facility, all yard waste would be composted, and all PET (polyethylene terephthalate) beverage bottles, and high-density polyethylene (HDPE) milk and water bottles would be recycled. Seventy-five percent of the glass, and about half of the newspaper, cardboard, ferrous metals, and aluminum, would be recycled also.

Smaller Boiler, Richer Fuel

The most obvious effect of the integrated approach would be to reduce boiler size. In our hypothetical case, a 35% reduction in the amount of waste burned would reduce boiler size by 26%—and construction costs by 16%. While a 16% reduction in cost may be less than hoped for, it more than covers the added cost of building the recycling and composting facilities.

But perhaps the most surprising result of the integrated approach is that, although fuel is cut by 35%, energy output (and earnings) decline by only 26%. That's because the waste that remains after recycling has a higher energy content. Table 1 shows how reducing inorganic noncombustibles increases carbon content and triggers other changes to produce a more uniform fuel having 13% more energy per pound.

Table 1

Btu Content Of Waste Before and After Recycling/Composting

Component	Before (lbs/100 lbs)	After (lbs/100 lbs)
Moisture	26.41%	24.16%
Inorganic	20.65%	16.31%
Carbon	27.62%	31.06%
Hydrogen	3.73%	4.25%
Oxygen	20.40%	22.74%
Nitrogen	.59%	.66%
Chlorine	.48%	.67%
Sulfur	.12%	.15%
% Weight	100.00%	65.34%
Btu/lb	4,880	5,539

Notes: The net effect of recycling and composting is to increase energy content of waste in the IRR burn unit by 13.5%.

Data source: Foster Wheeler Power Systems, Inc.

Easier on Boilers

Recycling removes or reduces waste elements that can damage boilers and interrupt their operation. One such component is chlorine, which in many municipal waste streams comes largely from waste paper and plastic. A U.S. Department of Commerce study of Baltimore County waste showed that 86% of the chlorine came from these two sources.

Chlorine forms chlorides that corrode metal tubes in the walls and superheater sections of the boiler. Recycling a significant amount of newspaper, cardboard, and plastic should reduce chlorine—and chloride attack.

Glass is another problem for boilers. Ash samples taken from waste-to-energy facilities typically contain 25% silica, the primary component of glass. If the temperature at which ash fuses into a coherent solid is at or below temperatures in the lower furnace, the ash can fuse onto the boiler walls, forming "slag." High silica content lowers ash fusion temperature. So, by removing glass from fuel, the silica fraction in the ash will decrease, the fusion temperature of the ash should rise, and the slagging phenomenon should be reduced. That should simplify maintenance, reduce hazards for boiler maintenance workers, and increase boiler efficiency.

Aluminum, also a worry, melts at 1,200 degrees Fahrenheit, a temperature far below those maintained in waste-to-energy boilers. Unless extracted from waste, the metal clogs boiler air circulation holes, causing uneven burning which reduces efficiency.

Easier on the Environment

An important result of integrated resource recovery, economically and environmentally, is reduction of ash and other landfill material. A computer comparison of our models shows that the amount going to a landfill from the integrated complex (178 tpd) is 39% less than the 290 tpd landfilled from the mass-burn unit.

Bypass waste, too, is less of a problem for the integrated complex. Nearly 50% more waste is processed in the summer and fall due to yard waste. Where yard waste is composted, deliveries of waste to the energy recovery facility are more uniform in amount throughout the year. Emergency storage capacity can be designed for a more predictable volume, reducing the likelihood of overflow (and the need to bypass during emergency shut-downs).

Getting the Lead Out

Spent batteries are a major source of heavy metals in the environment. Although automobile batteries have been recycled with commercial success since before World War II—and 80% are recycled today—they still add much lead to the waste stream. Recent marketing changes have made the problem worse by limiting their return to their place of purchase.

But because the market for spent auto batteries has been successful for so long, few states have thought it necessary to mandate recycling. This is changing, however.

In Connecticut, recycling of car batteries will be compulsory by January, 1991; other states are considering similar legislation. EPA's proposed standards for new and existing waste-to-energy plants would require removal of all lead-acid batteries from the waste stream before combustion.

EPA's standards would mandate some sort of program targeting removal of household batteries from the waste stream. The Hearing Aid Association and the American Watch Institute have mounted vigorous recycling programs in support of the market for spent "button" batteries. There is less of a market for spent nickel-cadmium batteries since most are found in small, cordless appliances not designed for battery extraction. According to the National Electrical Manufacturers Association, appliance manufacturers are now designing appliances from which batteries may be easily removed for recycling.

There is no market at all, to our knowledge, for used alkaline and zinc carbon dry cells, because, as yet, no commercially viable way exists to process them.

All dry cells—except the lithium type—contain small amounts of mercury.

In the absence of commercial recycling, battery manufacturers continue to reduce the amount of mercury used, while searching for nontoxic substitutes.

DISPOSAL ECONOMICS IMPROVED

Table 2 summarizes the capital cost, operating costs, and annual revenues of the 1,000-tpd mass-burn and integrated models. Both handle roughly 300,000 tons of waste per year and pay the same price for electricity, water, sewer use, and bypass and residue disposal. Both models assume that:

- electricity sells for six cents per kilowatt hour (.06/kWh);
- ash disposal costs $50/ton; and
- the state in which the facility operates does not have a beverage container redemption law or "bottle bill." (Since bottle-bill laws provide incentives for the return of aluminum cans to stores by consumers, the lack of such a law means more aluminum in the waste stream, from which it may be extracted to become a revenue.)

Table 2

Cost Comparisons

	IRR	Mass Burn
Thruput in Tons	300,000 (Burning: 196,000) (Recycling: 59,000) (Composting: 45,000)	300,000
Capital Cost	$102.5 m.	$110.5 m.
Operating Cost	$9.5 m.	$8.2 m.
Annual Revenue	$9.9 m.	$9.2 m.
Annual Landfill Costs	$2.7 m.	$4.3 m.
Net Tiping Fee	$47.36/ton	$55.02/ton

Data Source: Foster Wheeler Power Systems, Inc.

From the perspective of disposal economics alone, the cost advantage lies with the integrated model. Still, in some communities, the cost disadvantage of separate collection could easily outweigh cost-of-disposal advantages. The most sensitive variables are the amount of aluminum recovered and the cost of ash disposal.

Integrated resource recovery is more attractive economically wherever aluminum recovery can be maximized and residue disposal costs are high. In our models, by varying these assumptions within a reasonable range, the economic advantage of the integrated approach, in disposal costs only, ranges from a high of $11/ton to $0/ton.

Because it promises to benefit everyone in so many ways, and has already shown that it can keep its promises, integrated resource recovery is an idea whose time has come.

NO
Neil Seldman

WASTE MANAGEMENT: MASS BURN IS DYING

Public standards for solid waste management have been changing very rapidly. What was acceptable as recently as 1985 no longer meets with public approval. Furthermore, the new popular standards are being incorporated in state and federal regulations. This rapid change results from the fact that mass incineration technology, the solution proposed by federal and state agencies to the landfill disposal crisis, has proven to be technologically, environmentally, and financially unacceptable to concerned citizens, whose organizations now have a tight grip on local decisionmaking. After a mass incineration plan was unanimously rejected by the Prince George's County (Maryland) Council in favor of a comprehensive recycling and composting program, the *Washington Post* observed that citizens are involved, mobilized, and here to stay in the local political process. Roland Luedtke, the former mayor of Lincoln, Nebraska, put it more succinctly: "Garbage is an issue that can unseat an incumbent mayor."

Through the 1970s and the first half of the 1980s, mass incineration (burning garbage as is) was the technology of choice for most local planners. It was a convenient solution to the shortage of landfill space. Incinerators required essentially no change in the delivery system: Waste was simply delivered to an incinerator instead of to a landfill. Of course, the costs were much higher, but there appeared to be no alternative. The U.S. Environmental Protection Agency (EPA) and the U.S. Department of Energy (DOE) provided a decade of support worth perhaps $1 billion for planning, research, demonstration, and, finally, commercialization of incinerators. When the plants were unable to find steam customers, the 1979 Public Utilities Regulatory Policies Act (PURPA) provided a guaranteed market for electricity sales instead, despite the high cost of generating electricity from garbage. When incinerator ash was routinely found to be hazardous, it was exempted from state and federal hazardous waste laws. Also in 1979, DOE began a $300 million support program for plant construction. Program planners projected that, by 1992, they would be feeding 70 percent of the nation's waste into boilers. Recycling was considered merely a footnote. Don Walter, the head of the DOE Commercialization of Waste to Energy Program, glibly stated that he did not care if recycling saved more energy than was generated by incineration. "My job is to create energy from garbage, and that is it." Through 1979, the federal government spent only about $1 million on recycling. (After 1979, all solid waste program funds were transferred to hazardous waste programs.)

Today, about 10 percent of the country's municipal solid waste is incinerated and the same percentage is recycled. EPA programming for recycling is only now being planned.

Since 1985, some 40 mass burn plants, valued at about $4 billion, have been canceled, most before reaching the construction stage. In 1987, for the first time, more plant capacity was canceled than was ordered. Only 11 plants were ordered in 1988. Major cities such as Seattle, Philadelphia, Los Angeles, San Diego, and Portland, Oregon, have canceled plants, as have smaller cities such as Chattanooga, Tennessee; Lowell, Massachusetts; Saratoga, New York; and Gainesville, Florida. Of the 100 plants that remain in the planning stage, most face very stiff opposition and probably will not be built. Very few large incinerators (processing 1,000 tons or more of garbage per day) will be built. The most dramatic cancellation occurred in Austin, Texas, where the city had spent $23 million on a plant in construction. But when two businessmen were elected to the city council in early 1988, they voted with the environmentalists against the plant. The new council members foresaw a $150 million savings over a 20-year period if they invested in recycling. Florida cities in Broward and Pasco Counties have also pulled out of mass burn projects for financial reasons. The costs, it was concluded, would require steep property tax increases. Many jurisdictions fear repeating the experience of Warren County, New Jersey. A plant that was supposed to cost $35 per ton to operate came on line in 1988. By the end of the year, the cost was $98 per ton, and it is expected to rise to $140 by the end of 1989 because the ash was found to be hazardous and has to be shipped to Buffalo, New York. Now the county must build a $30 million ash landfill. In addition, the county was fined by the state for failing to meet ash regulations and continues to pay $59,000 per week to subsidize the plant.

Mass burn plants were advertised as a way to stabilize disposal costs. The record hardly bears out such a claim. In fact, the technology is so uncertain that each year reveals new difficulties. Before 1985, bag houses and scrubbers were not required; now they are. The need for monofills (landfills for only one type of waste) for toxic ash was not anticipated, but they also are now required. EPA recently ordered a scrubber called Thermal DeNox to be installed in plants to control oxides of nitrogen emissions.

Now, based on a grassroots appeal from citizens in Spokane, Washington, a whole new set of criteria is being established for mass burn plants. The EPA office for Region 10 (Alaska, Idaho, Oregon, ad Washington) has recommended to the national EPA office in Washington, D.C., that the mass burn permit given to Spokane be remanded because the plan for mass burn does not include comprehensive source reduction, recycling, and mechanical processing prior to incineration. Because these approaches are cost effective, commercially available, and known to reduce pollution, federal Best Available Control Technology (BACT) standards require that they be implemented before an air permit can be issued. Since June, the same arguments are being made in support of new Source Performance standards, which must be issued by EPA for national guidelines. In essence, mass burn is about to be declared unpermittable.

This development has not been lost on the industry. Already Westinghouse, Wheelabrator, and Waste Management, Inc., have acquired recycling and processing technologies to meet the new standards that have, in effect, been dictated by recycling and grassroots citizen groups. Further, the processing industry is now investing in ways to recover paper fiber for recycling instead of using it as a refuse-derived fuel. These developments seem to have caught the national environmental groups off guard. They never imagined that grassroots groups on their own could change federal policy. Many environmental organizations would have settled for less recycling and would have forgone requirements for mechanical processing.

What is the alternative to mass burn? The primary alternative is recycling that includes composting. Mechanical processing and some burning in small specialty boilers or existing coal-fired plants may be required. Depending upon source reduction activity, landfilling of 10 to 20 percent of the waste stream will still be necessary. Several key cities have already started implementation. Seattle has set a goal of 65 percent recycling/source reduction by 1992 and has already achieved half this goal. Philadelphia set a 50 percent goal, and Washington, D.C., a 45 percent goal. These recycling plans replace mass incineration; they are not simply afterthoughts to a mass burn system.

Throughout the United States, recycling is at the takeoff stage. This progress stems from the infinite patience and painstaking efforts of citizen groups. Cliff Humphrey, a recycling pioneer and organizer of the first environmental action organizations in the country, provides some insight:

> We knew early on [in 1968] that the tonnage was really in the commercial waste stream. But the houses were where the votes were. Even then we had to start with drop-off sites. This was a necessary first step, even though it was inefficient compared to curbside collection. It was the way that you established recycling values in the community.

The investment in time and effort has paid off. When Humphrey began his recycling odyssey in 1968, the word recycling was unfamiliar to the U.S. public. Today, it is the law of the land.

What are the mechanics of the new recycling paradigm that is replacing the burn and bury pattern of solid waste management? There are five key factors:

MANDATORY RECYCLING

While this may not be absolutely required, it certainly helps. When accompanied by proper education, mandatory recycling programs throughout the country have a 90 percent participation rate. Islip, New York, presents an interesting case study. Islip had a mandatory ordinance but never enforced it. Consequently, participation was at 25 percent. After the embarrassment of the garbage barge, the town's administration focused its attention on recycling. Now, according to Town supervisor Frank Jones, "We can't find a household that does not recycle."

ECONOMIC INCENTIVES TO HAULERS AND HOUSEHOLDS

In Seattle, the city pays haulers $48 per ton for recycled materials. This represents a shared savings payment, as the city would pay $96 per ton for new landfill space. Seattle households pay for garbage disposal by the bag. After the first bag, fees increase steeply, but collection of recyclables is free. In Perkasie, Pennsylvania, a similar system resulted in a 61 percent reduction in the amount of garbage going to landfills—49 percent through recycling and 13 percent through source reduction.

RECYCLING LITERACY

In-school education will be far more effective in the long run than general public-awareness raising. Children in daycare and elementary schools bring recycling home with them. Like reading and writing, once recycling is taught, it is never forgotten. Public schools must make sure that all graduates know why and how to recycle. This effort will make students better and less costly future citizens of the community. Thanks to the efforts of early environmental educators, a whole catalogue of lesson plans and curricula is available to teachers.

GOVERNMENT PROCUREMENT

If recycling programs are to succeed, local, state, and federal governments must start buying products made with recycled materials. Industry should be encouraged to do the same, perhaps with tax breaks in certain circumstances. Since 1976, EPA has been working on establishing government procurement guidelines, which are not yet mandatory. To sustain markets for recycled materials, EPA must act quickly. Such action would benefit trade as well as the environment because using recycled materials would make U.S. industry more competitive with other countries, which buy our exported recycled materials to outperform U.S. companies in the marketplace. Perhaps the U.S. Department of Commerce, and not EPA, should focus on procurement programs.

ECONOMIC INCENTIVES TO MANUFACTURERS

In most cases, all that is needed is a below-market rate loan to encourage manufacturers to locate in a given area and create markets for locally recovered materials. In this "everyone wins" scenario, there are tremendous payoffs for both public and private sectors. Recycled glass is worth $35 per ton as a raw material but from $4,000 to $7,000 as a finished glass product. Moreover, scrap-based manufacturing creates jobs and new skills, encourages investment, and enlarges the manufacturing taxbase of the local economy. The "Clean Michigan" program is a model for this policy, the program's low funding level notwithstanding. This five-year-old program provides grants to nonprofit groups, local governments, and private firms for planning, implementing, and maintaining recycling and market development projects. Along similar lines, the state of Minnesota lured a tire processor to locate there with an innovative loan package. As garbage becomes a larger and larger burden on local economies, the need to use economic development funds to attract users of recycled materials becomes more obvious.

IN MANY JURISDICTIONS, PROMOTERS OF mass burn have conceded 25 percent of the waste stream to recycling as a tactic to guarantee the remaining 75 percent to incinerators. While mass incineration advocates begrudgingly admit that 25 percent recycling may be possible, many cities and towns are moving beyond this arbitrary boundary, whose actual function is to limit the demonstrated success of recycling.

Economic realities and fears of environmental catastrophe have mobilized citizens on the garbage issue. Michigan is an excellent example. That state has three mass burn plants, representing a billion dollar commitment, all of which stand idle because there is no place to dispose the hazardous ash they produce. The state legislature is rushing through a law to exempt the ash from existing hazardous waste regulations so that the plants can be put back in use and others can be built. Michigan citizens are fighting for a far more balanced program that would include recycling, composting, and mechanical processing, as well as landfills, to relieve the state of the capital and operating costs of these mass burn plants. A recent study of Michigan recycling programs revealed that collection of curbside recyclables costs $9.61 per load as compared with $37.33 per load for regular garbage collection.

In addition to its own disposal problem, Michigan faces an influx of waste from the eastern states of Connecticut, New York, and New Jersey, which cannot find local landfills for their mass burn residues. Because Ohio, Illinois, and Indiana are stiffening landfill regulations while Michigan is lowering barriers, the eastern ash will soon be heading by rail for Michigan's private landfills, which cannot bar out-of-state waste: If Michigan exempts its own ash from hazardous waste regulation, it must do the same for imported ash.

Progress in recycling at the local level in the United States has international as well as national implications. As industrial energy use is reduced and as virgin ore and fiber is replaced with recycled materials, the U.S. contribution to global pollution should decline. Moreover, as U.S. recycling increases, imports of Third World raw materials should decrease, enabling Third World countries to use their own raw materials for manufactured products instead of buying finished products at high prices from the industrialized nations. Finally, if U. S. cities generated far less toxic ash, they would not need to ship these wastes to Third World countries. Recycling involves acting locally. It impacts globally.

POSTSCRIPT

Municipal Waste: Should Incineration Be a Part of Waste Disposal Methods?

Although the use of more effective devices for scrubbing and filtering the combustion gases has reduced the air pollution potential of new waste incinerators, these technologies have greatly increased the costs of waste disposal. The financial considerations include the expense of disposing of the ash in which toxic substances have been trapped. According to various estimates, this ash amounts to between 10 and 25 percent of the landfill volume required to dispose of the waste from which the ash is produced. The courts have interpreted federal legislation that exempts household garbage from the rules defining toxic waste to include the ash, even though tests show that much municipal waste ash would otherwise require disposal in a licensed hazardous waste landfill. Nevertheless, a growing number of states are defining the ash as a "special" waste that can be disposed of only in landfills that meet expensive leachate control specifications. Efforts are under way to explore possible productive, environmentally safe uses for municipal waste incinerator ash.

Even more effective than setting goals for the percentage of waste that should be recycled are regulations (such as those now in effect in New York) that require the recycling of all waste components for which the cost involved would be less than incineration. *Garbage in the Cities: Refuse, Reform, and Environment,* by Martin Melosi (Texas A&M University Press, 1981), is a historical analysis of the social, political, economic, and technological factors that led to our present waste disposal problems.

Two recent, informative reports on a variety of problems and controversies concerning the garbage disposal issue are "Trash Can Realities," by Jon Luoma, *Audubon* (March 1990) and "Wasting Away," by Howard Levenson, *Environment* (March 1990).

ISSUE 14

Nuclear Waste: Is Yucca Mountain an Appropriate Site for Nuclear Waste Disposal?

YES: Luther J. Carter, from "Siting the Nuclear Waste Repository: Last Stand at Yucca Mountain," *Environment* (October 1987)

NO: Charles R. Malone, from "The Yucca Mountain Project: Storage Problems of High-Level Radioactive Wastes," *Environmental Science & Technology* (December 1989)

ISSUE SUMMARY

YES: Science journalist Luther J. Carter describes a variety of technical and political considerations that support the choice of Yucca Mountain for the nuclear waste dump.
NO: Nuclear waste specialist Charles R. Malone cites several serious deficiencies in our ability to make the needed assessments of the proposed site, which make it impossible for the present effort to succeed.

The fission process by which the splitting of uranium and plutonium nuclei produces energy in commercial and military nuclear reactors also generates a large inventory of radioactive waste. This waste includes both the "high-level" radioactive by-products of the fission reaction, which are contained in the spent fuel rods removed from the reactors, and a larger volume of "low-level" material, which has been rendered radioactive through bombardment with neutrons—neutral subnuclear particles—emitted during the reaction. "Low-level" waste also includes the refuse produced during medical and research uses of radioactive chemicals.

The amount of highly radioactive material that builds up in the core of a commercial nuclear power plant during its operation far exceeds the radioactive release that results from the explosion of a high-yield nuclear weapon. Because radioactive emissions are lethal to all biological organisms—causing severe illness and death at high doses and inducing cancer at any dose level—it is necessary to make sure that the radioactive wastes are kept isolated from the biosphere. Since some of the nuclear products remain radioactive for hundreds of thousands of years, this is a formidable task.

The early proponents of nuclear reactor development recognized the need to solve this problem. Confident that scientists and engineers would find the solution, a decision was made to proceed with a program, sponsored and funded by the U.S. government, to promote nuclear power before the serious issue of permanent waste disposal had been resolved.

Thirty years later, with 100 commercial nuclear power plants licensed in this country, more than 300 in other countries around the world, and hundreds of additional military nuclear reactors piling up lethal wastes in temporary storage facilities every day, that early confidence that the disposal problem could be solved has long since disappeared. Nowhere in the world is there a proven, operating plan for permanent nuclear waste disposal. In the United States, several abortive plans and schedules have been mandated by Congress, only to be abandoned for a variety of technical and political reasons. The most recent Nuclear Waste Policy Act, legislated in 1982, set a step-by-step schedule to complete an operating permanent "high-level" waste repository by 1998. It is already clear that even this schedule will be impossible to meet. In December 1987, recognizing that serious problems were again developing in implementing the new plan, Congress short-circuited the process by designating Yucca Mountain, Nevada, as the only site to be considered for the first high-level repository.

A solution to the problem of "low-level" waste has been equally elusive. Amendments were passed in late 1985 in attempt to make the 1980 Low-Level Radioactive Waste Policy Act workable. But political and technological disputes continue unabated, and it seems certain that the timetable established in the legislation will not be met.

The history of the nuclear waste issue illustrates the folly of focusing on technological fixes without recognizing that solutions to real world problems must meet political, socioeconomic, and ecological criteria that are not revealed by isolating the results of laboratory investigations from the other aspects of the issue. The simplistic response of some nuclear scientists is to claim that the technological problems have been resolved and nuclear waste disposal is only a "political" issue. A careful examination of the situation reveals the inappropriateness of adopting such a perspective. The evaluation of a proposed technological solution to a problem is related to social values, which in turn affect the political position of the participants in the process. Serious differences exist as to the degree of isolation and period of time necessary for "high-level" waste containment. How certain need experts be about future geological processes before they can claim that burying wastes in a particular location will result in the required degree of isolation?

Luther J. Carter, a journalist who has written a book about nuclear issues, writing just before Congress chose the Yucca Mountain site, presents arguments in favor of that decision. Nevada nuclear waste agency scientist Charles Malone argues that the planned evaluation program cannot provide the needed assurances about the safety of the specified site.

YES

Luther J. Carter

SITING THE NUCLEAR WASTE REPOSITORY: LAST STAND AT YUCCA MOUNTAIN

The Great Basin, the immense desert region that extends from the Rockies across Utah and Nevada to the east wall of the Sierra Nevada in California, takes in some of the driest, most remote, most sparsely inhabited and hard-bitten country in the United States. The landscape is stark, severe, and often awesome: rugged north-south mountain ranges, forested only at higher elevations, are separated by long, flat valleys dominated by sagebrush. No rivers or streams flow out of the Great Basin; they either spread out and disappear in depressions such as the Humboldt Sink or Death Valley, or they discharge into lakes that are usually shrinking remnants of huge lakes of the wetter times that accompanied the last continental glaciations.

It is here in the Great Basin that the now politically desperate effort to establish a geologic repository for the highly radioactive spent fuel from nuclear power reactors may have to make its last stand. The repository site of primary interest is likely to be one in volcanic tuff near the far southwest corner of the Great Basin. This is at a place called Yucca Mountain, which straddles the southwestern boundary of the Nevada Test Site—used for the last 35 years as a testing grounds for nuclear weapons. . . .

The Yucca Mountain site has been rated the best of those identified to date, and it is clearly the one that presents the least environmental or land-use conflict. Moreover, if this site should in fact eventually turn out to be both technically acceptable and politically attainable, development of a repository and related spent fuel storage and packaging operations could provide a major alternative use for the Nevada Test Site in the event the United States and the Soviet Union agree to ban or sharply curtail the testing of nuclear weapons.

Luther J. Carter, "Siting the Nuclear Waste Repository: Last Stand at Yucca Mountain," *Environment*, vol. 29, no. 8 (October 1987).

POLITICAL PARALYSIS

The present repository-siting effort, mandated by Congress under the Nuclear Waste Policy Act (NWPA) of 1982, has been in a state of political paralysis since the U.S. Department of Energy's site-screening decisions of May 1986. At that time the department (DOE) announced that, in the search for a site for the first repository, three sites had been confirmed as the ones chosen for "characterization," or detailed exploration, from exploratory shafts and tunnels as well as from the surface. They were the tuff site at Yucca Mountain in Nevada; a basalt site on DOE's Hanford Reservation in Washington State; and a site in bedded salt in Deaf Smith County on the High Plains of west Texas. This was a major step in the screening process, for each of the characterization projects would cost on the order of a billion dollars (the estimates soared from $100 million or less per site only several years earlier) and, in the end, one of the three sites probably would be chosen by the president for the first repository.

The selection of these sites would have been strongly opposed in any case by the three "host states" of Nevada, Texas, and Washington. But the political problem was compounded and the resistance increased by DOE's announcement at the same time of another major decision—the indefinite postponement of further screening of sites for a second repository, a screening which had been undertaken in granite and other crystalline rocks in the eastern half of the country. Although defended by DOE on the grounds that a second repository would not be needed as soon as originally thought, the decision was in fact in response to political pressures. DOE arrived at it in close consultation with the Reagan White House, which was worried lest nuclear-waste-siting controversies contribute to a loss of Republican control of the Senate in the November 1986 elections.

Members of the House and Senate from the West were outraged by what they saw as a flouting by DOE of the regional balance and equity that Congress had sought to write into NWPA. . . .

Although DOE stands accused of undermining congressional support for the repository-siting effort by acts of gross political expedience, the record suggests that, whatever DOE had decided, this support (such as it was) was going to be lost one way or another. The site screening mandated by NWPA has proved to be confoundingly contentious and a veritable political and procedural marathon of such complexity and difficulty that neither DOE nor any successor agency is ever very likely to complete it successfully.

Moreover, while procedurally prescriptive and demanding, NWPA seems in some respects to have left DOE too much freedom, allowing the department to give preference even to sites that present major land-use or environmental conflicts, real or strongly perceived. For instance, the Hanford basalt site was selected for characterization despite its close proximity to the Columbia River, a resource of incalculable value to both Washington and Oregon. Similarly, the Deaf Smith salt site was chosen despite its location beneath the Ogallala aquifer, on which the abundant agriculture of the Texas High Plains depends for irrigation water.

These conflicts impose a heavy political burden on the Deaf Smith and Hanford sites and are probably sufficient in themselves to make the sites unattainable. But there are other problems, too. At Hanford the geohydrologic problems and uncertainties are so difficult that there is a question of whether a repository could be built at all. Indeed, in constructing the maze of tunnels deep in the basalt, there would be risk of catastrophic flooding. At Deaf Smith the tendency of salt to creep or flow plastically in response to pressure and heat promises to make impractical the waste retrieval option required by law and regulation.

Congress has been revisiting the NWPA since early this year. But with one exception, all of the principal bills have been proposals to declare a moratorium on siting activities for a geologic repository and a central spent fuel storage facility, pending a special commission study of what has gone wrong and (under two of the three moratorium bills) a congressional reauthorization of the siting effort. The exception is the Senate bill sponsored by J. Bennett Johnston (D-La.), chairman of the Energy and Natural Resources Committee, and James A. McClure (R-Idaho), the committee's ranking Republican minority member.

The Johnston-McClure bill, S. 1481, represents a clean break with past siting strategy. It calls on the secretary of energy to pick one of the three western sites for characterization by January 1, 1989. The affected state, if it chooses to cooperate, would be eligible for benefits of $50 million a year during site investigation and licensing and $100 million a year once the repository began receiving waste. If the state chose not to cooperate, the project would nevertheless proceed, with the prospect that the state could get the repository and no benefits. This bill has been favorably reported by both the Energy and Natural Resources Committee and the Senate Appropriations Committee.

YUCCA MOUNTAIN

The Johnston-McClure bill is obviously directed at Yucca Mountain, given that the technical and political burdens of this site appear lighter than those at either the Deaf Smith or the Hanford sites. This is not to say, however, that any effort to establish a repository at Yucca Mountain will not face formidable political hurdles.

In Nevada public discussion of the waste isolation issue is dominated by the governor, Richard H. Bryan. Bryan has declared that Nevada will not accept the "stigma" of being the "country's nuclear wasteland" and has denounced offers of generous benefits as bribery and "nuclear blackmail." The governor has carried public opinion with him. A University of Nevada poll in late 1986 found three-fourths of Nevadans agreeing that Nevada "should do everything in its power to prevent the locating of a high level nuclear waste site in the state."

Nevadans resent what they see as a stereotypical image of Nevada entertained by outsiders as "border to border sand dunes with occasional slot machines and brothels, and a fine spot for nuclear waste," to quote one Nevada environmentalist. Their opposition to a repository or "nuclear dump" stems partly from a belief that the coming of such a facility would confirm the stereotype.

Bryan has declared that the explanation for Nevada's opposition cannot be laid to "NIMBYism," or a predictable not-in-my-backyard response. There is, he indicates, besides fear of a wasteland label or stigma, a sense of gross inequity or unfairness. He points out that for many years Nevada has accommodated not only nuclear weapons testing but also disposal of low-level waste, the latter at a commercial burial ground near Beatty, some 15 miles or so to the west of Yucca Mountain. The governor has pointed out further that a number of technical problems and uncertainties are associated with the Yucca Mountain site, including the possibility of earthquakes and even renewed volcanism.

Tectonic History and Waste Isolation

Such uncertainties do in fact exist, but in considering them it is important to understand how they relate to the clearly positive attributes of the site and of the Great Basin for waste isolation. Lying not far to the east of where the Pacific and American tectonic plates meet in California, the Great Basin is tectonically active, especially along its west side. Indeed, the geologic history of this region has been marked by volcanic eruptions on an astonishing scale, with these eruptions followed by an equally remarkable period of earthquake activity, massive faulting, and mountain building.

The Yucca Mountain site itself appears to have been little visited, and certainly little affected, by earthquake activity during the Holocene, or last 10,000 years. But while this suggests that the next 10,000 years (the period of greatest hazard in waste isolation) also will be free of such activity, a recurrence at or near Yucca Mountain of volcanism and major earthquakes during this period cannot be entirely precluded.

Taken alone, these circumstances and this history would make Yucca Mountain unattractive for a nuclear waste repository. But they should not be taken alone, because the region's history of tectonic activity has lent to this site certain characteristics that appear ideal for waste isolation.

Yucca Mountain is tuffaceous, and the rock in which a repository would be built is known as welded tuff. The tuff formations originated in volcanic eruptions of 11 million to 15 million years ago, truly cataclysmic events in which the magma erupted explosively, ejecting ash, pumice, and other materials from the craters or calderas in enormous volume, sometimes measured in hundreds of cubic miles. The ejecta, charged with hot gases, flowed out over the existing landscape, filling in all the low places, finally coming to rest as a thick, hot mass of volcanic ash. Increasingly compressed and "welded" by its own weight and heat, the ash was transformed, especially in its deeper parts, into welded tuff.

To judge from the uniformity of the tuff that now forms the caprock at Yucca Mountain, the eruptions and the ash flow that created it settled over an area of low relief. But over the next several million years, tectonic forces were to break up what is now the Great Basin into structural blocks, creating the numerous mountain ranges and intervening basins that are this region's most salient characteristics. As is typical of the mountains and ridges that make up the ranges, Yucca Mountain—an unimposing flat-topped ridge that rises some 1,000 to 1,500 feet above the surrounding terrain—is bounded by faults on either side and is thus considered a "fault-block mountain." The escarpment on the west side of the mountain shows displacement of as much as 700 feet along the Solitario Canyon Fault. . . .

These extraordinary events have transformed the landscape and environment of the Great Basin, and given the Yucca

Mountain site both its positive and negative attributes as a site for nuclear waste isolation. First of all, the Great Basin owes it desert climate to the rain shadow created when the Sierra Nevada was uplifted as a towering barrier to eastward passage of moisture-laden clouds from the Pacific. Nevada has little agriculture and, apart from Las Vegas and Reno metropolitan areas, is a state of small towns, hamlets, and, more than anything, wide open spaces. The vast, dry basins that make up much of Nevada are known for an awesome solitude.

So while Nevada (now with an estimated 1 million inhabitants) is one of the smaller states in population, some 82 percent of its inhabitants live in its two metropolitan areas. Nevada's dryness and very limited, concentrated settlement offer, then, a major advantage for nuclear waste isolation, for in principle it should be possible to minimize conflicts, real or perceived, between waste isolation and people and their all-important water resources. Nye County, where Yucca Mountain is located, is one of the most sparsely inhabited counties in the United States, with an average of less than one person per square mile.

Moreover, at Yucca Mountain itself the Great Basin's tectonic history has made for the great depth to the water table, which lies on the order of 2,000 feet below ground surface, so deep that the repository could be built hundreds of feet above it. Emplacement of waste high above the water table is desirable because there, in the "unsaturated zone," the rock is relatively dry and there is little or no movement of groundwater, the principal mechanism for the leaching of waste and transport of radionuclides.

The deep-lying water table at Yucca Mountain is attributable in several ways to the region's tectonic history. First, owing to the Sierra Nevada rain shadow, only about 6 inches of rain a year falls at this site, so little that there is almost no local recharge of the tuff aquifer that begins at the water table. Second, the uptilting of the Yucca Mountian fault block itself has increased the depth to the water table. Finally, the highly permeable and transmissive tuff aquifer finds its point of ultimate discharge deep below sea level in Death Valley. This makes for a pronounced "drain tile" effect that may be unique to the Death Valley groundwater basin.

The Earthquake Risk

The advantages of the Yucca Mountain site are of course beside the point if the problems and uncertainties are severe and unresolvable, so it is important to examine some of the problems, especially the earthquake risk. Government scientists believe that the major faults that bound Yucca Mountain probably have not moved within at least the last few hundreds of thousands of years. The same is true, they think, of the Ghost Dance fault, within the mountain itself.

Further investigation must be carried out before these faults can be considered inactive, however. Some evidence of slight movement within the last 10,000 years has been found on the nearby Windy Wash fault in Crater Flat, just to the west of Yucca Mountain. The entire region is in any case heavily faulted, and a repository at Yucca Mountain would have to withstand ground motion from large earthquakes.

But in light of the advantages of building the repository above the water table in the unsaturated zone, the principal concern in the event of earthquakes would appear to be not loss of waste isolation but the safety of miners and repository workers. Earthquakes have far less significance for waste isolation than they do for reactor safety, for in the latter case the fuel assemblies give off immense amounts of radioactive decay heat, and pipe ruptures could cause a loss of cooling and a fuel core meltdown, possibly leading to a major release of radioactivity. By contrast, in relative terms the aged spent fuel emplaced in a geologic formation is not nearly so hot and needs no cooling beyond the natural diffusion of heat within the repository itself.

The uncertainties and the risks related to such phenomena as earthquakes and renewed volcanism or geothermal spring activity in the vicinity of Yucca Mountain are expected to be thoroughly studied. An extraordinary 7,000-page site characterization plan has been prepared by DOE at a cost of about $24 million. The site characterization would take at least five years to complete.

WEAPONS TESTING AND WASTE ISOLATION

Even as Governor Bryan has denounced consideration of Yucca Mountain as a possible waste repository site, he has opposed moves in Congress to drastically curtail nuclear weapons testing at the Nevada Test Site. Yet, unlike in the case of nuclear waste isolation, containment of the radioactivity left from the weapons detonations, or shots, must be left entirely to the geologic formation and cannot be redoubled by engineered barriers such as robust packaging.

This is not a trivial difference, even though the total amount of radioactivity in the approximately 550 to 600 shot holes and tunnels in which weapons have been detonated is equivalent to only a modest fraction of what would be present in a repository. According to the most recent unclassified figures, there were about 50 million curies of radioactivity in shot holes and tunnels as of January 1, 1983, contaminating some 700 million cubic feet of soil and rock in the immediate vicinity of the detonation cavities.

By contrast, a repository for commercial spent fuel would have several billion curies. On the other hand, there is more radioactivity in the weapons detonation cavities than there will be in DOE's Waste Isolation Pilot Plant repository at Carlsbad, New Mexico, for transuranic (or plutonium-contaminated) waste generated in the nuclear weapons program. Furthermore, even today, after years of underground testing, mishaps occasionally happen in the carrying out of the tests themselves. The last significant accidental venting to the atmosphere of an underground shot occurred in the Baneberry event in late 1970. But in the Mighty Oak tunnel shot of May 1986, high levels of radioactivity leaked beyond containment barriers within the tunnel complex; some $32 million worth of normally reusable equipment was lost, and delicate and complex problems of radiological safety confronted re-entry teams seeking to recover test data and to clean up or seal off the mess.

Nonetheless, despite the environmental and occupational safety risks (such as they are) associated with the weapons testing, only a tiny minority of Nevadans have taken part in the political demonstrations against the testing that have been mounted with increasing frequency at the Nevada Test Site. The protests are sustained chiefly by out-of-state religious and secular peace activists who see the weapon tests as contributing significantly to the nuclear arms race. (Since 1980 these tests have numbered from 13 to 19 a year.)

Bryan was no doubt quite right recently when he told a U.S. Senate committee, "Nevadans have been strongly supportive of the national defense efforts at the test site, and [of the continued] testing." One reason the testing is accepted by most Nevadans, and then goes largely unquestioned by them, is that it provides many jobs. DOE's Nevada Operations Office and other associated federal agencies and contractors employ about 8,000 people and spend a total of about $425 million annually on payroll and services. Most of the jobs and services are related to the weapons testing, making this activity a major force in the economy of metropolitan Las Vegas.

All things considered, there is a striking irony in Nevada's acceptance of weapons testing and its total, categorical rejection of waste isolation. Further investigation would very likely show that a repository at Yucca Mountain can certainly be every bit as safe as the weapons testing, and probably safer. Furthermore, waste isolation could serve as an important alternative use or mission for the Nevada Test Site.

Absent such a new mission, there is nothing hypothetical at all about the possibility that over the next few years jobs at the test site will be sharply reduced. The U.S. House of Representatives in April passed a measure to stop all but the lowest-yield tests, and a similar (although slightly more permissive) bill has been pending in the Senate with 33 cosponsors. . . .

The experience with the weapons testing actually bears positively in several ways on use of the test site and adjacent areas for waste isolation.

- First, as was recognized early on in the consideration of the test site for waste isolation, it is an area that must be policed and monitored indefinitely because of the contamination already present. The contamination is not limited to the radioactivity left in the shot holes and tunnels. There are more than 60 surface hot spots, each covering up to four or five acres, which have been fenced off and marked with radiation warning signs. Most are from either atmospheric testing of weapons in the 1950s or from the deliberate destruction of plutonium bombs to test safety systems designed to prevent a nuclear detonation in the event of accidents.
- Second, the testing experience suggests that the test site and the Nevada desert is in fact a suitable and forgiving area for waste isolation. Data from the extensive network of test wells monitored by the U.S. Environmental Protection Agency in and around the test site indicate that nearly all of the residual radioactivity from the underground tests has remained in or very near the detonation cavities. Even radioactive hydrogen, or tritium, which like stable hydrogen can join with oxygen to become part of water molecules and move with the groundwater, has not shown up at levels in excess of the drinking water standard at test wells, five of which are within a quarter-mile or so of the nearest shot holes.

In addition, tunnels excavated in tuff have been found to withstand shock from even very large tests conducted nearby. In particular, tunnel complexes in Rainier Mesa were undamaged by shots, some as large as 1.3 megatons, detonated 10 to 15 miles away on Pahute

Mesa. The ground motion at Rainier Mesa was in some instances similar to that associated with a major earthquake.

•Finally, despite all the talk by Bryan and others that a nuclear "dump" at Yucca Mountain would create a stigma, the experience with the weapons tests strongly suggests that there would be no such effect. The Las Vegas gambling and tourism industry has grown apace with the weapons testing for more than three decades. After the tests went underground in the early 1960s, the larger ones commonly produced noticeable effects in Las Vegas, sometimes causing high-rise buildings to sway, windows to break, swimming pools to crack, and so on, giving rise to thousands of complaints of damages. But if any of this has scared off tourists, it is not the least bit evident.

Bryan and his aides contend that a waste repository would alarm the public much more than weapons testing because for some 30 years commercial spent fuel and defense high-level waste would be constantly arriving, with perhaps a dozen or more such shipments on Nevada highways each day and all came by truck (rail transport is in fact expected to be emphasized). But it is relevant to note that Nevadans have for many years quietly accepted the presence of one of the world's largest ammunition plants at Hawthorne, Nevada, about 100 miles southeast of Reno. Last year about 5,000 truck shipments (or about 16 per working day) carrying bombs or other munitions—placarded "Explosive" or "Dangerous"—were rolling in and out of Hawthorne, some of them passing through Las Vegas. The Hawthorne depot's traffic manager recalls not a single complaint or protest about these shipments from anyone over the past decade. In today's climate nuclear waste cargoes would no doubt provoke protests, but Bryan could explain to his constituents that a spent fuel assembly, unlike a 1,000-pound bomb, cannot explode.

In sum, further technical studies probably will show that waste isolation can be accomplished safely at Yucca Mountain, and there is no reason to believe that a repository would frighten tourists away. Moreover, besides creating (directly and indirectly) up to several thousand jobs at periods of peak activity, nuclear waste storage and isolation at Yucca Mountain could be accompanied with large cash payments and other concessions for Nevada. . . .

THE BULLFROG CHARADE

The Nevada legislature's creation of the now infamous Bullfrog County shows the keen interest of many legislators, especially those from the Las Vegas area, in turning the Yucca Mountain project into a bonanza if, as most are quick to add, a waste repository is "shoved down our throat." By this bill, which was enacted at 3:30 in the morning just before the legislature adjourned on June 18, the legislature created a new county, named Bullfrog for an old mining district; it consists solely of that part of Nye County lying within the boundaries of the Yucca Mountain project. It would be in effect a special taxing district with no inhabitants and with the "county government" located at the state capital in Carson City, 270 miles away. Bullfrog County is but a device to permit the state to collect and distribute funds received from Washington in "grants equal to taxes" (GETT) under the Nuclear Waste Policy Act, funds which otherwise would have gone principally to Nye county. . . .

A special irony about the Bullfrog legislation, which Governor Bryan signed, was that it deals with Nye County in the same high-handed, patronizing way the federal government is accused of treating Nevada. The Bullfrog act contains no assurances of fair treatment for Nye and thus the county's officials have been left fearful (though the governor indicates their fears are groundless) that Nye will not get the GETT money properly due a county that is host to a repository. . . .

TOUGH BARGAINING CALLED FOR

To judge from the Bullfrog saga and the impressions one gets from talking with state legislators, the only politically acceptable way to consider the economic benefits that a repository could bring is to treat the matter as a contingency—in case, heaven forbid, Nevada is forced to take a repository. The reality is, however, that this contingency is one that is by no means unlikely to arise. The 34 states that have, or will have, operating nuclear power reactors cannot escape the fact that if there is no geologic repository, the reactor sites could become permanent spent fuel storage sites. Moreover, the fact that Yucca Mountain clearly appears to be the preferred site is not likely to be lost on the congressional delegations from these spent-fuel-generating states.

But whatever the contingencies, Nevada could have much to gain from tough bargaining. First, as for economic benefits, the state could seek not only large annual grants to help make ends meet but other concessions as well, such as transfer of sizable tracts of federal land to the state (85 percent of Nevada is in federal ownership). Second, as for project objectives, the state could—and in fact should—insist that the project be recast in a more experimental mode . . . and made to include new features that could increase public confidence in secure waste isolation and enhance the project's overall significance, nationally and internationally.

Robust waste canisters of a durability greater than is now required could be one such feature. Another could be an assured option for ready retrieval of waste canisters, either in case problems arise with the system of waste isolation or in case, on some distant tomorrow, the reprocessing of spent fuel for recovery of its plutonium makes more economic and political sense than it does today. Perfecting waste packaging and retrieval systems, together with the ongoing geologic investigation, could easily allow Yucca Mountain and the test site to become an important international center of waste isolation research.

Nevada would even do well to seek to have interim surface storage of spent fuel located at the Nevada Test Site instead of at a site in the eastern United States, such as the one proposed by DOE at Oak Ridge, Tennessee, in the "monitored retrievable storage" plan that the department has submitted to Congress. Such storage, besides easing the utilities' spent fuel storage problems and providing more jobs and contracts for Nevadans, would justify long-continuing federal benefits to Nevada even if the Yucca Mountain site were to prove unlicensable for a permanent repository. Equally important, in the event of a ban on nearly all underground weapons testing, establishing the spent fuel storage facility could be significant in the transition of the test site to a mission associated with peaceful uses of nuclear energy. The payoff for Nevada and the nation from farseeing and vigorous state leadership could be very large indeed.

For now, of course, the talk continues of the stigma that would come with a "nuclear dump." But this stigma, should it have any reality at all, would be mostly the product of the state officials' own insistence on treating as a curse something that would in fact be far more benign than the nuclear testing that has been tolerated and encouraged for better than three decades.

One is reminded of the parable of the wise man who was approached by a trickster who, motioning to his tightly clasped hands, said, "Sir, I hold in my hands a bird. Is the bird alive or dead?" If the wise man said the bird was dead, the trickster would open his hands and let the bird fly away; if he said it was alive, the trickster would crush the bird and kill it. But the wise man would not be thus fooled. He said, "The life of the bird is in your hands."

So it is too with Nevada, the test site, and Yucca Mountain. Governor Bryan and his colleagues in the state government can make a stigma or they can make a new industry and encourage a new and peaceful use for the Nevada Test Site.

NO

Charles R. Malone

THE YUCCA MOUNTAIN PROJECT: STORAGE PROBLEMS OF HIGH-LEVEL RADIOACTIVE WASTES

The U.S. Department of Energy has embarked on the unprecedented task of siting and developing a geologic repository that must safely contain high-level nuclear waste for at least 10,000 years. The site chosen for this endeavor is Yucca Mountain in southwestern Nevada, adjacent to the Nevada Test Site where the first nuclear bomb was tested. Plans for characterizing the site and determining its suitability have been prepared (*1*) based on probabilistic containment requirements that limit environmental releases of radionuclides (*2*). For Yucca Mountain to be an acceptable site for the repository there must be reasonable assurances that future geologic and hydrologic changes at the site will not result in transport of radionuclides to the accessible environment during the first 10,000 years following closure of the repository (*3, 4*). In order to meet such standards of performance, the site must become understood with a degree of confidence never before achieved. Natural geologic and hydrologic processes must be sufficiently understood to allow reliable predictions of radionuclide release rates and routes of transport to the accessible environment over thousands of years.

Characterizing a complex, heterogeneous environment like Yucca Mountain and designing an engineered system that complements the natural setting and complies with the 10,000-year performance standard implies the need to accept a degree of scientific and technical uncertainty. The environmental sciences lack adequate means for analyzing and quantifying the uncertainty associated with repository performance (*5–7*). These factors affect the efficacy of the Yucca Mountain project and limit the ability of science and technology to ensure the success of the repository program. For example, the regulations for siting and licensing a repository specify various criteria pertaining to potentially adverse site conditions and the long-term performance of a repository. Among them are requirements that concern the potential for volcano-tectonic and faulting processes to occur. However,

From Charles R. Malone, "The Yucca Mountain Project: Storage Problems of High-Level Radioactive Wastes," *Environmental Science & Technology*, vol. 23, no. 12 (1989), pp. 1452-1453.

uncertainties exist with accepted methods for predicting earthquakes and volcanic intrusions (*8, 9*) that must be resolved to meet the regulations. Available geophysical techniques for acquiring data from deep underground at Yucca Mountain have proven unsuccessful because of the complexity of the site, and new methods must be developed before data acceptable for repository licensing can be obtained (*10*).

Regulations also exist pertaining to repository performance criteria and the potential for adverse long-term geohydrologic changes that would transport radionuclides to the accessible environment. Because groundwater is one of the most likely routes for radioactive wastes to take from a geologic repository, it is essential that accurate and reliable geohydrologic models be developed (*5*). The heterogeneous, complex nature of the geohydrological system at Yucca Mountain and the absence of reliable methods for characterizing unsaturated zones in fractured media pose recalcitrant problems (*6, 10–12*).

Predictive geohydrologic modeling of the unsaturated zone is characterized by uncertainty regarding such questions as the applicability of Darcy's law to groundwater movement in unsaturated fractured rock (*5, 6, 11, 12*). These issues are critical for determining the suitability of Yucca Mountain for repository development and for assessing performance capabilities in accordance with regulatory requirements (2–4). Successfully carrying out performance assessment will involve selecting a range of scenarios within which must lie what really will occur at Yucca Mountain over the next 10,000 years. Studying and correctly measuring the proper parameters for predicting future conditions implies knowledge of existing environmental conditions and the availability of adequate scientific and technical tools for gathering reliable information.

In view of the known complexity of the Yucca Mountain environment, the limited ability to interrogate such a site nondestructively, and the lack of validated models for predicting geologic and hydrologic processes over 10,000 years, it is understandable that incorrect specification of repository performance scenarios is likely to be the most significant source of error in trying to assess how such a system will behave (*7*).

A performance assessment program has yet to be developed for the Yucca Mountain Project, so it is not known what data and information ultimately may be needed. Thus, there are no assurances that the ongoing site characterization program (*1*) will provide the appropriate information for the analyses required for addressing regulatory requirements and performance standards (*2, 3*).

In the face of these difficulties it appears that the limits of environmental science have been exceeded by the goals set for the nation's radioactive waste disposal program. There is so much uncertainty with respect to safely isolating nuclear waste for thousands of years that it seems unlikely that reasonable assurances can be provided that a nuclear waste repository at Yucca Mountain will meet health and safety standards.

For this reason it is prudent that the debate over national policy for dealing with high-level nuclear wastes be reopened and that alternatives to the present course of action receive further consideration. If this occurs, the state of the art and capabilities of environmental science and technology must be primary considerations so that a practicable solution to the problem can be reached.

REFERENCES

1. "Site Characterization Plan"; U.S. Department of Energy. U.S. Government Printing Office; Washington, DC, 1988; DOE/RW-0199.

2. "Environmental Standards for the Management of Spent Nuclear Fuel, High-Level, and Transuranic Radioactive Wastes"; U.S. Environmental Protection Agency, U.S. Government Printing Office; Washington, DC, September 19, 1985; 40 CFR, Part 191.

3. "Disposal of High-Level Radioactive Wastes in Geologic Repositories"; U.S. Nuclear Regulatory Commission, U.S. Government Printing Office: Washington, DC, July 31, 1986; 10 CFR, Part 60.

4. "Nuclear Waste Policy Act of 1982: General Guidelines for the Recommendation of Sites for Nuclear Waste Repositories"; U.S. Department of Energy. U.S. Government Printing Office: Washington, DC, December 6, 1984; 10 CFR, Part 960.

5. Djerrari, A. M.; Nguyen, V. V. In *Geostatistical, Sensitivity, and Uncertainty Methods for Ground-Water Flow and Radionuclide Transport Modeling;* Buxton, B. E., Ed.; Battelle Press: Columbus, OH, 1989; pp. 559–82.

6. Freeze, R. A. et al. In *Geostatistical, Sensitivity, and Uncertainty Methods for Ground-Water Flow and Radionuclide Transport Modeling;* Buxton, B. E., Ed.; Battelle Press: Columbus, OH, 1989; pp. 231–60.

7. Brandstetter, A.; Buxton, B. E. In *Geostatistical, Sensitivity, and Uncertainty Methods for Ground-Water Flow and Radionuclide Transport Modeling;* Buxton, B. E., Ed.; Battelle Press: Columbus, OH, 1989; pp. 89–110.

8. Crowe, B. M. In *Studies in Geophysics: Active Tectonics;* National Academy Press: Washington, DC, 1986; pp. 247–60.

9. National Academy of Sciences. *Probabilistic Seismic Hazard Analyses;* National Academy Press: Washington, DC, 1988.

10. Jones, G. M. et al. "Survey of Geophysical Techniques for Site Characterization in Basalt, Salt, and Tuff"; U.S. Nuclear Regulatory Commission: Washington, DC, 1987; NUREG/CR-4957.

11. National Water Works Association. *Proceedings of the NWWA Conference on Characterization and Monitoring of the Vadose (Unsaturated) Zone.* National Water Works Association: Dublin, OH, 1985; pp. 396–563.

12. National Academy of Sciences. *Ground-Water Models: Scientific and Regulatory Applications;* National Academy Press: Washington, DC, 1989.

POSTSCRIPT

Nuclear Waste: Is Yucca Mountain an Appropriate Site for Nuclear Waste Disposal?

Despite the present dry condition of the Yucca Mountain site, some geologists have questions about the geological history and structure of the area and the certainty that flooding will not occur. Some geologists have interpreted certain deposits as indicative of past flooding, which could reoccur and result in flooding of the waste storage chamber in the future. It is difficult for a non-expert to judge whether Carter is correct that the planned investigation will lay this and other concerns to rest. Likewise, it is difficult to know whether Malone is right in claiming that scientists have not done the basic research necessary to make the needed technical evaluations. If the mandated studies yield negative or inconclusive results, the timetable for resolving the nuclear waste issue will be set back, since Congress has focused on only this one site.

Carter cites Nevada's governor Richard Bryan as a key opponent of the decision made by Congress. For an elaboration of his arguments, read Bryan's article "The Politics and Promises of Nuclear Waste Disposal: The View From Nevada," *Environment* (October 1987).

Luther Carter's assessment of the entire nuclear waste controversy is elaborated in his book *Nuclear Imperatives and Public Trust* (Resources for the Future, 1987). Fred Shapiro's critical appraisal of the congressional decision is the subject of "Yucca Mountain," *The New Yorker* (May 23, 1988).

A very pessimistic assessment of the effort to resolve the low-level waste problem is contained in "The Deadliest Garbage of All," by Susan Q. Stranahan, *Science Digest* (April 1986).

On the optimistic side is a report by researchers George Wicks and Dennis Bickford entitled "Doing Something About High-Level Nuclear Waste," *Technology Review* (November/December 1989). They describe the process of classifying nuclear waste, scheduled to start in the United States in 1992, which they claim is the key to permanent safe disposal.

A scandal has recently erupted concerning nuclear waste issues related to the U.S. nuclear weapons program. For an account of this problem, which predates the publicity given to recent revelations, see William F. Lawless's article "Problems with Military Nuclear Waste," *Bulletin of the Atomic Scientists* (November 1985). In "Hanford Cleanup: Explosive Solution," published in the October 1990 issue of the same journal, environmental researchers Scott Saleska and Arjun Makhijani present a frightening assessment of ongoing nuclear waste storage and treatment practices at the large military nuclear reservation in Hanford, Washington.

New York State Department of Commerce

PART 4

The Environment and the Future

In addition to the many serious environmental problems of today, there are several potential future crises that might be averted or diminished if preventive measures are taken now. The destruction of Brazil's rain forest has emerged as a serious current problem with consequences for the future because rain forests play a life-sustaining role in the biosphere. Two other problems—pollution of the lower atmosphere by gases that could raise the average global temperature, and pollution of the stratosphere by gases that may destroy much of the protective ozone layer—are also international in terms of both cause and effect. The continued abuse of natural resources could also disastrously impact the world. The resolution of most of the problems debated in this section will require unprecedented cooperation among the nations of the world.

Is Brazil Serious About Preserving Its Environment?

Does Global Warming Require Immediate Action?

Is the Montreal Protocol Adequate for Solving the Ozone Depletion Problem?

Are Abundant Resources and an Improved Environment Likely Future Prospects for the World's People?

ISSUE 15

Is Brazil Serious About Preserving Its Environment?

YES: Nira Broner Worcman, from "Brazil's Thriving Environmental Movement," *Technology Review* (October 1990)

NO: Philip M. Fearnside, from "Deforestation in Brazilian Amazonia: The Rates and Causes of Forest Destruction," *The Ecologist* (November/December 1989)

ISSUE SUMMARY

YES: Brazilian environmental journalist Nira Broner Worcman is cautiously optimistic that new governmental policies that are a response to national environmental activism and global concern may halt Amazonian rain forest destruction.
NO: Brazilian ecologist Philip M. Fearnside fears that failure to deal with the economic and social root causes of deforestation will result in total destruction of the rain forests.

The area of the Earth that is covered by forests has decreased by approximately 33 percent as a result of human activity. The clearing of land for crop production or animal grazing, the harvesting of timber, and the gathering of wood for fuel have been the principal human activities that have shrunk the world's forests.

Rapid regional decimation of tree cover is not a new phenomenon. By the end of the eighteenth century, France had cleared almost 85 percent of its forested land over a period of less than two centuries. Other contemporary industrialized countries, including the United States, have engaged in wholesale deforestation. Until recently, however, these practices have had relatively minor deleterious effects on human welfare.

Today, hundreds of millions of poor people in South and Central America, Africa, and Asia find their principal source of fuel for cooking and home heating threatened by the accelerating clear-cutting of tropical forests. Natives of the forested areas are being uprooted and driven from their only available source of sustenance. The economic forces that promote this plague are large lumbering, cattle ranching, and land speculation entrepreneurs. The beneficiaries include foreign investors as well as local elites with strong ties to the entrenched minority who maintain political control in much of the developing world.

The problem is exacerbated by the displaced forest dwellers who contribute to Third World urbanization. In the cities charcoal, which is easier to transport, replaces wood as the predominant fuel. In the process of converting wood to charcoal, half of the fuel value of the wood is lost. Thus the movement of rural villagers to city slums results in a doubling of per capita fuel wood needs.

The recent recognition that as much as one-third of the annual contribution to the increase in atmospheric carbon dioxide (see the issue on global warming) comes from deforestation has greatly heightened worldwide concern about forest depletion. (Global warming is a result of the fact that the cutting and combustion of trees release carbon dioxide that would only be balanced by an equal number of new trees removing that same amount of carbon dioxide as they grew.) Since global warming is an issue of universal concern, there is hope among environmentalists that it will serve to mobilize the enormous number of people needed to save the forests.

Tropical forests supply many useful commercial products and are the source of a wide variety of chemicals, including natural products that are used in the pharmaceutical industry. Among the serious consequences of rain forest destruction would be the loss of a principal source of organic chemicals used in medical research.

Brazil has recently responded to the growing chorus of pleas to halt the destruction of its Amazonian rain forests by modifying some government policies that contributed to the economic climate supporting the tree-cutting entrepreneurs. Nira Broner Worcman, a journalist from Brazil, describes the little-known environmental movement within that country that has helped to raise consciousness and to pressure political leaders into adopting policies that she hopes will prevent ecological suicide. Philip M. Fearnside is a professor of ecology in the Amazon. He describes the factors that continue to promote forest devastation and he argues that present government action will be ineffective because it deals only with the symptoms rather than the root causes of the problem.

YES

Nira Broner Worcman

BRAZIL'S THRIVING ENVIRONMENTAL MOVEMENT

The impression has been widespread in developed nations that while the world fights to preserve Brazil's natural environment, Brazil simply doesn't want to be preserved. Granted, there is good reason for international concern about the South American environment. The burning of the Amazonian rainforest, which represents 30 percent of the world's tropical forest, contributes to the risk of global warming by releasing vast amounts of carbon dioxide. It also endangers numerous animal and plant species that are important to medicine, biotechnology, and agriculture. It has been estimated that the rainforests house 40 to 50 percent of the species on earth, some of them valuable sources of pharmaceuticals for glaucoma, cancer, and perhaps even AIDS.

Despite the impressions of outsiders, however, Brazilian ecologists started their own fight to preserve the rainforests two decades ago. In the mid-1970s ecologist and agronomist Jose Lutzenberger, now the Brazilian environmental secretary, started the first protests against World Bank financing of gigantic mining, hydroelectric, and agricultural projects in the Amazon. "When we noticed that a significant part of the forest devastation was caused by multinational companies and financed by the multilateral development banks with money from the First World taxpayers, we realized that it was necessary to make them aware of the problem," Lutzenberger says. "If today the First World governments, Margaret Thatcher, Francois Mitterrand, Helmut Kohl, and even George Bush show their concern with the destruction of the Brazilian rainforest, it is because the environmentalists from those countries reacted to the information we sent them."

An important victory came in 1987, when environmentalist Chico Mendes, head of a group of indigenous rubber tappers, persuaded the Inter-American Development Bank (IDB) to suspend a $150 million loan for paving highway BR-364 through the southwestern Amazon. The IDB and other development agencies also began in the late 1980s to demand assessments of environmental impact more consistently before loaning money to projects in Brazil.

These actions play an important role in measures adopted by the Brazilian government itself. The Our Nature Program, launched two years ago to halt deforestation in the Amazon, and IBAMA, the new Brazilian Institute of the Environment and Renewable Natural Resources, were both products of such international pressure, according to former IBAMA president Fernando Cesar Mesquita.

Still, there are also strong forces pulling in the other direction. Mendes himself was slain in 1988 for his efforts to slow the clearing of the rainforest, and the funds to complete highway BR-364 could still come from other sources. The Japanese, for instance, have shown an interest in funding the extension of highway BR-364 to Peru, allowing Pacific access to valuable Amazonian timber. Mesquita says the road must be constructed for its economic, political, and strategic importance to Brazil as the country's only connection with the Pacific Ocean. "We can build this road without damaging the environment, constructing it as a corridor passing through national reserves," he says.

But this is not what Lutzenberger has in mind. He accepted the new post of environmental secretary after President Fernando Collor de Melo agreed to halt plans for the highway. Collor also agreed to respect the rights of the indigenous people of the forest and to cut all economic incentives to destructive projects in the Amazon.

Choosing Lutzenberger as environmental secretary might be a sign that the Brazilian government is getting serious about preserving its environment, even if that means going against powerful economic interests. "We will have to face strong pressure from several economic groups that are interested in keeping the devastation going," Lutzenberger said in a visit to Washington last April.

Lutzenberger's appointment was at the very least a smart political move. A prominent ecologist in Brazil, Lutzenberger has won recognition worldwide for his work to preserve the environment, including his battle against pesticide use. In 1988 the Swedish government bestowed on him the Right Livelihood Award, considered the Nobel Prize for environmentalism.

For many environmentalists, however, it is odd to see Lutzenberger in a governmental position, since he has always assumed an opposing posture, calling the previous administration *desgoverno,* meaning nongovernment. Some are skeptical about his knowledge of the issues and his ability to implement environmental policy in Brazil, but they are reluctant to speak out, since he is part of their movement. They are doubtful about how much leeway he will have to reform old policies, and they believe that he will not remain in the government for very long.

President Collor has said that he considers the environmental question a priority on the national agenda. If Brazil does not face its environmental problems, he says, it will have trouble getting the external loans to implement new economic development projects. The money for environmental programs at least, is easy to borrow. A $117 million World Bank loan will pay three-quarters of the cost of environmental protection programs in Brazil over the next three years.

HOW BRAZIL'S ENVIRONMENTAL MOVEMENT EMERGED

The environment has long been a political issue in Brazil, serving for more than a decade as one of the only channels open for civilian activism. From 1964 to

1984 Brazil was ruled by a military dictatorship, which, like many regimes struggling to industrialize, dismissed ecological concerns. It focused instead on economic development at any cost, usually at the expense of the environment.

In 1972, during the first United Nations Environmental Conference in Stockholm, Brazilian Minister of the Interior Costa Cavalcanti made a remark that epitomized this attitude. He stated that Brazil would not change its economic policy, based on growing as quickly as possible. Gigantic and costly construction projects, including hydroelectric plants and roads, were the centerpieces of that model. Large loans from multilateral development agencies like the World Bank, whose policy was founded on the idea that high standards of living follow from industrialization, supported such an approach. . . .

THE OFFICIAL RESPONSE

Brazilian participation in the 1972 U.N. conference inspired at least a small official gesture toward environmentalism. In 1974 the government created SEMA—the Brazilian National Environment Agency—and appointed ecologist Paulo Nogueira Neto to head it.

In Nogueira Neto's opinion, the conference marked the emergence of environmental concern not only in Brazil but worldwide. Yet SEMA, a part of the Ministry of the Interior, was given virtually no resources to address that concern. "We had three employees and only two rooms to work in," says Nogueira Neto. He held the position for 12 years until 1986, working first with the military and then with the civilian government that succeeded it.

Environmentalists in Brazil say that Nogueira Neto's long coexistence with different regimes was possible only because SEMA was a puppet agency, intended to mask the government's inactivity on environmental problems. "SEMA did no good," says ecologist Azis Ab'Saber. "Nobody looked after the environment."

SEMA did not have the same political support that IBAMA has today for implementing environmental policy, but Nogueira Neto disagrees that it was a puppet. "I could keep my position in both regimes because I had a technical function, without any political or ideological guidance," he says. He was responsible for creating the Ecological Station Program, a project aimed at maintaining Brazil's extensive biological diversity. Twenty-one ecological stations have been established, monitoring a total area of some 7 million acres. At least 90 percent of the natural areas surrounding the ecological stations is to be left undisturbed, and up to 10 percent is to be dedicated to research on burning and other forms of human interference in natural systems.

The Anavilhana station, for instance, created in 1981 near the city of Manaus in Amazonas, covers 865,000 acres of seasonally flooded lowland forest, encompassing hundreds of river islands, many lakes, part of the Amazon rainforest, and marshy forests with many palms. At this station Michael Gouding, a researcher from the National Institute for Research in the Amazon, discovered that a species of fish called Tambaqui feeds itself with the fruits of trees when the area is flooded, spreading the seeds and aiding the trees' reproductive process in the same way that birds do. In Raso de Catarina, in the northeast state of Bahia, National Museum researcher Helmut Sick discovered a species of parrot that had been considered extinct for 50 years. Now this species is being protected.

Unfortunately, the stations and the research carried on there were practically abandoned in 1984, when SEMA became part of the Ministry of Habitation and the space for environmental issues on the Sarney administration's agenda began to shrink, culminating in Nogueria Neto's resignation in 1986.

Brazil's national forests and parks, including the ecological stations and other types of conservation areas, cover only 2.5 percent of the territory that, by law, ought to be preserved. Several reserves exist only on paper. "Those areas are nobody's land, where everything can happen, from fire, hunting, timber, and mineral exploitation, to roads and dam construction," says Maria Tereza Jorge Padua, president of the environmental organization Pró-Natureza Foundation, which is working to preserve several conservation areas in Brazil.

CHANGES UNDER DEMOCRATIC RULE

Brazil's transition to democracy in the early 1980s brought environmentalism to the fore, providing new leaders and im portant new laws to address environmental problems. In 1982 the first popular elections for state governments took place, in a process of democratization started by the military rulers in 1979. Several state governments recruited environmentalists to serve in appointed offices. In some cases, as in Sąo Paulo and Mato Grosso do Sul, the heads of the environmental agencies came directly from the environmental movement. Since then, not only has Lutzenberger been appointed to the nation's highest environmental post, but environmentalists have also become state deputies and members of congress.

Environmentalist Fabio Feldmann is president of OIKOS, the Union for the Defense of the Earth. Taking advantage of a 1985 law giving citizens the right to sue anyone damaging the environment, Feldmann's group is suing the entire industrial complex in the polluted city of Cubatao in Sąo Paulo State. In 1986 Feldmann became the first deputy to the Brazilian Congress to be elected on an exclusively environmentalist platform. His election surprised even himself. "We launched our campaign intending to draw attention to environmental problems," he says. "We didn't think we were going to win. There were so few of us that we all fit into a microbus with three seats left over."

The environmental chapter of the new Brazilian constitution, sponsored by Feldmann, will have an impact that lasts long after his term in office. It ensures the right to a balanced environment and imposes on the government and the community the duty to defend and preserve that balance. Under this chapter the government must also foster ecological education at all levels and promote public awareness of the need to preserve the environment.

The constitution declares that the Amazon forest, the Atlantic forest, and the coastal zone are national treasures, assigning responsibility to preserve them to the federal government. It imposes on the government the responsibility of maintaining essential ecological processes and ecosystems and the various species they host, as well as controlling the production and use of substances that may threaten public health and the environment. The new constitution also defines "ecological crime," providing penal sanctions against people and institutions who damage the environment.

Powerful lobbies, including the chemical, nuclear, and mining industries, opposed Feldmann in the fight for the environmental chapter. "Several times during the voting sessions, we had clauses that were mysteriously taken out of the article," he says. But he gained an important ally in the dispute: the Green Front, which united more than 80 members of congress from different parties determined to support the environmental cause in the constitutional assembly.

Constitutional protection for the environment represents a significant advance, but it alone cannot change Brazil. "It seems a paradox that we have such an advanced constitution and such a horrible reality," says Feldmann, who points out that much work remains to be done through ordinary legislation. He is also afraid that the laws will not be obeyed. "We have a tradition in Brazil of not respecting legislation," he says. "The applicability of the constitution will depend on the pressure that the civilian society places on the government."

The Our Nature Program, announced by Sarney when the new constitution was ratified, was intended to bolster enforcement efforts in the Amazon. Among other measures, the program includes an emergency plan to preserve the rainforests through the inspection and control of deforestation. It withdraws fiscal incentives for farming and cattle raising in the forests, strictly regulates the use and production of pesticides in Brazil, and creates a commission to coordinate research on the Amazon as well as a fund to channel donations into environmental projects. Under Our Nature, the rate of deforestation in the Amazon decreased 30 percent in 1989 as compared to 1988, according to Mesquita. (However, satellite data from INPE, the Brazilian Institute for Space Research, show that the rate of deforestation between January 1989 and July 1990 equaled the average rate of deforestation during the 1980s, about 9,250 square miles a year.)

The establishment of IBAMA in February 1989 was also part of Our Nature. IBAMA integrated the functions of the old agency SEMA with the duties of the fishery, rubber, and forestry development institutions in Brazil. To Feldmann, the creation of IBAMA represented "the acceptance of the idea that the management of natural resources is connected to the country's economic activities."

Like SEMA before it, IBAMA has a very small budget, but what the agency lacks in monetary support it makes up for in political resolve. Mesquita, the agency's first head, points out that IBAMA has helped ensure that previously unenforced environmental legislation will be followed. A 1965 code requiring logging companies to replant enough trees to match consumption, for instance, was virtually ignored until recently. "It became part of Our Nature, and with political will, we enforced it," Mesquita says. . . .

PACKAGING THE ENVIRONMENT

Environmentalism has become a popular cause in Brazil. A 1989 poll by the advertising firm Standard, Ogilvy & Mather showed that interest in deforestation among Brazilians was second only to concerns about salaries. Consumer prices, the federal budget, and state intervention in the economy were of lesser interest. Entrepreneurs have capitalized on this concern to market a wide array of new products. Fashion advertisements feature environmentalists such as Fabio Feldmann as models. T-shirts with environmental motifs sell well, and travel agents and environmental organizations are promoting "ecotour" theme travel packages. Large corporations produce luxurious ecology books for their best clients as the traditional New Year's gift.

The number of television programs, news reports, and ads concerning the environment is multiplying. The radio station at the University of Sąo Paulo, for instance, airs a weekly 50-minute program called "Nova Terra" (New Earth) dedicated exclusively to ecology. The Brazilian TV network Manchete took first place in the ratings after the premier this March of the environmentally oriented soap opera "Pantanal," named after the vast freshwater wetland system in the center of Brazil where the soap's action is located. . . .

Brazil now harbors some 1,000 environmental organizations, most of them concerned with local projects. S.O.S. Mata Atlantica, for instance, was founded in 1986 by a group of ecologists to save the remaining 5 percent of the Atlantic rainforest on Brazil's east coast. The group sponsors environmental education programs for teachers, instructs fishers on how to harvest oysters without depleting their supply, and reaches millions of Brazilians through the media with its campaign against deforestation.

INDIGENOUS EFFORTS TO SAVE THE AMAZON

Native people from the Amazon rainforest, like the Indians and the rubber tappers, are also organizing themselves to protect the area where they live. For these groups, preserving the natural environment is a struggle for survival, since the forest is their home and the source of their livelihood.

Indians have suffered the most from environmental damage to the rainforests. Their population, estimated at 1–5 million in the year 1500, has been reduced to about 220,000. The new constitution recognizes the Indians' right to preserve their culture and the land they have traditionally occupied or used in productive activities, but strong economic interests still push the Indians out of their homeland. For example, 50,000 gold miners invaded the mineral-rich Yanomami Indian territory in the northern Amazon last year, devastating much of the Yanomami land and bringing disease and death to the tribe. After a judicial order evacuated the miners, former president Sarney signed a decree opening 5 percent of the Indian territory to them. President Collor later reversed Sarney's decision and started an operation to destroy the 100 airstrips that were the miners' only access to the area. However, this operation was not effective. Federal attorneys who visited the area said in July that the exploded airstrips had already been reconstructed and that new ones were being built. The Yanomami Indians, they said, were in their terminal stage, decimated by malaria and other illnesses brought by the miners.

"For 400 years we have fled, but now we don't have anywhere to go," says Ailton Krenac, coordinator of the Union of Indigenous Nations of Brazil (UNI). "When almost the whole world is worried about the preservation of the rainforest, I believe that our community, which has always lived in harmony with nature, has a big contribution to make." UNI is working to identify sustainable economic activities that would enable Indian communities to participate directly in the economy without destroying the environment and their traditional way of life. Last year, UNI created the Indian Research Center for Resource Management in Goias state in the center of Brazil. There, Indians are being trained in such skills as identifying local fruits that can become marketable crops.

Rubber tappers, who banded together in the National Council of Rubber Tappers in 1985, have also lost much from the deforestation of the Amazon, including their former leader Chico Mendes. With the help of the Institute for Amazonian Studies, an environmental organization in Paraná state, the council has created and won governmental approval for the concept of extractive reserves. Such areas are set aside for extracting and processing rubber as well as for collecting nuts and fruits. Extractive reserves are a sustainable development alternative for the Amazon, since they do not cause deforestation. Several studies have shown that extractive activities can be more profitable than cattle raising or logging in the rainforest. Since January, 9.4 million acres have been set aside as extractive reserves.

OLD PROBLEMS, NEW THINKING

Despite some victories, the struggle in Brazil continues. The Amazon is still burning, destructive projects are still carried on, and environmentalists, Indians, and rubber tappers alike must still fight for the rights guaranteed to them by law.

Some Brazilians think the environmental movement is largely ineffective. Economist Henrique Rattner, from the University of Sąo Paulo, believes the movement sidesteps the political and economic factors that are the real cause of environmental abuse. The majority of environmental problems, he says, stem from the capitalist model of development, aimed at maximizing corporate profits and transferring the costs to society. The developed countries also have environmental problems, he points out, but since their societies are more mature and organized, they can take more efficient action to protect the environment.

Feldmann believes the environmental movement in Brazil has to adopt a less conservationist vision and start studying new systems of development. These should be based on democratized decision making, appropriate technologies, a more equal distribution of wealth, and a stronger domestic market to attend to the population's basic needs. Secretary of the Environment Lutzenberger says it is necessary to rethink the basic ideology of industrial society, which conceives of the earth only as a resource to keep society growing. The earth is a living organism, Lutzenberger asserts, and if humanity doesn't learn to live in harmony with nature, we will die along with it.

In response to the Brazilian government's invitation, the second United Nations Environmental Conference will take place in Brazil in 1992. The invitation could be a sign of Brazil's readiness to cooperate in a global effort to preserve the environment, which would be a step forward from the first conference in 1972. Ailton Krenac, from UNI, remains cautiously optimistic. "I think that humans have a very strong survival instinct. I don't think that we are going to destroy ourselves. But I want us to stop, and do it quickly, because what we are doing is leading toward suicide."

NO

Philip M. Fearnside

DEFORESTATION IN BRAZILIAN AMAZONIA: THE RATES AND CAUSES OF FOREST DESTRUCTION

The seemingly endless expanse of trees which covers most of Amazonia cannot delay the destruction of the forest by more than a brief moment in historical terms. It is of little importance whether 20 or 60 years pass before we come to the last tree. What is vital is whether or not future generations inherit a world with an Amazon forest.

Available figures indicate an apparently exponential rate of deforestation in the Brazilian states of Rondônia and Mato Groso. In other states, the increase may not be exponential; it is nevertheless very rapid. The deforestation rate in Rondônia declined slightly after 1985, partly as a result of migrants moving on to more distant frontiers in Acre and Roraima, and partly from a decrease in the number of migrants entering Rondônia from Mato Grosso. The increasing movement of migrants from Rondônia or Roraima—one of the areas with no recent data available—means deforestation there has progressed further than previously thought.

DISCREPANCIES BETWEEN DEFORESTATION ESTIMATES

A Brazilian National Institute for Space Studies (INPE) study of the areas burning in 1987 using data from a US National Oceanic and Atmospheric Administration (NOAA) satellite indicated that 204,000 square kilometres (20 million hectares) were burned in an area roughly corresponding to the Legal Amazon, of which 80,000 square kilometres (8 million hectares) were in the portion of the area classified as dense forest, the rest of the area being mostly cattle pasture, secondary forest and especially the *cerrado* (savanna) vegetation of Mato Grosso and Goiás. Our calculations, however, indicate that 35,000 square kilometres of dense forest were lost in 1987. The discrepancy between these and INPE's figures is probably in large part due to differences in assessing the degree to which the heat from the huge fires affected the sensitivity of the NOAA satellite.

From Philip M. Fearnside, "Deforestation in Brazilian Amazonia: The Rates and Causes of Forest Destruction," *The Ecologist,* vol. 19, no. 6 (November/December 1989). Reprinted by permission of *The Ecologist,* Worthyvale Manor, Camelford Cornwall, England. Notes omitted. *The Ecologist* is distributed in North America by The MIT Press, 55 Hayward St., Cambridge, MA 02142. Subscriptions are $30 for individuals, $65 for institutions, and $25 for students.

A World Bank estimate of deforestation for 1988 concluded that 12 per cent of Brazilian Amazonia had been cleared by that year, a figure four per cent higher than my own estimate. However, both the World Bank estimate and mine lead to the same conclusion: the deforested area of Brazil's Legal Amazon is still relatively small, but it is expanding explosively.

THE MOTOR OF DEFORESTATION

The process of deforestation in Amazonia is driven by both the expansion of already cleared areas and the appearance of new centres of deforestation. The formation of these new centres has been strongly influenced by governmental decisions over the past decades. The construction of the Belèm-Brasília Highway (BR-010) in 1960, its improvement for year-round traffic in 1967, and its paving in 1974, were significant milestones in creating the largest nucleus of deforestation in Amazonia. Within this nucleus, the area deforested in southern Pará and northern Mato Grosso has enlarged significantly in recent years.

The construction of the Cuiabá-Porto Velho Highway (BR-364) in 1967 initiated another focus for deforestation, and its paving in 1984, with financing from the World Bank, has allowed a wave of colonization into western Amazonia. The paving of the BR-364 from Porto Velho (Rondônia) to Rio Branco (Acre) began in 1986, with financing from the Interamerican Development Bank. The disbursement of funds was suspended because of public concern in North America and Europe over the project's potential environmental impacts, but was resumed in October 1988 when the Brazilian Government announced its *Nossa Natureza* ('Our Nature') programme, which established a series of committees and suspended for ninety days the export of logs and the approval of new ranching incentives. Opening Acre to rapid settlement can be expected to play a key role in accelerating deforestation throughout Amazonia. Discovery of oil and gas fields in the Juruá and Urucú River valleys has added to the pressure for road construction in western Amazonas, which could become the next destination for the influx of migrants no longer finding land in Rondônia and Acre.

INTERNAL MIGRATION

Deforestation has been indirectly stimulated by the government through programmes to attract new migrants from other parts of the country, along with the establishment of settlements and the improvement of access roads. These programmes have multiplied as a result of the increase in administrative units in Amazonia and the elevation of 'territories' to the status of 'states.' The proliferation of new political units results from interior areas of the Amazon almost always lending their support to incumbent governments, making it advantageous for any party in power to increase the political representation of these areas. Because the principal criterion for creating new territories and states is an increase in population, local politicians have been keen to attract colonists. In the early 1980s, for example, the governor of Rondônia launched a national media campaign to promote the "fertile land" in the region (which, in reality, represents under ten per cent of the area, and is mostly already occupied). The campaign was strongest just before the territory of Rondônia became a state in 1982.

In 1983, the Government of Roraima claimed in magazine advertisements that "thanks to its very rapid growth in the past four years, Roraima is almost ready to become the twenty-fourth state of Brazil . . . this dizzying expansion is due to the policy of attracting colonists. In four years—1979 to today—the government of Roraima distributed no less than one million hectares of land to ten thousand families. With this, the population has more than doubled in this period". In recent years the press has reported various government plans to create new federal territories in the southern, central and western parts of Pará and in the southwestern and western portions of Amazonas.

Migration to the Amazon has caused a rate of population growth far above the national average. The population of Brazil's Northern Region grew at 4.9 per cent per year between 1970 and 1980, compared with 2.5 per cent per year in Brazil as a whole and 14.9 per cent in Rondônia, where the deforested area increased at a rate of 37 per cent per year between 1975 and 1980, indicating that deforestation reached rates even higher than population growth. This suggests that the arrival of migrants explains only a part of the phenomenon of explosive deforestation.

Deforestation patterns in 100 hectare lots in the Ouro Preto Integrated Colonization Project (PIC) in Rondônia are being observed as part of the National Institute for Research in the Amazon (INPA)'s 'Carrying Capacity Estimation of Amazonian Agro-Ecosystems Project'. In 18 lots that had only one owner over a ten year period, the cumulative area deforested, on average, increased linearly until the sixth year of occupation, after which it increased much more slowly. The replacement of original colonists by new owners greatly increased the pace of deforestation. A comparison between 32 original colonists and 97 new colonists in the Ouro Preto PIC indicated that in the first four years after purchasing a lot, the new owner deforests, on the average, at an annual rate almost twice that of the original colonist. Therefore, the process of replacing original colonists with new owners, already common both in Rondônia and on the Transamazon Highway contributes to accelerating deforestation in these areas.

LAND USE AND DEFORESTATION

Pasture plays a central role in accelerating deforestation, both for small colonists and for large land owners and speculators. Even in official settlement areas in Rondônia—where almost all of the government effort in agricultural extension credit and advertising is focused on promoting perennial crops—it is pasture that occupies the greatest area.

Real estate speculation is a major force driving deforestation in the Brazilian Amazon and pasture has a central role in this system: besides increasing the value of legalized lots, deforestation followed by planting pasture is the method most often used both by small *posseiros* (squatters), not always thinking of speculation afterwards, and by large *grileiros* (land grabbers), attracted primarily by speculative opportunities. The granting of the right of possession to whoever deforests a piece of land is a centuries-old legal practice in the Brazilian Amazon. Such rights of possession are eventually transformed into full rights of ownership. Pasture represents the easiest way to occupy an extensive area, thus considerably increasing the impact of a small population on deforestation.

Financial incentives also continue to contribute to the deforestation, despite the myth that these incentives ceased to be important following the 1979 decision of the Superintendency for the Development of the Amazon (SUDAM) to stop approving incentives for new cattle projects in parts of the Amazon classified as 'dense forest'. New projects continue to be approved in the areas of 'transition forest', located between the Amazon forest and the *cerrado* (Central Brazilian savannah), contributing to intense deforestation in southern Pará and northern Mato Grosso. Old projects in dense forest continue to receive incentives for deforestation. The policy of denying new incentives to dense forest areas has not even always been followed: according to a member of the Consulting Council of SUDAM's renewable Resources Department, a large cattle project was approved for Acre, completely within the supposedly-protected dense forest zone. The *Nossa Natureza* programme, does little to stem the flow of inducements: the programme only suspends *new* incentives for ranching, and this only for a period of ninety days (later renewed for an additional ninety days). Generous governmental incentives make it possible for many projects to continue clearing pastures even after low beef production would have bankrupted any undertaking whose profits depended on agronomic results.

Deforestation for subsistence production is not a major cause of deforestation at present, but it may become more significant with increased population levels. Because settlement schemes are almost always unsustainable, more deforestation occurs as farmers and ranchers clear new areas when older clearings become exhausted. Increasing the output or the sustainability of agricultural systems would not necessarily decrease deforestation rates, however, because very little clearing now occurring in Brazilian Amazonia is done by traditional farmers who limit their activities when subsistence demands are satisfied.

THE DISTRIBUTION OF COSTS AND BENEFITS

An important factor preventing the control of deforestation is the current distribution of the costs and benefits of forest destruction. The groups and individuals profiting from deforestation are generally not the same ones that pay the resulting environmental, social and financial costs. Profits are often channelled to beneficiaries outside the Amazon region, and while the benefits are concentrated, the costs are widely distributed. Under these conditions, destruction continues to be completely rational in economic terms even if the total cost is much greater than the total benefits. On the other hand, some costs are concentrated, with the benefits accruing to larger, more influential groups, as in the case of land seized from indigenous tribes. Another important factor in the dynamic of deforestation is the monetary nature of the benefits, while the costs, being environmental and human, are more difficult to quantify. The non-monetary costs, unfortunately, are no less real than the monetary ones.

The fact that felling forest brings immediate profits—while many of the costs will only be paid by future generations—is one of the most fundamental aspects of the problem. In the middle of the economic crisis Brazil faced in July 1983,

Rondônia, Mato Grosso and Roraima were the only federative units whose monthly income from the Tax on Circulation of Merchandise (ICM) grew more than inflation. It is probably not a coincidence that the ICM, considered one of the best indices of economic activity, has increased most where deforestation is most explosive. This encouraging picture of immediate profits, however, should be evaluated taking into account the heavy costs following massive deforestation.

The discount rate—the speed future profits and costs have their weights diminished in calculating the net present value of an activity—is a part of the structure of decision-making that renders unviable many potentially renewable systems of resource management. The discount rate is an index that depends on the income that can be potentially earned in alternative investments. No logical connection exists between the discount rate and the biological rates (such as the rate of growth of a tree in the forest) limiting the rate of return from sustained exploitation of biological resources. Rational use of the Amazon forest would generate only a slow return.

POPULATION GROWTH IN AMAZONIA

Human population growth in the Amazon region could also frustrate any policy designed to control deforestation. The flow of new migrants now greatly surpasses the impact of the reproduction rate on population, but in the long term both must reach an equilibrium. The capacity of Amazonia to absorb growing numbers of people is very limited: the social problems motivating the rush of migrants to the region must be solved in the source areas themselves.

The expulsion of small agriculturalists by large landowners, both inside and outside Amazonia, together with the existence of a large landless rural population, makes finding a definitive solution to deforestation extremely difficult. The land tenure system in Amazonia, which is founded on deforestation, would have to be modified to make using the forest possible without clearing it. Since the tradition of legalizing land claims established by deforestation is an important factor in alleviating the impact of extreme social inequalities and the expulsion of rural populations, solutions for these problems would have to be implemented at the same time.

FUTURE PRESSURES FOR DEFORESTATION

Commercial logging, which until recently affected only a relatively small fraction of the region, is rapidly becoming a substantial source of disturbance.

At the moment, world markets for tropical hardwoods are being supplied principally by forest destruction in southeast Asia. Due to their more homogeneous character, the Asian forests are better suited to industrial uses than is the Amazon forest; however, at the present pace of destruction, virtually all of Asia's tropical forests will be destroyed before the end of the century, and, according to tropical wood merchants, commercial volumes of hardwood from Asia could be insignificant by the early 1990s. Large lumber firms are therefore likely to transfer their attention to Amazonia. Heavily logged forests have little chance of recuperation, even without having been clearcut or burnt. More advanced methods using a large number of species to make fuelwood chips, pulp, plywood, particle board or other wood products are likely to increase the devastation caused by commercial logging.

INDUSTRY IN THE AMAZON

Another potential cause of large-scale destruction is the making of charcoal. Wood is now being collected from native forest to supply a pig iron industry in conjunction with the Grande Carajás Programme. Recent statements by the Grande Carajás Interministerial Commission imply a charcoal demand which would consume 1000 square kilometres of surrounding forest per year. The Carajás iron deposit contains 18 billion tonnes of high grade ore—by far the world's largest and sufficient to sustain mining at current rates for at least 250 years. Only a tiny fraction is to be smelted in the area; the potential for expansion of smelting activity is limited only by the amount of available charcoal (that is by the amount of forest to be sacrificed). The first plant began operation on 8 January 1988. No environmental studies were done or impact statement prepared; it has not yet even been decided how much charcoal would be produced from plantations and how much harvested from native forest. Approval of the incentives, construction of the smelters and the beginning of operations all occurred after 23 January 1986, when environmental impact statements became a requirement in Brazil. The pig-iron programme also illustrates several ways potential environmental impacts of major development projects escape the environmental review processes of multilateral lending agencies such as the World Bank, which financed the Carajás railway and mine.

Mining activities are probably set to increase considerably in the future. The invasion of Amerindian reserves spearheaded by freelance gold prospectors (*garimpeiros*) is already a major concern, the continuing officially-condoned assault on Yanomami tribal areas in Roraima being the best known case. Another growing cause of disruption is the construction of military bases with roads and settlements, especially in the Calha Norte Programme. Yet another source of forest loss is hydroelectric development, plans for which imply flooding two per cent of Brazil's Legal Amazon.

ADDRESSING THE ROOT CAUSES

It is clear the range of problems that need to be solved to slow deforestation in the Amazon is enormous. Brazil must face all of these problems both present and future if destruction of the Amazon forest is to be avoided. Root causes of deforestation must be addressed, rather than restricting action to the more superficial symptoms.

Very little now stands in the way of massive increases in deforestation. Limited amounts of capital, especially in Brazil's current economic crisis, can temporarily slow the rate at which deforesters are able to realize their plans, but the deforestation process will run to completion unless fundamental changes are made in the structure of the system underlying clearing.

Many events in the process of Amazonian deforestation are beyond government control. Decrees prohibiting deforestation, such as Law 7511 of 7 July 1986, have minimal effect on land clearing decisions made by farmers or ranchers living many kilometres from major roads and cities, and spread over a region as vast as Amazonia. Some key points in the system, however, are subject to government control. The granting of land titles, with its associated criteria of land 'improvement' through deforestation, is entirely a government activity. The government is also responsible for the programmes granting special loans and tax incentives for agriculture and cattle ranching activities. Above all, only the government builds highways. Were the government to build and improve fewer highways in Amazonia, the vicious cycle of highway construction, population immigration, and deforestation would be broken.

Current deforestation rates indicate that such changes must be made without delay. In the face of such a daunting array of problems, paralysis is frequent: either accepting destruction as inevitable, or considering as useless any action less extreme than a complete restructuring of society. Paralysis, whatever its rationalization, is the most certain path to a future without an Amazon forest.

Acknowledgements
My research in Rondônia is funded by the Science and Technology Component of *Projeto POLONOROESTE*. J. G. Gunn, A. Setzer and S. Wilson provided useful comments on the manuscript. I thank the *Sociedade Brasileira para o Progresso da Ciencia* (SBPC) for permission to use portions of the text translated from *Ciencia Hoje*.

POSTSCRIPT

Is Brazil Serious About Preserving Its Environment?

The worldwide publicity following the murder of Chico Mendes in 1988 did much to popularize the plight of the Amazonian natives. Mendes, the leader of the rubber tappers, had established an international reputation as an effective nonviolent labor organizer and environmentalist before he was killed by ranchers opposed to his forest preservation efforts. "Whose Hands Will Shape the Future of the Amazon's Green Mansions?" is a very vivid and moving description of the struggle of the Brazilian rubber tappers by Michael Parfit, *Smithsonian* (November 1990). Unfortunately Parfit is pessimistic about the chances that the Brazilian government's policy of "extractive reserves" (described by Worcman) will succeed.

Greenpeace is one of the environmental organizations campaigning to save the forests. "Defenders of the Forest," by Andre Carothers in the July/August 1990 issue of the organization's magazine describes the political struggle of the native Penan tribe of Borneo against government-backed logging interests.

Boycotts of tropical forest products, including mahogany and teak, have been organized by some environmental activists. An argument against this strategy is the subject of "Soothing the Conscience—Tropical Forest Exploitation Revisited," by Richard Jagels, *Journal of Forestry* (October 1990).

Increasing the economic value of forest products other than wood has been suggested as a way to save the rain forests. This proposal is explored by Fred Pearce in "Brazil, Where the Ice Cream Comes From," *New Scientist*

(July 7, 1990). Other proposals are included in "The Peruvian Experiment," by Scott Landis, *Wilderness* (Spring 1990) and "Pardo's Law for Saving Tropical Forests" by Richard Pardo, *American Forests* (September/October 1990).

For a general discussion of the forest preservation crisis, see the chapter entitled "Reforesting the Earth," by Sandra Postel and Lori Heise in *State of the World—1988*, a Worldwatch Institute Report (Norton, 1988).

ISSUE 16

Does Global Warming Require Immediate Action?

YES: Claudine Schneider, from "Preventing Climate Change," *Issues in Science and Technology* (Summer 1989)

NO: Ari Patrinos, from "Greenhouse Effect: Should We *Really* Be Concerned?" *USA Today* (a publication of the Society for the Advancement of Education) (September 1990)

ISSUE SUMMARY

YES: Congresswoman Claudine Schneider acknowledges the uncertainties about the extent and consequences of global warming, but she outlines a series of preventative actions that make good sense even if little climate change occurs.
NO: Carbon dioxide research manager Ari Patrinos stresses the need for more research, arguing that there is no need for urgent action other than curtailing greenhouse gas emissions.

The likelihood of a major worldwide meteorological disturbance that would drastically affect climate and thus profoundly alter the balanced web of ecological cycles has provoked speculation by scientists as well as authors of science fiction. A currently prominent theory proposes that the age of dinosaurs was brought to a sudden and spectacular end as a result of just such an event. Supporters of this theory have pointed to evidence that an asteroid or comet may have struck the Earth, causing an enormous dust cloud to blanket the planet for many years, blocking much of the incident solar radiation. The lush tropical forests of the time would have been decimated by the reduction of sunlight. The resulting loss in the dinosaurs' primary food source, coupled with the climate change, could have produced sufficient stress to cause short-term massive extinctions.

Only recently, however, has there been serious scientific concern that significant changes in the average surface temperature or other worldwide meteorological effects could result from intentional or inadvertent human intervention. Although it is generally accepted that the per capita production of energy would have to increase by a factor of 100 or more before a direct, observable atmospheric heating could occur, there are other aspects of present industrial activity that are widely believed to be potential causes of calamitous atmospheric effects.

Many atmospheric scientists now predict our environment may be altered in a dramatic way as the result of the increase in carbon dioxide and other trace gas concentration in the air due to the burning of fossil fuels, destruction of forests, and other agricultural and industrial practices. Trace gases transmit visible sunlight but absorb the infrared radiation emitted by the Earth's surface, much as the glass covering of a greenhouse does. A continued increase in these gases could cause sufficient heat to be trapped in the lower atmosphere, raising the world's average temperature by several degrees over the next 50 to 100 years. This would result in major alterations in weather patterns and perhaps melt part of the polar ice cap. The rising ocean would then submerge vast low-lying coastal areas on all continents.

There is now little doubt that atmospheric carbon dioxide and other trace gas levels are rising. Most meteorologists now agree that there will be some resultant warming; but, due to the complexity of the many interacting phenomena that affect climatological patterns, there is much uncertainty on how large the temperature increases will be and whether they will actually be sufficient to cause the apocalyptic effects that some predict.

During the summer of 1988, unusually hot weather was experienced in many regions of the United States. This led to public speculation, and some by climatologists and meteorologists, that we were experiencing the beginning of the effects of the predicted greenhouse warming. These conjectures, and the scientific data offered to support them, are disputed by other scientists. One positive effect of this much publicized debate is that much more attention is now being paid to the causes and potential consequences of global warming, both nationally and internationally.

Rhode Island congresswoman Claudine Schneider is concerned that if we wait until scientists are more certain about its likely extent and consequences, it will be too late to prevent global warming. She advocates immediately initiating many programs designed to reduce greenhouse gas emissions that would make good economic and environmental sense even in the absence of a global warming threat. Ari Patrinos manages the U.S. Department of Energy's carbon dioxide research program. He stresses the uncertainties about the magnitude and the effects of greenhouse gas increases. He argues that since it will be decades before any major impact is felt, the most important task in the immediate future is to continue studying the problem.

YES

Claudine Schneider

PREVENTING CLIMATE CHANGE

Last summer [1988], American farms suffered multibillion dollar crop losses from a devastating drought. Thousands of square miles of dry forests went up in smoke as fires raged uncontrollably. Entire communities were left homeless and destitute in the wake of hurricanes Gilbert and Helene. And smog blanketed many cities as a mass of hot air hovered over the country. "It must be the greenhouse effect," was the comment made over and over again by members of Congress. Media coverage had sensitized most Americans to the peril posed by the unchecked growth of greenhouse gas emissions, and the unpleasant weather seemed to be a preview of what we can expect if global climate models are accurate in their predictions.

As scientists debate the validity of these climate models, policymakers must decide what to do. Two opposing schools of thought are emerging: One school promotes adaptation, the other prevention.

Advocates of the adaptation approach argue that we should wait until more scientific evidence is in, which will allow us to take more targeted actions. They worry that premature efforts to cut greenhouse gases may needlessly disrupt the world economy. But by the time we are sure that the greenhouse effect is with us, it will be too late to reverse it; the only recourse will be adaptation. To my mind, this approach is too much like the crisis management tactic of "Let's wait until the ship hits the sand, then figure out what to do" so prevalent in public policymaking today.

Luckily, many conclude that inaction with respect to global warming is foolhardy. When symptoms of an illness appear, they say, a remedy should be sought. Thus there is growing support for a preventive strategy, which means reducing the emissions of greenhouse gases—primarily carbon dioxide (CO_2), but also chlorofluorocarbons (CFCs) and methane. For example, the recommendations of the 1988 Toronto Conference on the Changing Atmosphere include a call for at least a 20 percent reduction in global CO_2 emissions below the 1987 level by the year 2005, and a 50 percent reduction by the year 2015. This goal has been endorsed by the Prime Ministers of Canada and Norway. The Global Warming Prevention Act (H.R. 1078) . . .

and . . . [the] National Energy Policy Act (S. 324) seek a 20 percent reduction by the end of the century.

Noting that society is at risk from the threat of global warming, the National Academy of Sciences recently called on President Bush to take a leadership role in gaining international consensus on this difficult issue and urged the president to adopt "prudent" policies to slow the pace of global climate change. Prudence dictates reducing our dependency on fossil fuels, which are the primary source of CO_2; phasing out most ozone-depleting chlorofluorocarbons and other halocarbons, which are also destroying the ozone layer; halting deforestation and promoting reforestation (trees remove CO_2 from the air whereas burning them releases it); and adopting agricultural and livestock practices that minimize forest conversion and reduce methane emissions. Slowing population growth will also have a profound impact.

Securing these changes will not be easy. Altering the ways of 240 million Americans, let alone 5 billion human beings worldwide, poses a formidable challenge. Confronting the vested economic interests adamantly opposed to change is an equally daunting task for political leaders. Moreover, the greenhouse peril must compete with other serious, and more immediate, environmental problems such as acid rain, urban ozone, indoor air pollution, the cleanup of contaminated nuclear weapons production facilities, and the disposal of toxic wastes.

The slow pace of climate change breeds complacency, but we must remember that the climate will also be slow to respond to after-the-fact solutions. We must begin now to adopt the good stewardship practices that will reduce the likelihood of human-induced climate disruption. Fortunately, congressional testimony received over the past several years indicates a wide range of actions to reduce greenhouse gases that are already cost-effective and many others that could be, given sufficient research and development. Rapid progress has already begun on reducing CFC emissions, and many of the suggestions for minimizing other threats to the climate have been incorporated into legislative proposals.

The path to pursue is one that not only reduces emissions of greenhouse gases but that makes economic sense as well. Each choice we make should spur multiple benefits. Stephen Schneider, a climate expert at the National Center for Atmospheric Research, calls this the "tie-in" strategy: Take those actions that reduce the trade deficit, free up capital, save consumers money, and enhance the competitiveness of U.S. industry, as well as reduce greenhouse gases.

In this spirit, the Global Warming Prevention Act builds upon existing federal programs and newly proposed legislation that can help slow climate change as well as achieve these multiple purposes. By linking these disparate efforts in a coordinated program and adding a few initiatives focused primarily on climate change, we can increase the likelihood that Congress will approve a plan sufficiently comprehensive to alter the conditions that threaten to disrupt the climate. In spite of the global nature of the problem, the emphasis is on domestic policy because the United States is the major producer of greenhouse gases—contributing approximately 20 percent of the total—and the place where we as Americans have the ability to take immediate action. In addition, we cannot expect other countries to listen to our advice until we lead by example.

MORE BANG FOR THE BTU

The most impressive tie-in advantages result from improving energy efficiency. We have already witnessed the power of this approach in responding to the energy crisis of the 1970s. Seeking to reduce oil imports and energy costs, scientists and engineers spawned a veritable revolution in product design and manufacturing techniques. The result was the emergence of an enormous new energy "resource." One can now "produce" energy by using light bulbs that are four times as efficient as conventional models or by living in a superinsulated home that requires only one-tenth as much energy for heating as a typical home.

Such innovations still await widespread adoption. But so far, even the modest improvements that have been made in America's stock of buildings, vehicles, factories, and appliances have secured dramatic results. Since 1973, energy efficiency gains have displaced the equivalent of 14 million barrels of oil per day, saving Americans more than $150 billion per year. In addition, CO_2 emissions are 50 percent lower than they otherwise would have been. Foreign oil imports are less than half of what they would have been, lowering the annual trade deficit by more than $50 billion. Efficiency gains worldwide have been instrumental in spawning a global oil glut, collapsing world oil prices, curtailing OPEC's power, and reducing inflation rates fanned by high energy costs.

How much efficiency is possible? A staggering amount, according to a study by the Global End-Use Oriented Energy Project based in part at Princeton University's Center for Energy and the Environment. For example, the international research team's 5-year energy scenario, based on extensive (but far from exhaustive) use of efficiency improvements, projects that it would be technologically feasible and economically compelling to hold energy use constant and reduce CO_2 emissions over the next half century even as world population doubled and gross world product quadrupled. . . .

The potential for energy savings is reflected in the success of the federally mandated Northwest Electric Power and Conservation Plan. Begun in 1980 when it became clear that the region had exhausted the supply of cheap federally produced hydropower, the plan established a rigorous least-cost planning process for Washington, Oregon, Idaho, and Montana. By focusing on low-cost energy efficiency measures, these states were able to indefinitely defer the need for all 16 planned generating plants, preventing the expenditure of billions of dollars in construction costs and keeping electric rates low. . . .

PRODUCTIVE R&D

Transportation, particularly automobiles and light trucks, is responsible for almost one-third of U.S. carbon emissions, as well as a large fraction of tropospheric ozone and acid rain pollutants. Market forces are not sufficient to spur optimal efficiency because the cost of gasoline comprises only about 20 percent of the total cost of operating a car. Even doubling the price of gasoline, as many countries have done, would not give consumers enough incentive to demand cars that get more than 35 miles per gallon (mpg).

Several policies to reduce fuel use in transportation are already in effect. Since 1975, the Corporate Average Fuel Economy (CAFE) standards have required U.S. manufacturers to achieve a minimum average efficiency among all new cars sold. The fuel economy of new U.S. cars doubled between 1974 and 1985 as the standard rose gradually to 27.5 mpg. Under pressure from manufacturers, the standard was relaxed to 26 mpg in 1986, 1987, and 1988. President Bush recently announced that the standard will be raised to 27.5 mpg in 1990. We should continue gradually increasing the standard over the next decade to 45 mpg for cars and 35 mpg for light trucks, thereby eliminating the need for 15 billion barrels of oil.

Some manufacturers argue that more demanding standards will make it impossible to produce the big cars consumers want and to provide adequate safety for drivers. But Department of Transportation studies indicate that the use of technologies such as continuously variable transmissions, multivalve engines, and advanced lightweight materials will make it possible to achieve 48 mpg in large cars. A few four- and five-passenger prototypes have done much better. As for safety, design is more important than weight. Chevrolet's 4,100-pound 1985 Astro minivan, for example, was one of the worst performers in crash tests; the high-mileage 2,600-pound Chevy Nova had the best crash test rating of any of the new cars tested that year.

Volvo's prototype mid-sized passenger sedan, the LCP 2000, indicates what is technologically feasible. Road tested at a combined city/highway mileage of 75 mpg, the car exceeds EPA's stringent safety standards, accelerates to 60 mph in 11 seconds, and is expected to cost no more than current models.

One drawback of the CAFE standards is that a manufacturer cannot force consumers to choose its more efficient models and therefore cannot guarantee that it will meet the specified average for all cars sold. Congress, therefore, should reinforce the efficiency mandate by strengthening the existing gas guzzler tax, which is levied on any car that gets less than 22 mpg and rises as the mileage falls. I propose increasing the amount of the tax, gradually increasing the kick-in mileage to 32 mpg by 2004, and extending it to light trucks, which are currently exempt. In addition, I propose a "gas-sippers" rebate of up to $2,000 on the purchase of the most fuel-efficient vehicles.

Industry should not have to bear the entire research burden for developing more efficient cars. The Department of Energy (DOE) spends a piddling $50 million per year on all transportation R&D, while the nation is spending $100 billion a year on gasoline. DOE needs to increase its research on auto efficiency, particularly in areas that will have near-term commercial application. To ensure that the money is spent wisely, Congress should ask the National Academy of Sciences to do a thorough review of fuel economy R&D opportunities and recommend priorities for the DOE program.

Underpinning all of these efforts must be a stable federal research and development effort, working jointly with private industry. A 50 percent cut since 1980 in the federal energy efficiency R&D budget has meant that there have been no new research projects begun this decade. These budget cuts seem particularly shortsighted in light of the spectacular success of federal energy-efficiency R&D. According to a 1987 analysis by the American Council for an Energy Efficient Economy, the $16 million that DOE spent on cooperative projects with industry to develop heat pumps, more efficient refrigerators, new ballasts to improve the efficiency of fluorescent lights, and glass coatings that control heat loss and gain through windows will help save the country billions of dollars through energy savings.

Federal R&D can be especially productive in the areas of building design and industrial processes. The annual utility bill for the nation's buildings is $160 billion, but the fragmented building industry spends almost nothing on energy efficiency R&D. The federal government can coordinate research with the 28,000 homebuilders and 150,000 special trade contractors to tap energy savings in virtually every part of the building, from the foundation to the roof and from the heating and cooling system to the windows.

Similar savings can be found in industry. Generic research in industrial process efficiency can lead to significant savings across many industries, but no individual company will reap sufficient benefit to justify funding the research itself. The Global Warming Prevention Act therefore calls for the creation of a number of industrial research centers to conduct joint projects with industry to improve the efficiency of manufacturing processes. This effort will not only reduce greenhouse gases and other pollutants but will enhance industrial competitiveness by cutting production costs.

REPLACING FOSSIL FUELS

Efficiency improvements alone will not be enough, but they will buy us time to develop renewable energy sources that do not produce greenhouse gases. The world's renewable resource base is enormous. According to a 1985 report by the Department of Energy (DOE), a 25-year research and development effort could make it possible to economically extract 85 quads of renewable energy in the United States. Combined with efficiency improvements, this amount could conceivably meet virtually the entire U.S. energy demand in 2010. But the necessary work is not being done. The federal R&D budget for renewable energy has fallen from more than $800 million in 1979 to a proposed $130 million in 1990—less than one percent of DOE's total budget.

With such a vast potential available and opinion polls reporting a high public regard for solar energy, why has there been so little progress? There are several reasons. In part, the outstanding success of energy efficiency has eclipsed the promise of sun power. In the wake of an oil glut and collapsed energy prices, the market for renewable energy technology virtually disappeared. Moreover, ill-conceived public policies did little to advance the infant solar industry toward maturity. R&D efforts were misplaced, emphasizing giant centralized projects, rather than capitalizing on the inherently dispersed and diffuse nature of solar energy flows. Tax incentives, while well-intentioned, were in some ways counterproductive. Pegged to system cost rather than energy production, the solar tax credits favored expensive "gold-plate" systems over low-cost options such as passive solar design.

Although some solar technologies will require more R&D before they are economically competitive, many untapped opportunities are now available. For example, large scale use of solar-generated electricity may still be a decade away, but the use of daylight in buildings is feasible now—and lighting accounts for 25 percent of all U.S. electricity use. DOE's Solar Energy Research Institute estimates that designing buildings to take advantage of daylight could provide one-fourth of lighting requirements. Daylighting should therefore be an integral part of all least-cost utility planning programs.

Several other renewable energy technologies are also in widespread practice or ready for use. Passive solar space heating has become an essential part of energy efficient building practice in northern climates, and solar water heating is being used in hundreds of thousands of buildings throughout the country. Although very cost-effective, these technologies are not being used to their full potential because many consumers lack information about their value, because renters have no incentive to invest in their homes, and because builders compete on home price rather than on operating costs. Least-cost planning programs need to include financial incentives or building standards that promote use of these established renewable-energy technologies.

Photovoltaic cells and high-temperature solar thermal systems offer economically attractive ways of meeting utility peak power demands during summer, and wind power offers the same potential during cold weather. Biomass-derived ethanol is already in use in blends with gasoline, and biogas from waste could be used to fuel new, highly efficient aircraft-derived gas turbines. In fact, Sweden plans to use biogas-fired turbines as one of the replacements for nuclear power, which it is phasing out over the coming decade.

Renewable energy sources, primarily hydropower and biomass, supply close to 10 percent of total U.S. energy needs. To tap the full potential of renewable energy, the federal government must reverse the decline in R&D funding. Government spending still reflects pre-greenhouse thinking. DOE will spend $1 billion on fossil fuel research this year, seven times what it will spend on all renewable energy technologies combined. At a time when we should be exploring every opportunity for replacing fossil fuels, this simply makes no sense.

In addition, the government itself can make a concerted effort to use renewable energy. Federal buildings, military bases, and public housing continue to ignore cost-effective investments in solar technology and energy-saving measures. Federal procurement guidelines should be made more stringent in requiring this use.

Likewise, national and multinational assistance to developing countries should include more attention to renewable energy technologies. For example, the World Bank, which currently underwrites the cost of diesel water pumps should also subsidize solar-powered pumps, which are just as cost-effective, more reliable, and also useful for operating filters to remove waterborne parasites. Stimulating use of photovoltaic cells would also enable the industry to reduce prices by scaling up production processes.

FOREST RELIEF

Between 10 and 20 percent of greenhouse gas emissions are due to the deforestation of some 11 million hectares annually. One "Yellowstone Park" is disappearing every month; a forested area the size of Pennsylvania is ravaged every year. Reforesting efforts are pathetically weak, with just one tree replanted for every 10 cut down. At the current rate of deforestation nearly half the world's remaining forests will be lost within the next several generations. . . .

Nevertheless, we cannot forget that preserving the tropical forests, which contain more than three-fourths of the world's biological diversity, has obvious tie-in appeal. The world relies heavily on the forests for a wide variety of products, including many important medicines. But as the forests disappear, plant and animal species are becoming extinct at a rate not witnessed since the demise of the dinosaurs 65 million years ago. We are in danger of losing an irreplaceable natural resource just as advances in molecular biology and genetic engineering are enhancing our ability to use it.

The importance of the tropical forests makes it incumbent on the United States to find a way to encourage their preservation. Self-righteous—and self-serving—homilies will not do. But through policy analysis, technological assistance, and financial aid, the United States can show Brazil and others how current practices are unsound and how alternatives can benefit them economically as well as environmentally. . . .

The developing countries will act to preserve the forests when they are convinced that it is in their interest to do so. The rural poor in developing countries need to grow food to survive, and the countries need to produce products for export. Cleared tropical forest land can be farmed (though only for a few years), and timber is a valuable export product. To reduce the incentive to destroy the forests, it is necessary to improve the productivity of agriculture on already cleared land and to find a way to generate wealth from the forests without destroying them.

The United States should expand on the pioneering work by the Agency for International Development (AID) to develop agroforestry, the nondestructive use of forest land to produce food and other products. For example, since 1965, AID forester Michael Benge has been demonstrating the advantages of leucaena, a fast-growing, nitrogen-fixing species that enriches the soil while serving as live fencing and producing forage for livestock and fuel wood. The National Academy of Sciences has published a series of handbooks on multiuse species for agroforestry in developing countries, and these could readily serve as the foundation for an expanded effort by AID and other development organizations.

Finally, we cannot forget that a rapidly growing population increases the pressure to burn fossil fuels and to use forests to produce energy and provide new land for farms and ranches. The United States pioneered family planning efforts to stabilize population growth several decades ago, long before most developing countries with high birth rates saw any need. Ironically, now that these same developing countries have reversed their positions and recognized the need to step up such efforts, the United States is steadily reducing its funding commitment to international family planning assistance.

Efforts to reduce CO_2 emissions through improving energy efficiency and tapping renewable energy sources can be overwhelmed by the increased energy needs that will accompany a growing world population. Every three years the world is adding a population the size of the United States, 95 percent of it in developing countries. Improving the standard of living in developing countries without increasing energy use is difficult enough without having to deal with the increased needs of a growing population. As a first step, the United States needs to double its international family planning funds. Developing countries need the means to control population growth—for their economic well-being and for the health of the planet.

The prevention approach to global climate change clearly entails an ambitious public policy agenda. Can we justify such a program with a few computer models? No, but we don't have to. Government initiatives to use energy more efficiently, to develop safe renewable alternatives to fossil fuels, to preserve tropical forests, and to slow population growth make economic and environmental sense even if climate disruption was not a danger.

If further research increases the certainty that we face rapid global warming, we will have to consider more dramatic measures. What I have proposed requires no sacrifice, and by setting in motion efforts that will steadily reduce CO_2 emissions, it will reduce the severity of measures needed if we find that the climate is changing more quickly than predicted.

NO

Ari Patrinos

GREENHOUSE EFFECT: SHOULD WE *REALLY* BE CONCERNED?

In confronting the problem of the greenhouse effect and its principal cause—the increase of atmospheric carbon dioxide (CO_2)—rational thinking and a balanced approach, not panic, are the best remedies. Climatic warming may cause stress and hardship, but could bring benefits as well. For example, while it is likely that average temperatures will rise a few degrees over the next half-century or so, laboratory experiments show crops yielding substantially larger harvests when grown in twice today's level of atmospheric carbon dioxide. Six to eight thousand years ago, when average summer temperatures in the Northern Hemisphere were, perhaps, two–three degrees warmer than they are now, rainfall in Africa and India seems to have been 50–100% greater than at present and the Sahara probably was a region of savannas or steppes. Warmer temperatures also allowed the Vikings to colonize Greenland and reach North America.

Nevertheless, it would be wise to curtail emission of gases that contribute to global warming. Increased conservation significantly could limit fossil fuel use, which releases CO_2 into the atmosphere, and international agreements may lessen the impact of chlorofluorocarbons, a second agent of the greenhouse effect and the leading cause of ozone loss in the upper atmosphere.

Although action is important, the most urgent need is to learn more. People have felt concerned about predictions that the midwestern grain belt will move north, but this change is by no means certain. The mathematical models used to project climate disagree about whether the Midwest will become dryer. More significantly, these models are constructed so that they give reliable estimates only for areas at least the size of a continent.

Researchers need to gather more data, refine present models, and construct new ones to answer these and other questions. Their efforts can improve the knowledge and estimates available as a foundation for making sound, practical decisions about how to cope effectively with the greenhouse effect. It also is important not to underestimate human resourcefulness.

People live in a variety of climates and, as the development of winter red wheat shows, can adapt and be productive even under inhospitable conditions.

The greenhouse effect is the warming of climate that results when the atmosphere traps heat radiating from Earth toward space. Certain gases in the atmosphere resemble glass in a greenhouse. They allow sunlight to pass into the "greenhouse," but block Earth's heat from escaping into space. The various atmospheric gases play a vital role in nurturing life. The oxygen humans and animals breathe makes up about 21% of the atmosphere by volume. Carbon dioxide, on the other hand, is essential to plant life, yet makes up only about 0.03% of the atmosphere.

Life also depends on energy coming from the sun. About half the light reaching Earth's atmosphere passes through the air and clouds to the planet's surface, where it is absorbed and then radiated upward in the form of infrared heat. About 90% of this heat then is absorbed by greenhouse gases and radiated back toward the surface, which is warmed to a life-supporting average of 59°F.

Human activities are increasing the amounts of "radiatively active" gases in the atmosphere and enhancing the natural greenhouse. Over the last century, use of fuels such as coal and petroleum has increased the concentration of atmospheric CO_2. To a lesser extent, so has the clearing of land for agriculture and cities. Agriculture, industry, and other human activities also have increased the concentrations of greenhouses gases like methane and chlorofluorocarbons.

What are the consequences of changing the natural atmospheric greenhouse? It is difficult to predict, but certain effects seem likely:

- On average, the Earth will become warmer. Some regions may welcome warmer temperatures, but others may not.
- Warmer conditions probably will lead to more evaporation and precipitation overall, but individual regions will vary, some becoming wetter and others dryer.
- A stronger greenhouse effect probably will warm the oceans and partially melt glaciers and other ice, increasing sea level. Ocean water also will expand if it warms, contributing to further sea-level rise.
- Meanwhile, crops and other plants may respond favorably to increased atmospheric CO_2, growing more vigorously and using water more efficiently. At the same time, higher temperatures and shifting climate patterns may change the areas where crops grow best and affect the makeup of natural plant communities. . . .

In contrast to acid rain and ozone loss, the greenhouse effect is not as immediately pressing a problem. Unlike acid rain, it probably will take decades and perhaps a century to make itself felt clearly. In contrast to ozone loss, increasing CO_2 is not directly harmful to human health, except in concentrations much greater than those expected over the next century. Nevertheless, the greenhouse effect poses a much greater challenge than acid rain or ozone loss. The factors contributing to an intensified greenhouse are much less amenable to controls, the consequences are likely to be much more far-reaching, and alternatives could demand a great increase in human resourcefulness and ingenuity.

INCREASES IN ATMOSPHERIC CARBON DIOXIDE

Atmospheric CO_2 has increased about 25%, scientists estimate, since the early

1800's. From 1958, when researchers began measuring it continuously, growth has been about 10%. Currently, the level is increasing at a rate of about 0.4% a year.

Human beings add CO_2 to the atmosphere primarily through burning fossil fuels, but also by clearing land (deforestation) and changing its uses. About three-quarters of CO_2 emissions come from fossil fuel consumption and one-quarter from deforestation. Ninety percent of the commercial energy used worldwide comes from fossil fuels, such as coal, petroleum, and natural gas. These fuels are rich in carbon because they consist of decomposed plant and animal matter. Burning oxidizes the carbon, converting it to CO_2 and releasing it into the atmosphere.

Through photosynthesis, living plants remove CO_2 from the atmosphere, convert it into plant tissue, and store the carbon in their roots, trunks, branches, and leaves. Deforestation limits this uptake and adds CO_2 directly to the atmosphere. Cut timber releases CO_2 immediately, if it is burned, or gradually, if it is left to rot. When agriculture replaces forests, plowing releases CO_2 from organic materials in the soils as well. Typically, farmers replace trees with annual crops, which give yearly harvests, rather than storing carbon for much longer periods. As populations expanded during the 19th century, so did the land devoted to agriculture, and continued industrialization in the 20th century greatly has accelerated the rate at which forests and other natural vegetation are converted into farms, residential areas, cities, roadways, and parking lots.

Human activities add appreciably to atmospheric CO_2, but natural processes are fundamental in regulating the amount in the atmosphere and elsewhere. Besides the atmosphere, the Earth has two major reservoirs for storing carbon dioxide—vegetation and soils and the oceans. Natural processes remove CO_2 from the atmosphere and transport it into these reservoirs.

The process primarily responsible for removing CO_2 from the atmosphere is absorption into the oceans. Atmospheric CO_2 dissolves into the surface water and natural mixing and circulation of ocean waters then transports it to greater depths and distant locations. Eventually, biochemical processes remove CO_2 from the water and "fix" it in various forms such as shells, which tend to drift toward the bottom of the ocean and form limestone deposits.

While these processes are storing carbon dioxide in oceans, plants, and soils, others are adding it to the atmosphere. The oceans, for example, give up CO_2 as well as absorb it. Respiration in humans, animals, and microorganisms takes in oxygen and release CO_2. Forest fires and volcanic eruptions also return it to the atmosphere. All of these processes together make up the global carbon cycle. Over extended periods, the carbon cycle tends toward equilibrium, with the mechanisms for absorbing CO_2 offsetting those that release it into the atmosphere. Because natural processes in the geologic past have caused the CO_2 concentration to vary, scientists are uncertain whether the natural carbon cycle actually is in balance now. Obviously, human activities are upsetting any natural balance that may exist.

Scientists estimate that the natural carbon cycle is removing and storing about half the excess CO_2 produced by human activities. Investigations are under way to determine more precisely how much is stored and whether the amount may change as the concentration of atmospheric CO_2 increases. Research shows that, under experimental greenhouse conditions, higher levels of atmospheric CO_2 enhance photosynthesis. If we can limit deforestation, perhaps Earth's vegetation will remove and store more carbon dioxide as the atmospheric concentration rises.

Pinpointing the rate of future increases is difficult, due partly to uncertainties about future human behavior. To provide a convenient, significant benchmark, scientists often focus on estimating when the concentration of atmospheric CO_2 is likely to double. When they set up experiments or equations, doubling the atmospheric level offers a handy reference point for assessing the impacts of increased carbon dioxide on climate, vegetation, and the like. When this doubling of CO_2 actually will occur is uncertain. Forecasts made several years ago anticipated a doubling between the years 2025 and 2040. More recent projections push the date back to the end of the 21st century and beyond.

Predictions become still more complicated when they take other greenhouse gases into account. While increasing CO_2 is expected to be the main factor in enhancing Earth's atmospheric greenhouse, gases like methane and the CFC's almost certainly will contribute to changing climate, but how quickly concentrations of these gases will increase is uncertain.

To project future CO_2 emissions, researchers sponsored by the Department of Energy and other agencies are developing techniques for making informed estimates about a host of complex social and technological questions:

- Will the world economy and industrial productivity grow and even thrive in the next century, thereby increasing energy demands?
- Will world population continue to expand, putting pressure on the economy and demanding still more energy?
- Will we continue our heavy reliance on fossil fuels; choose to pay more and depend to a greater extent than we do now on nuclear, hydro, wind, and solar power; or will there be technological breakthroughs that make these alternatives less costly?
- To what extent will we conserve energy, and how much will our technological ingenuity improve efficiency in energy use?
- Will we create and develop new sources of power like fusion, generating energy by fusing, rather than splitting, the nuclei of atoms?
- Will we be able to apply improved technologies and new discoveries, such as those in superconductivity, to increasing efficiency and making alternate energy sources more practical?

Despite present uncertainties, some steps seem advisable. Given the prospects for significantly increasing concentrations of CO_2 and other greenhouse gases, we would be prudent to adopt straightforward measures now—conserving energy and limiting or banning the use of CFC's. However, since major increases in the atmospheric concentrations of some greenhouse gases seem likely to take at least several decades to exert significant impacts on our lives, it probably would be wise to exercise caution about taking immediate steps that seriously might disrupt the economy and people's lives. Instead, we could use the next decade to learn more and gain greater certainty about these important questions as a basis for building sound strategies to prevent, lessen, or cope effectively with changes we can foresee.

CLIMATE

Globally, as the concentration of atmospheric CO_2 rises, temperatures will increase and other climatic characteristics probably will change. On average, surface air temperature around the world will increase. When the climate has adjusted to a doubling of the CO_2 concentration, researchers estimate a global average temperature increase of three-eight degrees. Over all, rain and perhaps snow may increase, but this trend may not appear everywhere. While some regions could experience increased precipitation, others might see decreases, and snow cover may recede toward the poles.

Changes in worldwide averages interest most of us less, however, than shifts in regional and local climate. Conditions locally may differ markedly from global ones in temperature, amounts of rain and snow, frequency or severity of major storms, and other respects. Farmers need to know, for example, whether the amounts of soil moisture available for their crops will increase or decrease and whether they should worry about frosts. In some regions, concerns for safety and property could require measures to deal with more frequent flooding and more severe storms.

The 1980's saw the four warmest years on record, and the summer of 1988 brought drought to large parts of the U.S. These events have led people to speculate that present conditions show what the greenhouse effect may bring in the future—or even that it has arrived.

There is evidence that average surface air temperature has increased worldwide by nearly one degree since 1850. Given the increase of about 25% in atmospheric CO_2 between the early 1800's and the present, we might conclude that the greenhouse effect indeed is producing a warming.

For several reasons, many scientists are not yet comfortable drawing this conclusion. Regional conditions often differ sharply from global averages. Heat and drought in the U.S. need not signal the advent of a worldwide greenhouse effect. In 1987, a record year for high temperatures, heat waves resulted primarily from very warm sea surface temperatures in the Southern Hemisphere; the Northern Hemisphere was only slightly warmer than normal. Unusually warm or cold years have occurred in strings five or 10 years long, then conditions have reverted to normal or even moved in the opposite direction. These are the kinds of reasons that persuade scientists to use 30-year averages when defining climate.

In addition, climate changes result from many factors, including natural forces, so it is hard to pinpoint a given cause, such as the greenhouse effect, with confidence. When volcanoes erupt, they can propel millions of tons of dust particles high into the atmosphere. The dust reflects solar energy back into space, cooling the Earth slightly for a few years. Over brief periods, energy output from the sun seems to vary somewhat, possibly warming or cooling the Earth. Longer-lasting climate changes sometimes have resulted from shifts in the Earth's position relative to the sun. About 6,000 to 8,000 years ago, the Earth was closer to the sun during Northern Hemisphere summers, and summertime in that hemisphere was roughly two-three degrees warmer than at present.

Warming a few degrees may not sound like much. We often experience changes of many degrees during a single day and at least a few degrees or more from one day to the next. These are fluctuations in daily weather, however. Changes of a few degrees in long-term global climate are a very different matter. Conditions in a given area could become considerably more or less hospitable to human life. We also need to remember that people live in many climates, and human beings can cope with and capitalize on change.

We should be aware that weather and climate differ significantly. Weather is the state of the atmosphere in a particular place and time: cloudy or clear; rainy, snowy, or sunny; windy or calm; hot or cold. These conditions can shift noticeably in a few hours, and changes in weather are commonplace. Changes in climate are not. Climate refers to averages for temperature, rain, snowfall, etc. over a given span of years. Weather services typically use 30 years as the period for establishing the averages.

Climate does vary, of course, from the tropics to the poles and over the passage of time, and these shifts in climatic averages change the centerline around which weather variations occur. A thousand years ago, climate in the North Atlantic region was perhaps one–two degrees warmer than now, and that difference was enough to allow the Vikings to colonize Greenland and use it as a staging area to discover America. Then, the Little Ice Age brought temperatures two degrees cooler than the present, and the Viking colonies disappeared.

The climatic optimum about 6,000 to 8,000 years ago also suggests how widespread the impact of a few degrees' warming can be on rainfall and vegetation. During this period, when summer temperatures in the Northern Hemisphere averaged about two–three degrees warmer than at present, prairies extended hundreds of miles further east in the U.S. than they do now. Meanwhile, higher water levels left their mark in the lakes of northern and eastern Africa, Saudi Arabia, India, and Australia.

As conditions during the climatic optimum suggest, greenhouse warming could create a climate more favorable to agriculture in some areas than we have now. Even so, changes might be stressful. In the past, climate variations occurred slowly, over thousands of years. Within a century, increasing atmospheric CO_2 may bring changes as great as those during the climatic optimum, and such a rapid shift would tax our resourcefulness.

In laboratory experiments with crops, higher concentrations of atmospheric CO_2 increase harvests. Experiments show that exposing plants to twice the concentration of carbon dioxide found in today's atmosphere produces dramatic results. Except for corn, whose yield rose by 10%, harvests for all the crops studied increased from 30% to more than 80%. Harvests studied come from different parts of the plants—stems, leaves, or seeds—so different kinds of crops all appear to benefit from increased levels of CO_2.

Laboratory experiments with increased atmospheric CO_2 also show improved water use efficiency, so crops would require less irrigation and be more likely to survive drought. In addition, plants may be less vulnerable to airborne pollution.

Unfortunately, public attention has generally ignored the implications of plant responses to carbon dioxide and focused almost entirely on the impacts of possible climate shifts, sometimes training the spotlight on only the most alarming possibilities. The public and legislators react with understandable anxiety to scenarios projecting climate conditions changing abruptly in the Midwest and the corn and wheat belts migrating into Canada. Even with respect to climate, this speculation is misleading. Estimates of future conditions in a given region are uncertain, and not all climate models predict drying in the Midwest.

Potential shifts in agriculture have touched off more anxiety than possible changes in water resources, but the effects on water supplies may be more important. Water resources also will depend on plant responses to increasing atmospheric CO_2 as well as changing climate. Other climatic factors like air temperature and cloudiness also help determine how much evaporation occurs and, hence, how much water remains in soils, lakes, and reservoirs.

While scientists know that elevated atmospheric CO_2 stimulates plant growth and improves water use efficiency under controlled conditions, they still aren't sure whether plants will sustain their present nutritional level. If they grow larger in response to rising concentrations of atmospheric CO_2, carbohydrate content may increase, upsetting the present balance of carbohydrates and protein.

One experiment with insects that feed on soybean leaves suggested such loss of nutrition. When soybeans grew under conditions of elevated CO_2, larvae of the soybean looper maintained their growth, but ate 25% more leaves than they do now. Although insect leaf consumption might seem irrelevant to human nutrition, the experiment may have implications for grazing animals. Cattle and sheep, for example, depend on a diet of leaves and provide one source of human food. The experiment also suggests broader questions about the future nutritional quality of grains, fruits, and vegetables that contribute directly to human diet.

Will weeds—such as kudzu—respond as vigorously as crops to increased atmospheric CO_2? Experiments support this possibility. Imported from Japan to control erosion in the southern U.S., this fast-growing vine now thrives prolifically along roadsides, in fields, and over treetops, choking other plants and trees. In experiments, it responds vigorously to increased atmospheric CO_2, suggesting that, in the future, it could spread considerably beyond the South. Greater concentrations of atmospheric CO_2 could strengthen the chances of weeds prevailing in their competition with crops. We also should remember that plants grown for food are not the only species important to our welfare. Natural vegetation in forests and grasslands also is vital to the economy, ecology, recreation, and aesthetics. Scientists know relatively little about how natural communities of vegetation will respond to increased atmospheric CO_2 and different climate conditions, so this entire area will be a field for research to determine how carbon dioxide would affect plant survival, growth, yield, and water use efficiency.

THE RISING SEA LEVEL

The sea level is rising, and the possibilities for future increases probably have caused more concern than any other aspect of the greenhouse effect. Many factors can affect sea level. The Earth's tectonic plates shift and subside, for example, and the continents are still "rebounding" from glacial pressure during the last ice age. While some of these forces tend to lower sea level, others cause it to rise.

The greenhouse effect may increase sea level because warmer temperatures could lead to partial melting of three bodies of ice that rest on land: glaciers and small ice caps around the world; the Greenland ice cap; and Antarctica. Since North Pole ice floats on ocean water, any melting there would not affect sea level—just as melting ice cubes have no effect on the level of a cold drink in a glass. Melting ice from other sources, however, would add to ocean waters, perhaps raising sea level. In addition, because water expands as it warms, a global temperature increase could contribute to sea level rise.

Over the past century, the sea level is estimated to have risen about four–eight inches worldwide. About one-third, scientists agree, has come from partial melting of small glaciers and other ice caps. Some probably has come from expansion of warming ocean waters. According to most specialists in ice studies, however, very little of the rise so far has come from any melting of Greenland or Antarctica.

What about the future? Sea level almost certainly will continue to rise—the question is how much? Artists have depicted the Statue of Liberty waist-deep in water and the TransAmerica tower surrounded by San Francisco Bay. Pushing possible conditions to extremes—and thus presenting worst-case scenarios—some research studies have speculated that sea level would rise as much as 12 feet by the year 2100. On the other hand, scientists who study ice—glaciologists—estimate that melting ice could raise sea level perhaps as much as two feet by 2100.

They expect some of this increase—perhaps two–eight inches—to come from mountain glaciers and small ice caps and about the same amount from Greenland. More uncertain is the contribution of the Antarctic, one of the least understood regions of the planet despite 30 years of accelerated exploration.

The most dramatic question is whether the West Antarctic Ice Sheet will disintegrate. Glaciologists point out that a series of complicated events would have to occur to produce disintegration. If a breakup occurred at all, they estimate it would take a century to trigger and hundreds, if not thousands, of years for collapse to occur.

Antarctica certainly has the potential to exhibit quite contrary tendencies. Ice at its edges could melt, raising sea level. At the other extreme, this surprising region could cause sea level to fall. The key factor is the extreme Antarctic cold. While air temperatures probably will warm, they are unlikely to do so enough to melt ice, except at the edges. In most places, warming air probably will carry more moisture, which then would precipitate as snow. Moisture to form the new snow deposited on this enormous continent would come from the oceans and thus could lower sea level.

In addition to melting ice, expansion of ocean waters probably will cause further sea level rise, but how much is uncertain. In the laboratory, it is easy to show that water expands when heated, but in the oceans, complex circulation makes the effects of this simple phenomenon very difficult to assess. One calculation often cited in scientific literature makes some assumptions about the uncertainties and estimates that ocean expansion will add 16 inches to sea levels by the year 2070.

Whatever the size of the increase, rising sea level may have considerable impact on coastal areas. Besides changing coastlines, it could drive storm surges further inland. Residents of places like Florida and Louisiana already are experienced in dealing with such problems, and they no doubt will need all their skill to cope with these challenges.

POSTSCRIPT

Does Global Warming Require Immediate Action?

A careful reading of Schneider's and Patrinos's essays reveals no significant disagreements about the facts or the uncertainties concerning global warming. They draw very different conclusions, however, about what political, industrial, and scientific leaders should do in response to the present understanding of the problem. Patrinos supports the "wait and see" philosophy that is the official position of the administration for which he works. Although he makes passing reference to the wisdom of implementing conservation strategies, the Department of Energy has done little to promote such initiatives. In fact, key energy and budget officials in the Bush administration have recently deleted many of the specific congressional proposals for energy efficiency legislation or policy advocated by Congresswoman Schneider.

The repeated failures of the oil-producing nations to adopt policies that economists and scientists consider to be in their long-term self-interest demonstrates the extraordinary difficulty of achieving international cooperation in industrial planning and development. At a minimum it will require the emergence of more enlightened political leadership than is now evident before a complex program of global energy planning has any chance of being implemented. Such a program must take into account the disparity between the industrial and developing nations in terms of energy needs, available resources, and economic capabilities. This will require more than the multinational assistance in developing renewable energy technologies proposed by Schneider.

Nuclear power advocates (see the issue on nuclear power) are proposing that revival of this technology is justified by the need to reduce the amount of carbon dioxide released by fossil fuel combustion. Articles critiquing this controversial suggestion are listed in the postscript to that issue.

A serious proposal to develop massive tree plantations as a means of controlling carbon dioxide is discussed in "Forests—A Tool to Moderate Global Warming?" by Roger Sedjo, *Environment* (January/February 1989).

Other recent articles that contain useful information related to global warming are: "The Great Climate Debate," by Robert White, *Scientific American* (July 1990); "Slowing Global Warming," by Christopher Flavin, *American Forests* (May/June 1990); "Swept Away," by Jodi Jacobson, *World Watch* (January/February 1989); "Practical Responses to Climate Change," by Jill Jager, *Environment* (July/August 1990); and "Murder in Acre—Global Warming and Sustainable Development," by Mason Willrich, *The Amicus Journal* (Spring 1989).

ISSUE 17

Is the Montreal Protocol Adequate for Solving the Ozone Depletion Problem?

YES: Richard Elliot Benedick, from "Ozone Diplomacy," *Issues in Science and Technology* (Fall 1989)

NO: Arjun Makhijani, Amanda Bickel, and Annie Makhijani, from "Beyond the Montreal Protocol: Still Working on the Ozone Hole," *Technology Review* (May/June 1990)

ISSUE SUMMARY

YES: Deputy Assistant Secretary of State Richard Elliot Benedick describes the technical background and political history of the Montreal Protocol designed to protect the planet from ozone depletion.
NO: Energy researchers Arjun Makhijani, Amanda Bickel, and Annie Makhijani acknowledge the significance of the Protocol but argue that it must be strengthened and its outreach broadened if the ozone problem is to be solved.

In 1974 a short paper by M. J. Molina and F. S. Rowland was published in *Nature*. Based on laboratory experiments, the paper warned of a potential threat to the stratospheric ozone layer resulting from the rapidly expanding use and release of a family of synthetic chemicals. This speculative prediction gave rise to an immediate controversy among the scientific, environmental, and industrial communities because of the essential role played by atmospheric ozone in shielding human beings, as well as other terrestrial flora and fauna, from the harmful effects of the high-energy ultraviolet radiation emitted by the sun.

The molecules in question are chlorofluorocarbons (CFCs), which were developed by the chemical industry as inert, volatile nontoxic fluids. They have been used as the working fluids in refrigeration and air conditioning systems, as the propellants in pressurized spray cans for a variety of industrial and consumer products, as foaming agents for plastic foams, and as cleansing solvents in the high technology electronics industry. Ultimately they find their way into the atmosphere where, because of their chemical inertness and insolubility in water, they neither decompose nor get washed out in rainfall. What Molina and Rowland proposed was that chloroflurocarbons would, over a period of about a decade, rise to levels in the stratosphere where they would be exposed to the high-energy ultraviolet radiation that fails to reach the lower atmosphere because of ozone absorption. They would then photodecompose, producing atomic chlorine, which laboratory studies showed would act as an effective catalytic agent in destroying ozone.

As might be expected, the predictions of Molina and Rowland met with much skepticism and stimulated intense activity on the part of scientists, particularly those associated with chemical companies and industries having a vested interest in selling or using CFCs, who tried to disprove the dire predictions of serious ozone depletion. Dr. Rowland has devoted much of his professional and private life to defending and substantiating his prediction as well as to seeking an international response to the need to phase out the use of CFCs. Gradually, through both laboratory work and environmental sampling, data accumulated that tended to support the ozone depletion prediction and to refute proposed alternative mechanisms for the removal of CFCs from the atmosphere.

Predictions of the extent of ultimate ozone loss from continued use of CFCs have tended to rise and fall as new data accumulated. Then in 1985 came definitive reports of large scale seasonal losses of ozone over the Antarctic. Follow-up studies have now resulted in a growing consensus that this effect is indeed a result of the increasing atmospheric concentrations of CFCs. Furthermore, estimates of the ultimate steady-state reduction in stratospheric ozone and the health and environmental consequences (including increased incidence of human skin cancers and phytoplankton destruction) if CFC use is not drastically curtailed are again on the rise.

This problem, like the climatic greenhouse warming discussed in the preceding issue, is a worldwide environmental problem that requires a concerted international response. Until recently, the only regulatory action was the banning of CFC pressurizers in aerosol spray cans (the least essential use of these chemicals) by the United States, Canada, and Scandinavia during 1976–78. Beginning in 1985, stimulated by the dramatic Antarctic findings, a series of yearly international conferences have been held and one result has been a landmark agreement reached in September 1987 in Montreal.

Richard Elliot Benedick was the chief U.S. negotiator at the 1987 meeting in Montreal. He understandably takes great pride in describing the process by which the many controversies were overcome in arriving at a broad protocol, with provisions for further modification, that he thinks provides the basis for solving the ozone depletion problem. Arjun and Annie Makhijani and Amanda Bickel are researchers at the Institute for Energy and Environmental Research. They acknowledge the significance of the Montreal Protocol, but they stress the need to extend its coverage to other important ozone-depleting chemicals and to assure the participation of Third World countries. They argue that if this is not accomplished, substantial ozone layer destruction will occur.

YES

Richard Elliot Benedick

OZONE DIPLOMACY

On September 16, 1987, representatives of countries from every region of the world reached an agreement unique in the annals of international diplomacy. In the Montreal Protocol on Substances that Deplete the Ozone Layer, nations agreed to significantly reduce production of chemicals that can destroy the stratospheric ozone layer (which protects life on earth from harmful ultraviolet radiation) and can also change global climate.

The protocol was not a response to an environmental disaster such as Chernobyl, but rather *preventive* action on a global scale. That action, based at the time not on measurable evidence of ozone depletion or increased radiation but rather on scientific hypotheses, required an unprecedented amount of foresight. The links between causes and effects were not obvious: a perfume spray in Paris helps to destroy an invisible gas 6 to 30 miles above the earth, and thereby contributes to deaths from skin cancer and extinction of species half a world and several generations away.

The ozone protocol was only possible through an intimate collaboration between scientists and policymakers. Based as it was on continually evolving theories of atmospheric processes, on state-of-the-art computer models simulating the results of intricate chemical and physical reactions for decades into the future, and on satellite-, land- and rocket-based monitoring of remote gases measured in parts per trillion, the ozone treaty could not have occurred at an earlier point in human history.

Another noteworthy aspect of the Montreal Protocol was the negotiators' decision not to take the timid path of controlling through "best available technology"—the traditional accommodation to economic interests. Instead, the treaty boldly established firm target dates for emissions reductions, even though the technologies for accomplishing these goals did not yet exist.

The ozone protocol sounded a death knell for an important part of the international chemical industry, with implications for billions of dollars in investment and hundreds of thousands of jobs in related industries such as food, transportation, plastics, electronics, cosmetics, and health care. Here, as in many other areas, international economic competition clashed with the

Richard Elliot Benedick, "Ozone Diplomacy," *Issues in Science and Technology*, vol. 6, no. 1 (Fall 1989).

need for international environmental cooperation, but in this case concerns about the environment eventually carried the day.

Similar conflicts between economic and environmental imperatives are bound to arise in the future, as more and more environmental problems cross national boundaries and require international solutions. Furthermore, there will be a growing number of threats to the environment that, although not obvious or immediate, pose serious long-term dangers. So it is worth considering what factors contributed to the Montreal Protocol's success, and what lessons the negotiations might hold for future attempts to deal with similar situations.

ENVIRONMENTAL BOMBSHELLS

In 1974, two theories were advanced that suggested potentially grave damage to the ozone layer. According to the first, chlorine in the atmosphere could continually destroy ozone for a period of decades: A single chlorine atom was capable, through a catalytic chain reaction, of eliminating tens of thousands of ozone molecules. The other theory postulated that man-made chlorofluorocarbons (CFCs) would break down in the presence of radiation in the stratosphere and release dangerously large quantities of chlorine.

These hypotheses were environmental bombshells. Production of CFCs had soared from 150,000 metric tons in 1960 to over 800,000 metric tons in 1974, reflecting their broad usefulness: CFCs are chemically stable and vaporize at low temperatures, which make them excellent coolants in refrigerators and air conditioners,and ideal as propellant gases in spray cans; they are good insulators; they are standard ingredients in the manufacture of such ubiquitous materials as styrofoam; and they are generally inexpensive to produce. The stability of CFCs means that, unlike other man-made gases, they are not chemically destroyed or rained-out quickly in the lower atmosphere. Rather, they migrate slowly upward, remaining intact for decades. The halons, a related family of chemicals used in fire protection, were found to have similar properties.

Theories about the relationship between the ozone layer, chlorine, and CFCs stimulated tremendous activity in scientific and industrial circles. Although the chemical industry vigorously denied the validity of any linkage between the state of the ozone layer and their growing sales of CFCs, the U.S. scientific community mounted a major research campaign that confirmed the fundamental validity of the chlorine-ozone hypotheses.

But although the theory was sound, making precise measurements of effects was not so easy—especially since growing concentrations of carbon dioxide and methane (originating at least in part from human activities) could greatly offset the projected chlorine impact, and nitrogen compounds could influence the reaction in either direction. Thus, in the years following the initial hypotheses, there were wide fluctuations in the predicted results of CFC emissions. Theoretical-model projections of global average ozone depletion 50 to 100 years in the future began at about 13 percent in 1974, increased to 19 percent in 1979, and dropped to less than 5 percent in 1982–83. For a time, these swings tended to diminish public concern over the urgency of the problem.

But consensus was soon to develop. In 1986, an assessment spearheaded by NASA and sponsored by the United Nations Environment Programme, the World Meteorological Organization, and other agencies concluded that continued CFC emissions at the 1980 rate would reduce global average ozone by about 9 percent by the latter half of the next century, with much larger seasonal and latitudinal decreases. New measurements also indicated that accumulations of CFCs in the atmosphere had nearly doubled between 1975 and 1985, even though production of these chemicals had stagnated over the same period, illustrating the potential long-term danger from these substances.

On the basis of these figures and of projections of continuing, though moderate, CFC emissions, the U.S. Environmental Protection Agency estimated that in the United States alone there could be over 150 million new cases of skin cancer among people currently alive and born by the year 2075, resulting in over 3 million deaths. EPA also projected 18 million additional eye cataract cases in the United States for the same population. Other possible results of CFC emissions included damage to the human immune system, serious impacts on agriculture and fisheries, increased formation of urban smog, and warming of the global climate.

THE GREAT ATLANTIC DIVIDE

As scientists analyzed the effects of CFCs on the ozone layer and the resultant implications for human health and the environment, the United States and the European Community (EC)—comprising 12 sovereign nations—emerged as the principal protagonists in the diplomatic process that culminated in the Montreal Protocol. Despite their shared political, economic, and environmental orientations, the United States and the European Community, which together accounted for 84 percent of world CFC output in 1974, differed over almost every issue at every step along the route to Montreal.

The U.S. Congress held formal hearings on the ozone layer soon after the theories were published, which led in 1977 to ozone protection legislation that banned use of CFCs as aerosol propellants in all but essential applications. This affected nearly $3 billion worth of sales in a wide range of household and cosmetic products, and rapidly reduced U.S. production of CFCs for aerosols by 95 percent. The U.S. action was paralleled by Canada (a small producer), and by Sweden, Norway, Denmark and Finland (all nonproducing, importing countries).

In contrast, European parliaments (except for the German Bundestag) showed scant interest in CFCs. The European Community delayed until 1980, and then enacted a 30 percent cutback in CFC aerosol use from 1976 levels and announced a decision not to increase production capacity.

These EC actions, however, were feeble compared to the U.S. regulation. With respect to the 30 percent aerosol reduction, European sales of CFCs for this purpose had, by 1980, already declined by over 28 percent from the 1976 peak year. Moreover, the European Community two years later defined "production capacity" in a manner that would enable current output to increase by over 60 percent. The capacity cap was therefore a painless move, supported by European industry, which gave the appearance of control while in reality permitting undiminished rates of expansion for at least two more decades.

Relative to gross national product, EC production of CFCs was over 50 percent higher than that of the United States. Aerosols, which had virtually disappeared in the U.S., still comprised during the 1980s over half of CFC sales within the European Community. The European Community was also the CFC supplier to the rest of the world, particularly the growing markets in developing countries. EC exports rose by 43 percent from 1976 to 1985 and averaged almost one-third of its production, whereas the United States consumed virtually all it produced.

These developments were reflected in growing differences in attitude between the chemical industries on the two sides of the Atlantic. Shaken by the force of public reaction in the 1970s over the threat to the ozone layer, American producers had quickly developed substitutes for CFCs in spray cans. U.S. chemical companies were also constantly aware of their vulnerability in the environmentally charged domestic atmosphere. In their public pronouncements, industry spokespersons took the ozone problem with growing seriousness, and appeared increasingly concerned about its effect on companies' reputations. The threat of a patchwork of state laws—legislation against CFCs was actually introduced or passed by Oregon, New York, California, Michigan, Minnesota, and others—made U.S. industry not only resigned to but even publicly in favor of federal controls, which would at least be uniform and therefore less disruptive.

There was also resentment among American producers that their European rivals had escaped meaningful controls. A constant theme in the U.S. chemical industry during the 1980s was the need to have a "level playing field"—to avoid recurrence of unilateral U.S. regulatory action that was not followed by the other major producers.

In September 1986, the Alliance for Responsible CFC Policy, a coalition of about 500 U.S. producer and user companies, issued a pivotal statement. Following the obligatory reiteration of industry's position that CFCs posed no immediate threat to human health or the environment, the Alliance spokesman declared that "large future increases . . . in CFCs . . . would be unacceptable to future generations," and that it would be "inconsistent with [industry] goals . . . to ignore the potential for risk to those future generations." Thus, only three months before the protocol negotiations began, U.S. industry announced its support for new international controls on CFCs.

This unexpected policy change, which came after much soul-searching within U.S. industry, aroused consternation in Europe. The British and French had been suspicious all along that the United States was using an environmental scare to cloak commercial motivations. Now, some Europeans surmised (incorrectly) that the United States wanted CFC controls because they had substitute products on the shelf with which to enter the profitable EC export markets.

For its part, EC industry's primary objective was to preserve its dominance and to avoid the costs of switching to alternative products for as long as possible. Taking advantage of public indifference and political skepticism, European industrial leaders were able to persuade most EC governments that substitutes for CFCs were neither feasible nor necessary—despite the demonstrated U.S. success in marketing alternative spray propellants. Industry statements were echoed in official EC pronouncements that continually stressed the scientific uncertainties, the impossibility of finding effective substitutes, and the adverse effects of regulations on European living standards.

SELF-SERVING POSITIONS

Although the United States and the European Community were the major CFC producers, the ozone problem threatened the entire world and therefore could be solved only by international agreement. Filling a catalytic role for such an agreement became the mission of a small and hitherto little-publicized United Nations agency, the UN Environment Programme (UNEP). Under the dynamic leadership of its executive director, Mostafa Tolba, an Egyptian scientist, UNEP soon made ozone protection a top priority. The agency worked to inform governments and world public opinion about the danger to the ozone layer, it provided a nonpoliticized international forum for the negotiations, and it was a driving force behind the consensus that was eventually reached.

In January 1982, representatives of 24 countries met in Stockholm under UNEP auspices to decide on a "Global Framework Convention for the Protection of the Ozone Layer." The following year a group of countries, including the United States, Canada, the Nordic nations and Switzerland, proposed a worldwide ban on "nonessential" uses of CFCs in spray cans, pointing out that the United States and others had already demonstrated that alternatives to CFC sprays were technically and economically feasible. In late 1984, the European Community countered with a proposal for alternative controls that would prohibit new additions to CFC production capacity.

Each side was backing a protocol that would require no new controls for itself, but considerable adjustment for the other. The United States had already imposed a ban on nonessential uses of CFCs, but U.S. chemical companies were operating at close to capacity and thus would suffer under a production cap. Their European counterparts, on the other hand, had substantial underutilized capacity and could expand CFC production at current rates for another 20 years before hitting the cap.

Despite these disagreements, by March 1985 the negotiators had drafted all elements of a protocol for CFC reductions except the crucial control provisions. Meeting in Vienna, all major producers except Japan signed an interim agreement—the Vienna Convention on Protection of the Ozone Layer—which promoted international monitoring, research, and exchange of data and provided the framework for eventual protocols to control ozone-modifying substances. Over strong objections from European industry representatives, the Vienna Conference passed a separate resolution that called upon UNEP to continue work on a CFC protocol with a target for adoption in 1987. . . .

DEEP CUTS

One of the central disputes of the negotiations was whether restrictions would be placed on the production or consumption of the substances covered by the agreement. This issue, though seemingly arcane, was one of the most important and most difficult to resolve.

The European Community pushed for controls on production, arguing that it was simpler to control output since there were only a small number of producing countries, whereas there were thousands of consuming industries and countless points of consumption. But the United States, Canada, and others who favored a consumption-related formula pointed out that controlling production would confer unusual advantages on the European Community while particularly prejudicing importing nations, including the developing countries. Since about a third of EC output was exported and there were no other exporters in the picture, a production limit essentially locked in the EC export markets. The only way the United States or others could supply those markets would be to decrease their domestic consumption.

The European Community, with no viable competitors, would thus have a virtual monopoly. If European domestic consumption should rise, the European Community could cut back its exports, leaving the current importing countries, with no recourse to other suppliers, to bear the brunt of CFC reductions. Because of this vulnerability, there would be incentives for importing countries to remain outside the treaty and build their own CFC facilities.

To meet the valid EC argument about controlling multiple consumption points, the United States and its allies came up with an ingenious solution: A limit would be placed on production *plus* imports *minus* exports to other Montreal Protocol signatories. This "adjusted production" formula eliminated any monopoly based on current export positions, in that producing countries could raise production for exports to protocol parties without having to cut their own domestic consumption. Only exports to nonparties would have to come out of domestic consumption, and this would be an added incentive for importing countries to join the protocol, lest they lose access to supplies. Additionally, an importing signatory whose traditional supplier raised prices excessively or refused to export could either produce on its own or turn to another producer country.

The single most contentious issue was the timing and extent of reductions. Again, the European Community and the United States were the principal opponents. The United States originally called for a freeze to be followed by three phases of progressively more stringent reductions, all the way up to a possible 95 percent cut. But even late into the negotiations, the European Community was reluctant to consider reductions beyond 10 to 20 percent.

The United States and others rejected this as inadequate. In fact, Germany, which had become increasingly concerned over the ozone problem, was already planning an independent 50 percent reduction and early in 1987 it made urgent appeals to the other EC members also to accept deep reductions. Meanwhile, new scientific research was demonstrating that all of the control strategies under consideration would result in some degree of ozone depletion, the extent of which would depend on the stringency of international regulation. These developments helped garner support for deep cuts.

(An interesting sideshow during the negotiations was the attempt by some antiregulatory ideologues within the Reagan administration to overturn or weaken the U.S. position by reopening basic questions about the science and possible impacts of ozone depletion. These efforts, which did not even have support from most of American industry, were effectively countered by the combined efforts of the Department of State, EPA, and the Council of Economic Advisors, among other agencies. The issue went for decision to the President, who reaffirmed the basic elements of the U.S. drive for strong international regulation of CFCs and halons.)

The turning point in the negotiations came when Mostafa Tolba, head of UNEP, began to play a central role. He personally proposed a freeze by 1990, followed by successive 20 percent reductions every two years down to a complete phaseout, and he pressed for deep cuts during informal consultations with heads of key delegations.

Ultimately, even the most reluctant parties—the European Community, Japan, and the Soviet Union—agreed to a 50 percent decrease. The treaty as signed stipulated an initial 20 percent reduction from the 1986 level of CFCs, followed by 30 percent. Halons were frozen at 1986 levels, pending further research. And one innovative provision—that these reductions were to be made on specific dates regardless of when the treaty should enter into force—removed any temptation to stall enactment of the protocol in the hopes of delaying cuts, and also provided industry with dates upon which to base its planning.

Negotiators at Montreal also faced the difficult task of encouraging developing countries to participate in the treaty. Per capita consumption of CFCs in those countries was tiny in comparison to that of the industrialized world, but their domestic consumption requirements—for refrigeration, for example—were growing, and CFC technology is relatively easy to obtain. The protocol thus had to meet their needs during a transition period while substitutes were being developed, and it had to discourage them from becoming major new sources of CFC emissions.

A formula was developed whereby developing countries would be permitted a 10-year grace period before they had to comply with the control provisions. During this time they could increase their consumption up to an annual level of 0.3 kilogram per capita—approximately one-third of the 1986 level prevailing in industrialized countries. It was felt that the realistic prospects of growth in CFC use to these levels in the developing countries was not great, as they would not want to invest in a technology that was environmentally detrimental and would soon be obsolete.

BASIS FOR OPTIMISM

Science is demonstrating that this planet is more vulnerable than had previously been thought. For example, the Antarctic ozone hole discovered in 1985 made it clear that the atmosphere, upon which all life depends, is capable of surprises: There is a potential for large and unexpected change. The international community can no longer pretend that nothing is happening, or that the planet will somehow automatically adjust itself to the billions of tons of man-made pollutants to which it is annually being subjected.

But now there is some basis for optimism. In September 1987, 24 countries signed the Montreal Protocol on Substances That Deplete the Ozone Layer; many other countries added their signatures over the ensuing months. Six months later, in a rare display of unanimity, the U.S. Senate approved the protocol by a vote of 83–0, and President

Reagan promptly signed the ratification instrument, making the United States the second nation to ratify (after Mexico).

The treaty entered into force on January 1, 1989. By the time of the First Meeting of the Parties, held in Helsinki May 2–5, 1989, 36 countries, accounting for about 85 percent of global consumption of CFCs and halons, had ratified it. Even now, under the farsighted process established by the negotiators, governments are actively considering whether to strengthen the protocol's control provisions on the basis of more recent scientific evidence concerning the impact of CFCs and halons on the ozone layer.

The Montreal Protocol stands as a landmark—a symbol both of fundamental changes in the kinds of problems facing the modern world and of the way the international community can address those problems. Mostafa Tolba has described it as "the beginning of a new era of environmental statesmanship." But the protocol may also have relevance for dealing with other common dangers, including national rivalries and war. The ozone treaty reflects a realization that nations must work together in the face of global threats, and that if some major actors do not participate, the efforts of others will be vitiated.

In the realm of international relations, there will always be uncertainties—political, economic, scientific, psychological. The protocol's greatest significance may be its demonstration that the international community is capable of undertaking complicated cooperative actions in the real world of ambiguity and imperfect knowledge. The Montreal Protocol can be a hopeful paradigm of an evolving global diplomacy, one wherein sovereign nations find ways to accept common responsibility for stewardship of the planet and for the security of generations to come.

NO

Arjun Makhijani, Amanda Bickel, and Annie Makhijani

BEYOND THE MONTREAL PROTOCOL: STILL WORKING ON THE OZONE HOLE

In June, environmental officials from around the world will meet in London to determine what further steps to take to protect the stratospheric ozone layer. The London meeting will allow the more than 50 nations ratifying the 1987 Montreal Protocol, which mandated substantial cuts in the use of chemicals that can damage the earth's protective ozone layer, to make the first changes since the protocol went into force in January 1989. The changes are likely to be significant.

Until recently, debate on chlorine, the most important element that catalyzes destruction of stratospheric ozone, focused almost exclusively on chlorofluorocarbons (CFCs). These compounds are used in refrigeration and air-conditioning, as aerosol propellants and solvents, and to form foams, including those used in fast-food packaging and as rigid insulation. The Montreal agreement requires industrialized countries to halve their use of five CFCs by 1998 and to freeze their use of three halons, chemicals used in firefighting that contain ozone-depleting bromine.

However, the protocol makes no mention of two other ozone-depleting compounds: methyl chloroform, one of the most widely used organic solvents, and carbon tetrachloride, perhaps the cheapest and most toxic organic solvent. Carbon tetrachloride is pound for pound more ozone-depleting than any of the five regulated CFCs, and both it and methyl chloroform contribute more to ozone-threatening chlorine levels than all but two of the eight regulated CFCs and halons.

Emissions from all these compounds make human industrial activity the overwhelming source of stratospheric chlorine. Chlorine released from the oceans, large volcanic eruptions, and biomass burning, including fuel wood as well as natural and anthropogenic forest fires, create an average level of .7

part per billion by volume (ppbv) in the stratosphere. Chlorine emissions from industrially produced chemicals add about four times that amount.

The Montreal Protocol and legislation in some countries have made inroads on reducing these anthropogenic emissions. But to get chlorine concentrations anywhere near their natural levels—and below those that threaten the ozone layer—will require much more stringent controls. Proposals to enact such controls will be on the table in London.

The parties will also begin considering limits on HCFCs, newly created CFC substitutes that themselves damage the ozone layer. And the signatories will have to confront the lack of broad participation by Third World countries, which threatens progress on all fronts. Even a phaseout of the use of methyl chloroform and carbon tetrachloride, a complete CFC and halon phaseout by 2000, and regulation of HCFCs will not be enough to stop ozone destruction if large Third World countries such as China and India do not join the agreement.

CONVINCING EVIDENCE APPEARS

Recent results from ongoing studies of the stratosphere lend urgency to the June meeting.

The 1985 report by the British Antarctic Survey—which helped propel the protocol forward—showed that massive ozone depletion occurs over the Antarctic each spring. However, the report never formally recognized the relationship between the huge seasonal depletion and the use of CFCs. This link came into sharper focus soon after the protocol was signed in September 1987. The second American National Ozone Expedition (NOZE II) and the international Airborne Antarctic Ozone Experiment, which sent planes into the ozone hole that October, found strong correlations between levels of chlorine monoxide and ozone depletion. (Chlorine monoxide, a chemical that reacts readily with ozone, forms when CFCs disintegrate.) These observations also helped reveal why high chlorine levels might become especially damaging over the Arctic and Antarctic: ice particles in stratospheric clouds provide surfaces on which longer-lived chlorine molecules are converted into forms like chlorine monoxide. The ice particles can also "lock up" forms of nitrogen that would normally react with chlorine to form more stable compounds.

Then in March 1988, the prestigious Ozone Trends Panel released an analysis showing that ozone loss was considerably greater over middle latitudes than computer models had predicted. The panel concluded that from 1969 to 1986, ozone levels had dropped 1.7 to 3 percent in the latitude band 30° to 64°N, which covers most of the United States, Europe, the Soviet Union, and China. Wintertime depletion in the northern portion of this region was even more severe: 5 to 6 percent.

These decreases mean that ever higher levels of dangerous ultraviolet light are probably penetrating to the earth's surface. According to the U.N. Environment Programme, every 1 percent drop in ozone levels will lead to a 3 percent increase in non-melanoma skin cancers in light-skinned people, as well as dramatic increases in cataracts, lethal melanoma cancers, and damage to the human immune system. Higher levels of ultraviolet light may also worsen ground-level pollution and hurt plants, animals, and ecosystems, especially light-sensitive single-celled aquatic organisms.

Although recent findings provide conclusive evidence that human activities have been damaging the ozone layer, the political situation, as much as the scientific evidence, has changed significantly since 1987. International opinion was divided only three years ago, making even modest reductions in CFC production and use difficult to attain. Most CFC producers, who had bene fighting any kind of controls since the 1970s, argued that scientific evidence was too inconclusive to warrant regulation, although some were willing to support limited international regulation. These producers had only recently resumed testing of alternative chemicals, having drastically cut back on research in the early 1980s.

The outlook changed in March 1988 when DuPont, the largest producer of CFCs, reversed its position and said it would phase out its production. (It has now said this will occur by the year 2000.) Although not all corporations have agreed to change as readily as DuPont, the trend is in that direction. Both producers and users have adjusted to the idea that CFCs are not likely to be on the market much longer.

In March 1989, the European Economic Community announced support for a complete phaseout of CFCs by 2000. President Bush also announced support for a 2000 phaseout, contingent on the availability of substitutes. A phaseout will probably become a binding amendment to the protocol this year, though there may be some controversy over the schedule for intermediate cutbacks. Participants in a preliminary meeting in Helsinki last year also supported an undetermined date for a halon phaseout, and a cutback in other unspecified substances that deplete ozone.

Those "other" substances have been much slower to receive high-profile attention than CFCs. This is largely because no one had systematically reviewed compounds such as methyl chloroform and carbon tetrachloride before the Montreal Protocol, although some evidence that they cause damage was available. Not till many months after the Ozone Trends Panel announcement did the U.S. EPA and some environmental groups show that chlorine levels could double, triple, or even quadruple because of these chemicals, even if CFCs were completely phased out. Models now reveal that overall chlorine concentrations could expand from around 3 parts per billion by volume (ppbv) in 1985 to around 12 ppbv in 2100 under the protocol's current limits.

Of course, actual stratospheric loading and ozone destruction could be less—or more. Other changes in the atmosphere, including global warming, may have unpredictable effects on the stratosphere. For example, the stratosphere could cool as greenhouse gases trap heat near the earth's surface. Or sulfuric-acid aerosols from industrial pollution and volcanos could act like the icy particles in polar clouds, speeding up ozone loss at middle latitudes.

To allow any significant increase in ozone-depleting chlorine, much less a rise to 12 ppbv, is to court catastrophe. The sudden occurrence of the Antarctic ozone hole in the late 1970s, as well as the unexpected effects over mid-latitudes, show that growing chlorine concentrations are producing effects much greater than scientists believed only a few years ago.

DEALING WITH METHYL CHLOROFORM

If CFCs are phased out, methyl chloroform will be the single greatest source of ozone-depleting chlorine emissions. Industry uses this solvent—1,1,1 Trichloroethane, or TCA—extensively to clean metal. It is also used in aerosols, coatings, and adhesives, as well as in cleaning electronic equipment. Molecule for molecule, methyl chloroform is not as ozone depleting as the five regulated CFCs, since most of it degrades in the lower atmosphere. Its punch is only about 15 percent that of CFC-11, one of the two most widely used CFCs, for example. But methyl chloroform emissions pose a serious threat to the ozone layer because industries use almost as much of it as all the CFCs combined.

EPA figures show that this compound accounts for about 15 percent of today's atmospheric chlorine buildup—almost four times more than CFC-113, the primary CFC solvent. Although methyl chloroform emissions have stabilized in the United States, regulation of other chlorinated solvents could boost releases once again. Besides cutting back the use of CFC-113, EPA is restricting the use of volatile organic solvents (VOCs) because they contribute to ground-level pollution. EPA has also lowered exposure limits for perchloroethylene and trichloroethylene, which could substitute for CFC-113 if they weren't so toxic. These changes could lead to significant increases in methyl chloroform's contribution to stratospheric chlorine.

A phaseout of methyl chloroform, in contrast, would cut chlorine levels rapidly and dramatically in a way that a CFC ban cannot match. That's because the former has a half-life of 6 years, while CFC-11 and CFC-12, the longest-lived CFCs, have half-lives of about 60 years and 120 years, respectively. If methyl chloroform emissions are not halted, declines in ozone destruction will be much slower, since the CFCs and halons already released will continue to cause damage for decades.

Fortunately, phasing out methyl chloroform should be relatively easy. Replacements are available for 90 to 95 percent of the chemical's uses, according to a recent evaluation prepared for the U.N. Environment Programme. Several corporations, including General Dynamics, have already eliminated it or are in the process of doing so. And, as is true of CFC-113, most of the alternatives, such as water-based cleaning, are less damaging to the environment. (In water-based cleaning, slightly acidic or alkaline soapy water is sprayed onto a surface. The wastewater can be recycled.) Until recently, however, most businesses were not aware of the threat that methyl chloroform poses to the ozone layer, and many were in fact planning to substitute it for CFC-113. With better information and stricter controls, the transition from both CFC-113 and methyl chloroform to alternatives could proceed rapidly.

CUTTING CARBON TETRACHLORIDE

The potential for phasing out most uses of carbon tetrachloride is also good, and there are similarly compelling reasons to do so. Carbon tetrachloride is used as a solvent in a range of common metal-cleaning applications, and in manufacturing CFCs. The 1987 protocol overlooked this chemical despite the fact that it, like the more well-known CFCs, does not break down in the lower atmosphere. Pound for pound, it is 10 to 20 percent more ozone-depleting than CFC-11, the most potent of the CFCs.

Since carbon tetrachloride is highly toxic, the United States and many other Western countries have largely restricted its use to the manufacture of CFCs. But Eastern Europe, the Soviet Union, and the Third World still use carbon tetrachloride as a solvent in considerable quantities. It is not uncommon in some parts of the Third World to see auto mechanics dipping parts into open vats of carbon tetrachloride with their bare hands.

Carbon tetrachloride is attractive to Third World countries because it is much cheaper—costing around 25 cents per pound—than many other chemicals, including CFC-113, which costs around $1 per pound. The Third World's tendency to use carbon tetrachloride could be even greater when the alternative is no longer CFC-113 but hydrochlorofluorocarbons (HCFCs) and hydrofluorocarbons (HFCs), replacements for CFCs that will probably cost several times as much as carbon tet. And if a revised Montreal Protocol does not cover that compound, production capacity idled by a CFC phaseout in industrialized countries could be redirected to provide solvents to the Third World.

Growing use of carbon tetrachloride in developing nations would lead not only to significant increases in ozone-depleting emissions but would also contribute to severe worker health problems and water pollution. Since less toxic replacements are available for virtually all solvents, carbon tetrachloride could be confined to use as a feedstock in making some CFC substitutes.

RISKY REPLACEMENTS

The new HCFCs and HFCs created to replace CFCs are more controversial targets for regulation than methyl chloroform or carbon tetrachloride. HFCs do not contain chlorine and thus do not deplete ozone. HCFCs, which contain hydrogen in addition to the chlorine, fluorine, and carbon normally present in CFCs, are much less ozone depleting because they largely break down in the lower atmosphere.

HCFC-22, the only HCFC widely available, is used in some refrigeration systems as well as in aerosols and foam blowing. (It is now the primary gas used to make styrofoam food packaging in the United States.) According to the EPA, HCFC-22 contributes about 4 percent of the chlorine buildup in the stratosphere. But as CFCs are phased out, HCFC-22 and other HCFCs being developed will undoubtedly account for more of the buildup.

Both HCFCs and HFCs pose another problem: they contribute to global warming. Like their relatives the CFCs, HCFCs and HFCs trap infrared radiation refracted from the earth's surface in wavelengths that other atmospheric constituents do not. Although the new chemicals have less of an impact than CFCs because they break down faster, many have global-warming effects as high as one-third—and some as much as two-thirds—that of CFC-11. Since CFCs are expected to contribute from 15 to 25 percent of the greenhouse problem, the possibility that HCFCs and HFCs may take their place is worrisome.

HCFCs and HFCs are unlikely to replace CFCs completely. In many cases, alternatives such as water-based cleaning will be the best substitutes for CFC solvents. However, HCFC use could also expand into new areas. For example, the United States, Canada, and Sweden have banned most uses of CFCs in aerosol sprays, although those products continue to account for about one-third of CFC use in the rest of the world. The U.S. ban alone has prevented the release of about 1 billion tons of CFCs over the past decade. But the 1977 U.S. legislation mandating this ban does not include HCFCs. If they became the standard propellant in aerosol sprays, HCFCs could constitute the world's single most significant source of ozone-depleting emissions.

In the long run, replacing 30 percent of CFCs with HCFCs could keep stratospheric chlorine levels above those that caused the Antarctic ozone hole—even *with* a ban on CFCs, halons, methyl chloroform, and carbon tetrachloride. And some scientists hypothesize that damage from HCFCs and other chemicals could increase substantially if the atmosphere becomes overburdened with pollutants and loses some of its capacity to cleanse itself.

Some parties at the upcoming London meeting will propose regulating or at least monitoring these compounds. A few participants have suggested phasing them out in 20 to 30 years, while others want to prohibit their use in non-essential areas such as aerosol sprays and foam not used for insulation.

But chemical companies such as DuPont and Allied-Signal are naturally encouraging a transition to the compounds they raced to design when regulation of CFCs became imminent. Some of the largest international producers have forged an unusual collaboration, known as the Program for Alternative Fluorocarbon Toxicity Testing (PAFT), to test the new compounds as quickly as possible. And some industries are counting almost exclusively on them to replace CFCs. For example, auto companies are expecting to market HFC-134a air-conditioners within just a few years, while refrigeration firms are looking hard at various mixtures of HCFCs as compressor-gas substitutes. These chemicals will undoubtedly fill the gap in vital areas where no other substitutes are available. But unfortunately, little research is apparently being done on longer-term options.

Helium refrigeration may be one such option. Although helium is commonly used for very low temperature cooling, it has generally been too expensive for home refrigeration, since critical parts must be machined to within one ten-thousandth of an inch. However, Cryodynamics Corp., a New Jersey-based company, claims to have developed a helium-cooling technology suitable for domestic refrigerator and freezer temperatures. A team at Oak Ridge National Laboratory thinks the efficiency of these freezers and refrigerators could be equal to or better than that of the best CFC-based technology within a few years, although some reliability and efficiency questions remain. The potential of the Cryodynamics model is not yet clear, but this kind of system clearly merits further study. A group at Oxford University has apparently run a helium-based cooler at domestic temperatures for 40,000 hours.

INCREASING THIRD WORLD PARTICIPATION

In many respects, progress on eliminating ozone-depleting chemicals has been phenomenal: science and technology have been on a fast track, and the international policymaking process has encouraged these changes. Yet lack of participation in the Montreal Protocol by many Third World countries promises to be a serious obstacle to cutting stratospheric chlorine. As environmentalists pointed out when the protocol was signed, Third World exemptions mean that real consumption of CFCs will likely drop by only 35 percent, rather than the targeted 50 percent.

The protocol allows participating Third World countries to increase CFC use for 10 years before they, like their industrialized counterparts, must reduce consumption by half. Developing countries can take this temporary exemption as long as per capita use remains no higher than .3 kilogram per year—six times their current levels. But many of the most industrialized Third World countries, including India and China, have chosen not to participate in the protocol at all.

Liu Ming-Pu, the Chinese environment commissioner, pointed out at the 1989 Saving the Ozone Layer conference in Britain that developing countries would suffer unduly from the protocol's restrictions. He said that China would not participate unless wealthy nations help poorer nations develop and buy new technologies that do not rely on ozone-depleting compounds. Many large Third World countries, including India and China, have been making substantial investments in CFC-based refrigeration, for example—investments that they now stand to lose.

Third World countries are in a difficult position. CFC replacement chemicals are expected to be expensive. A pound of CFC-11 cost about 50 cents in 1987. The projected 1994 price for HFC-134a, a substitute refrigerant, is $3. And developing and adopting technologies that are less ozone-depleting promises to be a major undertaking. Some substitutes may involve equipment that is less energy-efficient and less durable.

As China has pointed out, the industrialized countries have been the source of the vast majority of ozone-depleting releases. In fact, OECD and Soviet-bloc countries accounted for more than 85 percent of all CFC emissions in 1985. And these countries will continue to be the main source of releases for at least the next decade.

However, failing to obtain broader Third World participation in a stronger protocol would be a mistake. If China and India were to use CFCs at the full .3 kilogram per capita each year, worldwide consumption would rise half again above 1985 levels. If Third World nations consumed at the level of industrialized countries, world use would roughly triple. Such enormous increases would clearly cripple all attempts to limit stratospheric chlorine.

The original protocol, which simply urges technology transfer between rich and poor nations without providing concrete funding, is clearly inadequate. A well-funded research program, perhaps relying on collaboration between scientists from rich and poor countries, could tailor new technologies for Third World use and even foster specialized industries in countries like India and China. The U.N. Environment Programme has begun holding informal discussions with industrialized countries that could provide the needed funding.

We are in the midst of both dangerous pollution of the earth and dramatic efforts to change course. The Montreal Protocol has proven a flexible instrument of international diplomacy. But it needs to be amended, and its international reach broadened. The ability to deal equitably with the needs of the Third World will be a serious test for tackling the far more difficult task of controlling CO^2 emissions and global climate change. Closing the ozone hole is only the beginning.

POSTSCRIPT

Is the Montreal Protocol Adequate for Solving the Ozone Depletion Problem?

The article by Makhijani, Bickel, and Makhijani was written prior to the June 1990 international meeting that considered modifications to the Montreal Protocol. Several of the concerns that they raise were discussed and were incorporated into amendments to the protocol. These include target dates for the complete phase-out of the use of methyl chloroform and CFCs. A fund of $240 million will be established to aid developing nations in meeting the cost of replacing CFCs by other technologies. China and India have now signed the treaty. Nevertheless, environmental organizations still argue that the scope and timetables of the agreement are not adequate since they will allow up to 20 billion pounds of ozone-depleting chemicals to be emitted over the next 25 years. World leaders are scheduled to meet again in 1992 to renegotiate the treaty.

Anyone familiar with the history of international agreements will know that simply negotiating a treaty does not assure compliance. Greenpeace and other environmental organizations are continuing to mobilize public pressure to guarantee enforcement of the provisions of the agreement. In "The World Can't Wait for DuPont," *Greenpeace* (July/August 1990), Judy Christrup accuses the principal U.S. producer of CFCs of hedging on its commitment to cooperate in efforts to end ozone depletion.

F. S. Rowland (coauthor of the research paper that first predicted the effect of CFCs on the ozone layer) has published a detailed summary of the scientific, technological, and regulatory aspects of the problem in an article in the Summer 1988 issue of *Environmental Conservation.*

For an in-depth analysis of the provisions of the original Montreal Protocol, see "The Montreal Protocol: A Dynamic Agreement for Protecting the Ozone Layer," by Jamison Koehler and Scott Hajost, *Ambio* (April 1990). Dr. Mostafa Tolba, executive director of the United Nations Environment Program, is widely acknowledged to have played a key role in gaining the necessary consensus at the Montreal meeting that produced the protocol. For his own assessment of the significance of this achievement, see "The Ozone Agreement—and Beyond," *Environmental Conservation* (Winter 1987).

The "ozone hole" poses a specific threat to the Antarctic food chain. For a discussion of this local, but significant aspect of the problem, see "Fragile Life Under the Ozone Hole," by S. Z. El-Sayed, *Natural History* (October 1986).

ISSUE 18

Are Abundant Resources and an Improved Environment Likely Future Prospects for the World's People?

YES: Julian L. Simon, from "Life on Earth Is Getting Better, Not Worse," *The Futurist* (August 1983)

NO: Lindsey Grant, from "The Cornucopian Fallacies: The Myth of Perpetual Growth," *The Futurist* (August 1983)

ISSUE SUMMARY

YES: Economist Julian L. Simon is optimistic about the likelihood that human minds and muscle will overcome resource and environmental problems.
NO: Environmental consultant Lindsey Grant fears that, unless a "sustainable relationship between people and earth" is developed, the future may bring famine and ecological disaster.

In 1972, the results of a study by a Massachusetts Institute of Technology computer modeling team triggered an avalanche of controversy about the future course of worldwide economic growth. The results appeared in a book entitled *The Limits to Growth* (Universe Books, 1972). The book's authors—Donella Meadows, Dennis Meadows, Jorgen Randers, and William Behrens—predicted that exponential growth in population and capital, accompanied by increasing pollution, would culminate in sudden resource depletion and economic collapse before the middle of the next century. The sponsors of the study, a group of rich European and American industrialists called the Club of Rome, popularized its conclusions by distributing 12,000 copies of the book to prominent government, business, and labor leaders.

Critiques of the study emerged from all sectors of the political spectrum. Conservatives rejected the implication that international controls on industrial development were necessary to prevent disaster. Liberals asserted that no-growth policies would hurt the poor more than the affluent. Radicals contended the results were only applicable to the type of profit-motivated growth that occurs under capitalism. Among the universal criticisms of the study were the simplicity of the computer models used and the questionable practice of making long-term extrapolations based on present increasing growth rates. The book's authors admitted that no attempt was made to incorporate the complex sociopolitical interactions that can profoundly affect the type and level of international industrial activities.

Although the debate about the specific catastrophic predictions of Meadows et al. has died down, the questions raised during that controversy continue to receive attention. In 1980 a three-volume publication entitled *The Global 2000 Report to the President* was released by the U.S. government. This report, which has sold over half a million copies, is the result of a joint study by the Department of State and the Council on Environmental Quality under President Carter of trends in population growth, natural resource development, and environmental quality through the end of the century. The dire projections of this study include increased environmental degradation, continued abuse of natural resources, and a widening of the gap between the rich and the poor.

This study, like its predecessors, has had its share of methodological criticism. For example, anticipated changes in energy use during the period of the study are not taken into account. Despite these flaws, the *Global 2000 Report* has contributed to the growing consensus that present patterns and rates of worldwide industrial growth are likely to cause intolerable environmental stress. One of the principal opponents of this pessimistic view has been the energetic, outspoken economist Julian L. Simon. Simon reviews the facts behind the *Global 2000 Report* and finds reason to believe that the quality of life of the world's peoples and the environment will continue to improve in the foreseeable future. Lindsey Grant, a consultant to the Environmental Fund, takes issue with Simon and the late futurologist Herman Kahn (founder and former director of the conservative Hudson Institute), both of whom he refers to as "cornucopians." Grant argues that the analytical methods they use are seriously flawed and the laissez-faire approach they advocate with respect to population growth, resource depletion, and environmental degradation will indeed lead to the gloomy forecast of the *Global 2000 Report,* which depicts an overpopulated world with widespread famine, insufficient natural resources and monumental ecological distress. Grant does not think this pessimistic prospect is inevitable—but avoiding it will require increased governmental planning and regulation.

YES

Julian L. Simon

LIFE ON EARTH IS GETTING BETTER, NOT WORSE

If we lift our gaze from the frightening daily headlines and look instead at wide-ranging scientific data as well as the evidence of our senses, we shall see that economic life in the United States and the rest of the world has been getting better rather than worse during recent centuries and decades. There is, moreover, no persuasive reason to believe that these trends will not continue indefinitely.

But first: I am *not* saying that all is well everywhere, and I do not predict that all will be rosy in the future. Children are hungry and sick; people live out lives of physical or intellectual poverty, with little opportunity for improvement; war or some new pollution may finish us. What I *am* saying is that for most relevant economic matters I have checked, aggregate trends are improving rather than deteriorating. Also, I do not say that a better future will happen automatically or without effort. It will happen because men and women will use muscle and mind to struggle with problems that they will probably overcome, as they have in the past.

LONGER AND HEALTHIER LIVES

Life cannot be good unless you are alive. Plentiful resources and a clean environment have little value unless we and others are alive to enjoy them. The fact that your chances of living through any given age now are much better than in earlier times must therefore mean that life has gotten better. In France, for example, female life expectancy at birth rose from under 30 years in the 1740s to 75 years in the 1960s. And this trend has not yet run its course. The increases have been rapid in recent years in the United States: a 2.1-year gain between 1970 and 1976 versus a 0.8-year gain in the entire decade of the 1960s. This pattern is now being repeated in the poorer countries of the world as they improve their economic lot. Life expectancy at birth in low-income countries rose from an average of 35.2 years in 1950 to 49.9 years in 1978, a much bigger jump than the rise from 66.0 to 73.5 years in the industrialized countries.

The threat of our loved ones dying greatly affects our assessment of the quality of our lives. Infant mortality is a reasonable measure of child mortality generally. In Europe in the eighteenth and nineteenth centuries, 200 or more children of each thousand died during their first year. As late as 1900, infant mortality was 200 per 1000 or higher in Spain, Russia, Hungary, and even Germany. Now it is about 15 per 1000 or less in a great many countries.

Health has improved, too. The incidence of both chronic and acute conditions has declined. While a perceived "epidemic" of cancer indicates to some a drop in the quality of life, the data show no increase in cancer except for deaths due to smoking-caused lung cancer. As Philip Handler, president of the National Academy of Sciences, said:

> The United States is not suffering an "epidemic of cancer," it is experiencing an "epidemic of life"—in that an even greater fraction of the population survives to the advanced ages at which cancer has always been prevalent. The overall, age-corrected incidence of cancer has not been increasing; it has been declining slowly for some years.

ABATING POLLUTION

About pollution now: The main air pollutants—particulates and sulfur dioxide—have declined since 1960 and 1970 respectively, the periods for which there is data in the U.S. The Environmental Protection Agency's Pollutant Standard Index, which takes into account all the most important air pollutants, shows that the number of days rated "unhealthful" has declined steadily since the index's inauguration in 1974. And the proportion of monitoring sites in the U.S. having good drinking water has greatly increased since record-keeping began in 1961.

Pollution in the less-developed countries is a different, though not necessarily discouraging, story. No worldwide pollution data are available. Nevertheless, it is reasonable to assume that pollution of various kinds has increased as poor countries have gotten somewhat less poor. Industrial pollution rises along with new factories. The same is true of consumer pollution—junked cars, plastic wrappers, and such oddments as the hundreds of discarded antibiotics vials I saw on the ground in an isolated Iranian village. Such industrial wastes do not exist in the poorest preindustrial countries. And in the early stages of development, countries and people are not ready to pay for clean-up operations. But further increases in income almost surely will bring about pollution abatement, just as increases in income in the United States have provided the wherewithal for better garbage collection and cleaner air and water.

THE MYTH OF FINITE RESOURCES

Though natural resources are a smaller part of the economy with every succeeding year, they are still important, and their availability causes grave concern to many. Yet, measured by cost or price, the scarcity of all raw materials except lumber and oil has been *decreasing* rather than increasing over the long run. . . .

Perhaps surprisingly, oil also shows downward cost trend in the long run. The price rise in the 1970s was purely political; the cost of producing a barrel of oil in the Persian Gulf is still only perhaps 15 to 25 cents.

There is no reason to believe that the supply of energy is finite, or that the price will not continue its long-run decrease. This statement may sound less preposterous if you consider that for a quantity to be finite it must be measurable. The future supply of oil includes what we usually think of as oil, plus the oil that can be produced from shale, tar sands, and coal. It also includes the oil from plants that we grow, whose key input is sunlight. So the measure of the future oil supply must therefore be at least as large as the sun's 7 billion or so years of future life. And it may include other suns whose energy might be exploited in the future. Even if you believe that one can in principle measure the energy from suns that will be available in the future—a belief that requires a lot of confidence that the knowledge of the physical world we have developed in the past century will not be superseded in the next 7 billion years, plus the belief that the universe is not expanding—this measurement would hardly be relevant for any practical contemporary decision-making.

Energy provides a good example of the process by which resources become more abundant and hence cheaper. Seventeenth-century England was full of alarm at an impending energy shortage due to the country's deforestation for firewood. People feared a scarcity of fuel for both heating and the vital iron industry. This impending scarcity led inventors and businessmen to develop coal.

Then, in the mid-1800s, the English came to worry about an impending coal crisis. The great English economist William Stanley Jevons calculated then that a shortage of coal would surely bring England's industry to a standstill by 1900; he carefully assessed that oil could never make a decisive difference. But spurred by the impending scarcity of coal (and of whale oil, whose story comes next), ingenious and profit-minded people developed oil into a more desirable fuel than coal ever was. And today England exports both coal and oil.

Another strand in the story: Because of increased demand due to population growth and increased income, the price of whale oil used in lamps jumped in the 1840s. Then the Civil War pushed it even higher, leading to a whale oil "crisis." The resulting high price provided an incentive for imaginative and enterprising people to discover and produce substitutes. First came oil from rapeseed, olives, linseed, and pine trees. Then inventors learned how to get coal oil from coal, which became a flourishing industry. Other ingenious persons produced kerosene from the rock oil that seeped to the surface. Kerosene was so desirable a product that its price rose from 75 cents to $2 a gallon, which stimulated enterprisers to increase its supply. Finally, Edwin L. Drake sunk his famous oil well in Titusville, Pennsylvania. Learning how to refine the oil took a while, but in a few years there were hundreds of small refiners in the U.S. Soon the bottom dropped out of the whale oil market: the price feel from $2.50 or more a gallon at its peak around 1866 to well below a dollar.

Lumber has been cited as an exception to the general resource story of falling costs. For decades in the U.S., farmers clearing land disposed of trees as a nuisance. As lumber came to be more a commercial crop and a good for builders and railroad men, its price rose. For some time, resource economists expected the price to hit a plateau and then follow the course of other raw materials as the transition to a commercial crop would be completed. There was evidence consistent with this view in the increase, rather than the popularly supposed decrease, in the tree stock in the U.S., yet for some time the price did not fall. But now that expectation seems finally to have been realized as prices of lumber have fallen to a fourth of their peak in the late 1970s.

MORE FOOD FOR MORE PEOPLE

Food is an especially important resource, and the evidence indicates that its supply is increasing despite rising population. The long-run prices of food relative to wages, and even relative to consumer goods, are down. Famine deaths have decreased in the past century even in absolute terms, let alone relative to the much larger population, a special boon for poor countries. Per person food production in the world is up over the last 30 years and more. And there are no data showing that the people at the bottom of the income distribution have fared worse, or have failed to share in the general improvement, as the average has improved. Africa's food production per capita is down, but that clearly stems from governmental blunders with price controls, subsidies, farm collectivization, and other institutional problems.

There is, of course, a food-production problem in the U.S. today: too much production. Prices are falling due to high productivity, falling consumer demand for meat in the U.S., and increased foreign competition in such crops as soybeans. In response to the farmers' complaints, the government will now foot an unprecedentedly heavy bill for keeping vast amounts of acreage out of production.

THE DISAPPEARING-SPECIES SCARE

Many are alarmed that the earth is losing large numbers of its species. For example, the *Global 2000 Report to the President* says: "Extinctions of plant and animal species will increase dramatically. Hundreds of thousands of species—perhaps as many as 20 percent of all species on earth—will be irretrievably lost as their habitats vanish, especially in tropical forests," by the year 2000.

The available facts, however, are not consistent with the level of concern expressed in *Global 2000*, nor do they warrant the various policies suggested to deal with the purported dangers.

The *Global 2000* projection is based upon a report by contributor Thomas Lovejoy, who estimates that between 437,000 and 1,875,000 extinctions will occur out of a present estimated total of 3 to 10 million species. Lovejoy's estimate is based on a linear relationship running from 0% species extinguished at 0% tropical forest cleared, to about 95% extinguished at 100% tropical forest cleared. (The main source of differences in the range of estimated losses is the range of 3 to 10 million species in the overall estimate.)

The basis of any useful projection must be a body of experience collected under a range of conditions that encompass the expected conditions, or that can reasonably be extrapolated to the expected conditions. But none of Lovejoy's references seems to contain any scientifically impressive body of experience.

A projected drop in the amount of tropical forests underlines Lovejoy's projection of species losses in the future. Yet to connect these two events as Lovejoy has done requires systematic evidence relating an amount of tropical forest removed to a rate of species reduction. Neither *Global 2000* nor any of the other sources I checked give such empirical evidence. If there is no better evidence for Lovejoy's projected rates, one could extrapolate almost any rate one chooses for the year 2000. Until more of the facts are in, we need not undertake alarmist protection policies. Rather, we need other sorts of data to estimate extinction rates and decide on policy. None of this is to say that we need not worry about endangered species. The planet's flora and fauna constitute a valuable natural endowment; we must guard them as we do our other physical and social assets. But we should also strive for a clear, unbiased view of this set of assets in order to make the best possible judgments about how much time and money to spend guarding them, in a world where this valuable activity must compete with other valuable activities, including the preservation of other assets and human life.

MORE WEALTH FROM LESS WORK

One of the great trends of economic history is the shortening of the workweek coupled with increasing income. A shorter workweek represents an increase in one's freedom to dispose of that most treasured possession—time—as one wishes. In the U.S., the decline was from about 60 hours per week in 1870 to less than 40 hours at present. This benign trend is true for an array of countries in which the length of the workweek shows an inverse relationship with income.

With respect to progress in income generally, the most straightforward and meaningful index is the proportion of persons in the labor force working in agriculture. In 1800, the percentage in the U.S. was 73.6%, whereas in 1980 the proportion was 2.7%. That is, relative to population size, only 1/25 as many persons today are working in agriculture as in 1800. This suggests that the effort that produced one bushel of grain or one loaf of bread in 1800 will now produce the bushel of grain plus what 24 other bushels will buy in other goods, which is equivalent to an increase in income by a factor of 25.

Income in less-developed countries has not reached nearly so high a level as in the more-developed countries, by definition. But it would be utterly wrong to think that income in less-developed countries has stagnated rather than risen. In fact, income per person has increased at a proportional rate at least as fast, or faster, in less-developed than in more-developed countries since World War II.

THE ULTIMATE RESOURCE

What explains the enhancement of our material life in the face of supposed limits to growth? I offer an extended answer in my recent book, *The Ultimate Resource* (1981). In short, the source of our increased economic blessings is the human mind, and, all other things being equal, when there are more people, there are more productive minds. Productivity increases come directly from the additional minds that develop productive new ideas, as well as indirectly from the impact upon industrial productivity of the additional demand for goods. That is, population growth in the form of babies or immigrants helps in the long run to raise the standard of living because it brings increased productivity. Immigrants are the best deal of all because they usually migrate when they are young and strong; in the U.S., they contribute more in taxes to the public coffers than they take out in welfare services.

In the short run, of course, additional people mean lower income for other people because children must be fed and housed by their parents, and educated and equipped partly by the community. Even immigrants are a burden for a brief time until they find jobs. But after the children grow up and enter the work force, and contribute to the support of others as well as increasing productivity, their net effect upon others becomes positive. Over their lifetimes they are a boon to others.

I hope you will now agree that the long-run outlook is for a more abundant material life rather than for increased scarcity, in the U.S. and in the world as a whole. Of course, such progress does not come about automatically. And my message certainly is not one of complacency. In this I agree with the doomsayers—that our world needs the best efforts of all humanity to improve our lot. I part company with them in that they expect us to come to a bad end despite the efforts we make, whereas I expect a continuation of successful efforts. Their message is self-fulfilling because if you expect inexorable natural limits to stymie your efforts you are likely to feel resigned and give up. But if you recognize the possibility—indeed, the probability—of success, you can tap large reserves of energy and enthusiasm. Energy and enthusiasm, together with the human mind and spirit, constitute our solid hope for the economic future, just as they have been our salvation in ages past. With these forces at work, we will leave a richer, safer, and more beautiful world to our descendants, just as our ancestors improved the world that they bestowed upon us.

NO

Lindsey Grant

THE CORNUCOPIAN FALLACIES: THE MYTH OF PERPETUAL GROWTH

An intense if intermittent debate is under way between environmentalists and a pair of traveling "cornucopians," Julian Simon and Herman Kahn, who manage to appear in a remarkable number of forums to press their case. The environmentalists, drawing extensively upon the 1980 *Global 2000 Report to the President*, warn of threats to the ecosystem and to renewable resources such as cropland and forests generated by population growth and exploitative economic activities. The cornucopians say that population growth is good, not bad (Simon), or that it will solve itself (Kahn), that shortages are mythical or can be made good by technology and substitution, and generally that we may expect a glorious future.

The debate has strong political overtones. If things are going well, we don't need to do anything about them—a useful argument for *laissez-faire*. If something is going wrong, the environmentalists usually want the government to do something about it. The debate thus gets mixed up in the current reaction against "petty government interference" and a generalized yearning to return to earlier, more permissive economic and political practices.

One could hardly object to having a couple of cornucopians urging people to be of good cheer and stout heart, were it not for the danger that they may convince some citizens and policy makers not to worry about some pressing problems that urgently need attention.

The cornucopians' argumentation, however, is seriously flawed as a tool for identifying the real and important present trends.

There is an asymmetry in the nature of the arguments of the environmentalists and the cornucopians. The environmentalist—the proponent of corrective action—is (or should be) simply warning of consequences if trends or problems are ignored; he does not need to *predict*. The cornucopian, on the other hand, must predict to make his case. He must argue that problems will be solved and good things will happen if we let nature take its course. Since nobody has yet been able to predict the future, they are asking their listeners to take a lot on faith. They say, in effect, "Believe as I do, and you will feel better." Simon says explicitly that his conversion to his present viewpoint improved his state of mind.

The cornucopians have made assumptions and chosen methodologies that simply ignore or dismiss the most critical issues that have led the environmentalists to their concerns:

- The cornucopians pay little attention to causation, and they project past economic trends mechanically.
- They casually dismiss the evidence that doesn't "fit."
- They employ a static analysis that makes no provision for feedback from one sector to another.
- They understate the implications of geometric growth.
- They base their predictions on an extraordinary faith in uninterrupted technological progress.

Let us look into some of these cornucopian fallacies—the reasoning processes and omissions that characterize Simon's and Kahn's analyses.

EXTRAPOLATING PAST GROWTH: THE WRONG METHODOLOGY

Simon argues that the past is the best guide to the future. Perhaps, but much depends on what part of the past you look at. He devotes most of his effort to demonstrating in various ways that mankind's economic lot has improved in the past century or so, which is not an issue.

Simon bases much of his argument on an econometric study of past correlations between the number of children and economic growth. This approach leaves unanswered the question: Which, if either, phenomenon caused the other one? Or is this simply a process of using complex mathematical relationships to obscure the commonsense proposition that the children shared in a period of prosperity?

Cited in increasingly simplistic terms, that study remains the basis for Simon's views, but his subsequent efforts have been directed almost exclusively to a search for errors in the statistics of *Global 2000* or indeed of any environmentalist, on the assumption that a shaky statistic undermines the credibility of the method.

Kahn is more nimble polemically, but there is less evidence of any systematic undergirding for his projections. He extrapolates mid-twentieth-century growth trends with a line of reasoning that comes very close to economic vitalism. His specialty is impressive graphic presentations of the future, but examination suggests there is more of the airbrush than of intellectual discipline in those graphs.

In one of his major works, *The Next 200 Years*, he projects per capita "gross world product" at $20,000 in 2176 A.D., but his evidence raises a doubt whether these are constant dollars, current dollars, or imaginary ones. As best one can gather from the text, this projection is based on a freehand plot of "S-curves" (slow/fast/slow) of GNP growth for different categories of countries, drawn roughly from the U.S. and European experience.

There are two problems with this method. First, analogy can be a dangerous process. To predict the future performance of the poor countries based upon the past performance of the rich countries may involve too loose an analogy to justify the faith put in it. The analogy assumes that the underlying factors are substantially similar. They are not. In contrast to Europe when it industrialized, poor countries today tend to have faster population growth rates, no colonies where capital can be mobilized, lower incomes (probably), extreme foreign exchange problems, no technological lead over the rest of the world, and no empty new worlds to absorb their emigrants.

Second and even more important, gross national product (GNP)—or "gross world product"—is neither tangible nor real except in people's minds. It has no life of its own. It is simply a way of giving numerical abbreviation to a sum of economic activities. It is determined by underlying realities: the availability and quality of land, water, industrial raw materials, and energy; technological change; the impact of population change on production and consumption; the productivity of the supporting ecosystems; labor productivity; and so on. Kahn simply projects GNP without analyzing the forces that generate it.

Proof of past success is no assurance of future well-being, and the mechanical projection of economic curves is hardly a reliable guide to the future.

Most of us would agree that the general condition of mankind has been improving for a sustained period, at least until the past decade. Indeed, the scale of the growth is a new thing on earth; and the very magnitude of the growth of population and of economic activity is the source of the issue. For the first time, population and economic activities have grown so sharply as to bring them into a new relationship with the scale of the earth itself.

The hallmark of recent history has been this explosive growth, supported by and supporting an extraordinary burst of technological change and mankind's first intensive exploitation of fossil fuels. The central question for the future is not "Did it happen?" but rather: "Can such growth be sustained, or does it itself generate dynamics that will bring the era to an end? If the latter, what will the changes be, and what if anything should mankind be doing to forestall them or shape them in beneficial directions?"

IGNORING CLIMATE CHANGE

As a single example, let us take the question of carbon dioxide in the atmosphere. It takes little imagination to recognize that CO_2-induced rainfall and temperature changes, rising sea levels, and perhaps the necessity to curtail fossil fuel use could influence future economic activities. To most of us, the fact that human activity is changing the very chemical composition of the air we live in would seem adequate justification to bring the issue into any consideration of current trends affecting the well-being of mankind.

Global 2000 devoted 14 pages to man-induced effects on the climate, focusing primarily on CO_2 but dealing with other issues as well. It concluded that agreed climate projections are not currently possible to make, but it called attention to the problem: "The energy, food, water and forestry projections [in the report] all assume implicitly a continuation of the nearly ideal climate of the 1950s and 1960s. . . . The scenarios are reported here to indicate the range of climatic change that should be analyzed in a study of this sort."

Simon seems to have ignored the carbon dioxide issue.

Kahn discusses the problem along with other possible causes of a warming trend. He concedes that a warming trend might raise the level of the oceans, but argues that "this would hardly mean the end of human society. Major shifts, might be forced in agricultural areas and in coastal cities." He concludes, "It seems unlikely now that the carbon dioxide content will ever double unless mankind wants it to happen." Thus, he cheerfully dismisses this problem and thereby illustrates the curious inversion of his logic.

Why does he dismiss substance for "prediction"? He dismisses the real issues for fear they would lead his readers to lose faith in the future he has promised them. Would he not better join the environmentalists and concentrate on telling his audience that, if they want that future, they may need to take the carbon dioxide problem seriously?

Which intellectual approach is the more valid way of attempting to understand current trends affecting human welfare?

If you seek a sense of what will shape the future, examine the issues generated by population and economic growth; do not simply extrapolate the growth. Economic changes cannot be studied in a vacuum.

DOCTORING THE NEWS

One cannot escape the feeling that some of the Simon/Kahn rebuttals of "bad news" are directed more by polemical ends than by an effort to get at the truth. Such casual hip-shots are more likely to generate doubts about the writer's credentials than to convince readers that bad news is false.

For example, Simon makes points by using gross totals rather than per capita figures and shifts sources to manufacture trends. On world population, for example, to show that "U.N. and other standard estimates" have been steadily lowering their projections of anticipated population in 2000, he starts with a 1969 U.N. worst-case scenario, higher than their "high" series, then moves down to a later U.N. "low" series projection, and winds up with a 1977 Worldwatch Institute figure, justifying his inclusion of the Worldwatch figure by saying that Worldwatch is U.N.-supported. Through these devices, he manages to show the projection declining from 7.5 billion to 5.4 billion. In fact the U.N. projection for 2000 has remained remarkably constant, the median projection having fluctuated between 6.1 and 6.5 billion since 1957. And Lester Brown points out that his Worldwatch figure was not a projection but a proposed timetable.

Also, to prove that *world* air quality is improving, Simon, in *The Ultimate Resource,* cites statistics on *U.S.* air quality in the 1970s. His data are dated and limited, but they are nevertheless gratifying. He pays the environmentalists whom he excoriates the ultimate compliment of appropriating their work. If U.S. air quality has stabilized or improved in some ways in the past decade, it is at least in some measure the product of environmental efforts such as the Clean Air Act.

Kahn, generally takes a subtler line. He, too, points to improvements in air quality, but he is quite willing to accept the need for some expenditures on air quality and other environmental measures and he includes "possible damage to earth because of complicated, complex and subtle ecological and environmental effects" among eight "real issues of the future." In effect, his technique is to admit the possibility of environmental problems but to avoid focusing on them or attempting to measure their importance; he moves quickly on to extolling the brightness of the future and attacking those he deems pessimistic.

The cornucopians slight the resource and environmental issues that the environmentalists consider the most important questions to be examined.

THE LACK OF FEEDBACK

The cornucopians stand breathless on the edge of wonderful new expectations. Simon writes: "Energy . . . is the 'master resource'; energy is the key constraint on the availability of all other resources. Even so, our energy supply is nonfinite. . . ."

Certainly there are remarkable possibilities implicit in our growing awareness of what can be done with energy. But energy does not solve all problems.

The Sorcerer's Apprentice learned that immense power is not always benign to those who set it in motion. All of us have learned many sobering things about nuclear power since 1945.

Any projection for continued expansion in the use of energy must ask the question: What are the implications of developing the energy for the environment and for resources, and what are the consequences of its use likely to be? The same question should be asked about projections calling for continuing expansion in the use of chemicals, or indeed of any physical resource.

Global 2000 undertook to carry out as much as it could of this kind of interactive analysis and found that the state of current knowledge did not permit it to be carried very far. Nevertheless, it undertook to examine literally hundreds of such interactions.

The agricultural projections, for instance—themselves central to other major projections such as population and GNP—require certain assumptions about intensification of agriculture, a doubling or trebling of chemical fertilizer inputs (to a point where man-made introduction of nitrogen compounds into the biosphere will exceed the natural production), parallel increases in herbicides and pesticides, and reliance upon monocultures. These assumptions in turn generate questions concerning desertification, the conversion of forest and loss of forest cover, the effect of intensive agriculture on soil productivity, the impact of increased fertilizer application on watercourses and fisheries and perhaps on climate, and risks associated with pesticides and monocultures—all of which relate back to the initial assumptions about agricultural productivity and eventually to GNP and population assumptions. The degree of confidence concerning different interrelationships is made clear, and reference is made to the technologies that can help forestall or mitigate the harmful interactions foreseen.

There is nothing remotely approaching this sort of interactive analysis in the works of the cornucopians. Kahn simply projects economic growth and assumes that the necessary inputs will be available and that the environmental problems will be surmounted. Simon does not address these questions in any integrated fashion. One may question whether they are even addressing themselves to the real issue.

The speed with which technology is changing, the demands for economic growth posed by population growth, and the effort to raise living standards in developing countries are combining to force change at an unprecedented rate, which makes the study of the future more important than ever. The principal purpose of future studies should be to look as far ahead as possible, to study the implications of current and projected activity, to see how different sectors and issues interrelate. This process is anything but static. It should be a continuing process of probing and testing the potential consequences of different activities and directions of growth, of identifying the issues that need attention and the potential directions for beneficial change.

It was the lack of and the need for this capability that *Global 2000* highlighted. A follow-up study made specific recommendations as to how the capability might be improved within the U.S. government. Simon and Kahn, standing aside and reassuring everybody that the future looks good, seem strangely irrelevant to this entire process.

THE INFINITE-EARTH FALLACY

Neither Kahn nor Simon successfully deals with the simple facts that the earth is finite and that no physical growth can be indefinitely sustained. Let us cite three mathematical examples of the power of geometric growth, and preface them with the warning that they are not predictions:

- Even if the entire mass of the earth were petroleum, it would have been exhausted in 342 years if pre-1973 rates of increase in consumption had been maintained.
- Assume that we have one million years' supply of something—anything with a fixed supply—at current rates of consumption. Then let us increase the rate by just 2% per year (very roughly the current world population growth rate). Now, how long would the supply last? Answer: 501 years.
- At current growth rates, how long would it take for the world's human population to reach the absurdity of one person on each square meter of ice-free land? Answer: about 600 years.

These things won't happen. Resource use won't rise in a geometric curve until a resource is exhausted, then plunge suddenly to zero. There will be changes in real prices, adjustments, and substitutions—the whole pattern of constantly shifting realities that makes prediction impossible. The population will never remotely approach such a level. Long before then, birthrates will fall sharply, death rates will rise, or both.

However, the examples dramatize that the outer limits to current growth patterns are not so very far away. Populations have exceeded the carrying capacity of local environments many times and have sometimes paid the price of a population collapse, but human geometry for the first time requires that we think in terms of the relationship of population and economic activities to the entire earth.

World population has risen from about 1 billion to about 4.5 billion in about six generations. The demand for resources and the environmental impacts have been more than proportional, because per capita consumption has risen. This is not a mathematical fantasy or a projection for the future. It is a description of current reality. What the mathematical examples above suggest is that there are real limits, and not so very far away, that lead inescapably to this conclusion: Indefinitely sustained growth is mathematically impossible on a finite earth.

Kahn and Simon offer several responses to this point, none of them satisfactory.

• They fudge the problem by shifting the calculations. They project the potential longevity of supply of raw materials based on *current* demand rather than on increasing demand. Kahn and Simon have both used this technique. Since they are also assuming rising populations and rising per capita consumption, this is not an argument. It is a moonbeam. The calculations above should have disposed of it permanently.

A more sophisticated variant is to say that GNP will rise, but not resource consumption, because we will be more efficient and we will be consuming more intangibles such as culture. Very likely, within limits. However, nobody has yet drawn a model of sustained growth relying upon the consumption of operas to feed the multitudes.

• They suggest that the problem is so far away as to be irrelevant to those living now. Simon, in a bit of sophistry that he has probably come to regret by now, says: "The length of a one-inch line is finite in the sense that it is bounded at both ends. But the line within the end points contains an infinite number of points. . . . Therefore the number of points in that one-inch segment is not finite." He then extends the analogy to copper and oil. He argues that we cannot know the size of the resource "or its economic equivalent," and concludes, "Hence, resources are not 'finite' in any meaningful sense."

This kind of argument is really pretty shocking. An inch of string is finite, even if it can theoretically be cut into infinitesimal pieces. The earth is finite, even though we may differ endlessly about how much of a given resource may be available.

• Kahn says that population and consumption levels will stabilize in two centuries. If he paints with an airbrush, it is a broad one. He projects population stabilization at 15 billion, but allows himself a margin of error of two—i.e., the population may be somewhere between 7.5 and 30 billion, or a rise of something between 67% and 567%. Most of us suspect that population will stop growing *somewhere* within that range.

He does not attempt to explore whether the resource base would support the 15 billion population he posits, or what the ecological and environmental effects of such population and consumption levels would be; he simply announces that we can handle them. He thinks that prosperity will lead to lower fertility, but he does not ask whether the population growth itself will in some countries preclude the prosperity he expects. He offers no capital/output analysis to suggest how world consumption levels will progress from where they are to where he hopes they will be. In short, he states a dream without attempting to explore how it will be realized or what the effects of its achievement will be.

• Simon says different things at different times. Sometimes, he advocates population growth without limits of time or circumstance, and he speaks of resource availability and population growth "forever" without recognizing the crudest of barriers: lack of space. Elsewhere, he advocates "moderate" population growth. Still elsewhere, he urges that we not worry about the effects of geometric population growth, since it has never been sustained in the past, and he documents his remark by showing how population growth has been periodically reversed by pestilence, invasion, and famine. Is this the man who professes such warm feelings toward his fellow humans?

Most of us agree that population growth will eventually stop, if only through the operation of the Four Horsemen. Most of us hope that it will be stabilized by limiting fertility rather than through hunger and rising mortality. It is this goal that leads many environmentalists to advocate conscious efforts to limit fertility. Kahn thinks it will happen automatically (but does not know how). Simon, apparently, isn't dismayed at the alternative.

TECHNOLOGY AS A FAITH

Technology is knowledge. It is very difficult to predict knowledge if you don't have it yet, and technological trends are among the least predictable of the forces that will shape our future. The cornucopians are justified in reminding us forcefully of technology. A lot of people from Malthus on have underestimated it, and some environmentalists still ignore it.

Let us agree on one point: The world has been experiencing a burst of remarkable technological growth.

Although Kahn and Simon seem to have missed this point, *Global 2000* assumes that this rate will continue for the next 20 years. This approach may be faulted as too sanguine, but it is perhaps the safest projection given the relatively short time frame.

From here, however, we move to an article of faith among the cornucopians that the more pragmatic among us do not share: that the recent high rate of technological growth will continue *indefinitely*.

Yet Simon's advice to use the past as a guide argues against too much faith. Human history has been characterized by spurts of technological growth alternating with periods of slow growth, dormancy, or retrogression.

Technology may continue its recent phenomenal growth. It may not. It is an act of faith to assume that it will.

In addition, technology is not necessarily benign. It shapes us, as we shape it. Right now, it may be making communications cheaper, while it makes unemployment worse. It helped to generate the spurt in population growth that now concerns the environmentalists. New industrial and agricultural technologies have created many of our present environmental problems. Other technology will almost certainly help us to correct our mistakes. A sensible observer with a feeling for history would be justified in assuming that those solutions will in turn generate new problems to be addressed.

If one chooses *not* to stake human welfare on unsupported faith in technology, a certain caution seems in order. Mankind will not have suffered if population growth is less than the advance of technology makes possible, but it may suffer very seriously if hopes for technology prove too high and if populations outrun the ability of science to support them.

NO LIMITS TO DEBATING

If this article reflects a jaundiced view of the cornucopians' methods, it is not meant to discourage the debate. We can learn from each other.

We are all—cornucopians and environmentalists alike—trying to understand and describe a world in vast change. The technological growth on which the cornucopians pin their hopes is itself part of that change, as are the population growth and the environmental by-products of technological growth that concern the environmentalists.

We are all—except perhaps for a few nuts who enjoy human misery—interested in seeing the modern improvement in human welfare continue.

Cornucopians by their nature tend to emphasize *solutions* where environmentalists emphasize *problems*. An interchange can be useful. Do the environmentalists overstate difficulties and fail to recognize new directions that can be helpful? Have we explored the opportunities presented by the oceans, by recent breakthroughs in biology, and by electronics and data processing as thoroughly as we have explored the dangers from desertification, deforestation, and acid rain? Have we pressed for the elimination of legal and administrative impediments to beneficial change as eagerly as we have pressed for restrictive legislation?

If we urge the cornucopians to recognize the problems, we should share their interest in promoting technological change that will help to address the problems.

Those of both persuasions should remember that this is no single battle at Armageddon. Solutions will create their own problems, and problems their solutions. We should perhaps all recognize that only change is constant. And change is very fast right now.

We should ask of the cornucopians that they accept as much. Growth such as we have witnessed cannot be indefinitely extended. We must all seek a sustainable relationship between people and the earth. Most particularly, we must work out the implications of population growth. The issue cannot be: Should it stop? The questions are only: When should it stop? And how?

POSTSCRIPT

Are Abundant Resources and an Improved Environment Likely Prospects for the World's People?

It is certainly tempting to accept Simon's rosy predictions for the future and his faith in the ability of human beings to solve whatever problems they confront. Simon's minority view that increasing world population is positive, rather than problematic, is fully explicated in his book *The Ultimate Resource* (Princeton University Press, 1981). For a recent look at Simon and his world view, his controversial standing among ecologists, and additional background to this debate on the Earth's future, see John Tierney's "Betting the Planet," *The New York Times Magazine* (December 2, 1990).

It is important to distinguish Simon's views from those of environmentalists like Barry Commoner (Issue 5), who accept the premise that population explosion is a problem but argue against population control as the principal mode of controlling environmental degradation. Commoner, unlike Simon, sees the need for very active governmental involvement in technological planning and regulation in order to promote the economic conditions under which decreased population growth can be achieved.

As Grant points out, the analysis of data and extrapolation of trends used by "cornucopians" like Simon and Kahn betray a lack of understanding of ecological principles. For example, they ignore the fact that there may not be clearly observable signals of future environmental problems before it is too late to avert disaster.

The detailed response of Julian Simon and Herman Kahn to the *Global 2000 Report* is the subject of their book *The Resourceful Earth* (Basil Blackwell, 1984).

The issue of resource depletion has recently been incorporated into the more general debate about the concept of "sustainable development." Very few experts share Simon's optimism that the ten billion people that may inhabit the Earth by the end of the next century, if present trends continue, will be able to thrive if current industrial practices and materialistic goals continue to prevail. The World Commission on Environment and Development published a much publicized report, entitled "Our Common Future," on many aspects of this issue in 1987. Commission chairperson Gro Harlem Bruntland, prime minister of Norway, has actively publicized its findings and recommendations, which include unprecedented efforts to obtain international cooperation in achieving a more equitable distribution of the world's wealth and new systems of sustainable developmental practices. Her keynote address at the 1989 Forum on Global Change, "Global Change and Our Common Future," was published in *Environment* (June 1989).

CONTRIBUTORS TO THIS VOLUME

EDITOR

THEODORE D. GOLDFARB is an associate professor of chemistry at the State University of New York at Stony Brook. He received his B.A. from Cornell University and his Ph.D. from the University of California at Berkeley. He is the author of twenty research papers, which have appeared in scientific journals. He is also the coauthor of *A Search for Order in the Physical Universe* (W. H. Freeman, 1974).

Professor Goldfarb is a member of the American Association for the Advancement of Science, Science for the People, the American Chemical Society, and several other professional and community organizations. His present research interests include the environmental effects of energy-related technology and the use of agricultural chemicals. He has served as an advisor to local governments on environmental matters.

STAFF

Marguerite L. Egan Program Manager
Brenda S. Filley Production Manager
Whit Vye Designer
Libra Ann Cusack Typesetting Supervisor
Juliana Arbo Typesetter
David Brackley Copy Editor
David Dean Administrative Assistant
Diane Barker Editorial Assistant
James and David Filley Graphics

AUTHORS

RICHARD ELLIOT BENEDICK, as deputy assistant secretary of state for environment, health, and natural resources, was the chief U.S. negotiator for both the Vienna Convention on the Protection of the Ozone Layer and the Montreal Protocol on Substances That Deplete the Ozone Layer. He is currently on assignment as senior fellow of the Conservation Foundation and World Wildlife Fund in Washington, D.C.

AMANDA BICKEL is a staff researcher at the Institute for Energy and Environmental Research (IEER) in Takoma Park, Maryland.

SHIRLEY A. BRIGGS is executive director of the Rachel Carson Council, Inc., in Chevy Chase, Maryland. She is the coeditor, with Irston R. Barnes and Gilbert Gude, of *Landscaping for Birds* (Audubon Naturalist, 1973).

LUTHER J. CARTER is an independent Washington, D.C.-based journalist and the author of *The Florida Experience: Land and Water Policy in a Growth State* (Johns Hopkins University Press, 1976) and *Nuclear Imperatives and the Public Trust: Dealing with Radioactive Waste* (Resources for the Future, 1987).

BARRY COMMONER is a professor of earth and environmental sciences and director of the Center for the Biology of Natural Systems at Queens College in Flushing, New York, where he has taught since 1981. He is the author of *The Poverty of Power: Energy and the Economic Crisis* (Alfred A. Knopf, 1976) and *The Politics of Energy* (Alfred A. Knopf, 1979).

T. ALLAN COMP is the director of program planning for the American Forestry Association. He is the editor of *Blueprint for the Environment* (Howe Brothers, 1989).

MICHAEL CORR has served on the Committee on Environmental Alterations of the American Society for the Advancement of Science in Washington, D.C. He is the author of *To Leave the Standing Grain* (Copper Canyon, 1977) and *Cape Alava* (White Pine, 1981) and the editor of *Power Consumption and Human Welfare* (Macmillan, 1975).

PAUL R. EHRLICH is the Bing Professor of Population Studies at Stanford University, where he has taught since 1976. He is a member of the National Academy of Sciences and a fellow of the American Academy of Arts and Sciences. His publications include *New World, New Mind: Moving Toward Conscious Evolution* (Doubleday, 1989) and, with Anne H. Ehrlich, *The Population Explosion* (Simon & Schuster, 1990).

A. DENNY ELLERMAN is executive vice president of the National Coal Association. Founded in 1917 and currently comprised of 150 members, it is an organization of producers and sellers of coal, equipment suppliers, other energy suppliers, consultants, and coal transporters. The National Coal Association serves as a liaison between the coal industry and federal branches and agencies of government.

SAMUEL S. EPSTEIN is a professor of occupational and environmental medicine in the School of Public Health at the University of Illinois Medical Center in Chicago, president of the Rachel Carson Trust, and chairperson of the Commission for the Advancement of Public Interest Organization. He is an associate editor of *The Ecologist* magazine, the author of *The Politics of Cancer* (Sierra, 1978), and the coauthor of *Hazardous Waste in America: Our Number One Environmental Crisis* (Sierra, 1983).

PHILIP M. FEARNSIDE is a research professor in the Department of Ecology at the National Institute for Research in the Amazon. He is the author of *Human Carrying Capacity of the Brazilian Rainforest* (Columbia University Press, 1986) and has spent 13 years doing research in the Amazon.

HUGH M. FINNERAN is the senior labor counsel for PPG Industries, Inc., a *Fortune 500* company based in Pittsburgh, Pennsylvania, that manufactures paints, glass, printing inks, paper coatings, varnishes, adhesives, and many other such products.

HILARY F. FRENCH is a senior researcher at the Worldwatch Institute in Washington, D.C. and the author of Worldwatch Paper 94, *Clearing the Air: A Global Agenda.* Through its magazine and other publications, Worldwatch's goal "is to help reverse the environmental trends that are undermining the human prospective."

WILLIAM R. FURTICK was chief of the Crop Protection Service of the Food and Agriculture Organization of the United Nations.

BIL GILBERT has been writing about conservation matters for more than 30 years. His publications include *In God's Countries* (University of Nebraska Press, 1984), *Our Nature* (University of Nebraska Press, 1987), and *God Gave Us This Country: Tecumseh and the First American Civil War* (Macmillan, 1989).

LINDSEY GRANT is a consultant to the Environmental Fund and the author of *Foresight & National Decisions: The Horseman & the Bureaucrat* (University Press of America, 1988). He was the U.S. State Department coordinator for the *Global 2000* report.

DENIS HAYES, the 1985 recipient of the Sierra Club's John Muir Award and the American Solar Energy Society's Certificate of Outstanding Achievement, is a professor at Stanford University and a lawyer with the firm of Cooley, Godward, Castro, Huddleston & Tatum, located in

San Francisco, California. Hayes was the executive director of Earth Day 1970 and national chairman of Earth Day 1990. He is the author of *Rays of Hope: A Global Energy Strategy* (Norton, 1977).

JOHN P. HOLDREN is a professor in the energy and resources program at the University of California at Berkeley and a fellow of the American Academy of Arts and Sciences. He is a member of the executive committee of the Pugwash Conference on Science and World Affairs, London and Geneva, and coeditor, with Joseph Rotblat, of *Strategic Defense & the Future of the Arms Race: A Pugwash Symposium* (St. Martin's Press, 1987).

JON R. LUOMA is a widely published science writer and a regular contributor to magazines and newspapers, including *Audubon*. He is a member of the World Wildlife Fund, the Nature Conservancy, and the National Audubon Society. His publications include *Troubled Shores, Troubled Waters* (Viking Press, 1984); *The Air Around Us: An Air Pollution Primer* (Acid Rain Foundation, 1987); and *A Crowded Ark: The Role of Zoos in Wildlife Conservation* (Houghton Mifflin, 1988).

ARJUN MAKHIJANI directs the Institute for Energy and Environmental Research (IEER) in Takoma Park, Maryland. He holds a Ph.D. in electrical engineering from the University of California at Berkeley.

ANNIE MAKHIJANI is a staff researcher at the Institute for Energy and Environmental Research (IEER) in Takoma Park, Maryland.

CHARLES R. MALONE is an environmental scientist with the Nevada Agency for Nuclear Projects/Nuclear Waste Project Office in Carson City, Nevada. He has a Ph.D. in ecology from Rutgers University and has been involved with nuclear waste management since 1980.

CAROLYN MARSHALL is director of the Reproductive Hazards Project, a journalism project of the Tides Foundation.

JOHN G. MCDONALD is president of BP Oil Company and executive vice president of BP America, Inc., located in Cleveland, Ohio.

ROBERT J. MENTZINGER is an assistant researcher at Public Citizen's Congress Watch in Washington, D.C.

ROBERT K. OLSON, a former career officer in the Foreign Service, is a member of the U.S. Association for the Club of Rome and an active supporter of the Global Tomorrow Coalition. He is the author of *U.S. Foreign Policy and the New International Economic Order* (Westview, 1981).

ARI PATRINOS is program manager for the Carbon Dioxide Research Program.

LEWIS REGENSTEIN is a conservationist, author, and lecturer. He is the vice president of the Fund for Animals in Atlanta, Georgia, and serves on the board of directors for the Washington Humane Society, The Monitor Consortium, and the Interfaith Council for the Protection of Animals and Nature. His publications include *America the Poisoned: How Deadly Chemicals Are Destroying Our Environment, Our Wildlife, Ourselves, and How We Can Survive!* (Acropolis, 1982) and *How to Survive in America the Poisoned* (Acropolis, 1986).

ROBERT ROCHE is the manager of proposals at Foster Wheeler Power Systems, Inc.

WILLIAM D. RUCKELSHAUS, former head of the Environmental Protection Agency, is currently chairman of the board and chief executive officer of Browning-Ferris Industries, Inc., a waste disposal firm.

CLAUDINE SCHNEIDER served as congresswoman for the second district of Rhode Island. She is founder of the Rhode Island Committee on Energy and was chosen Woman of the Year by Rhode Island's Woman's Political Caucus in 1978.

NEIL SELDMAN is director of waste utilization at the Institute for Local Self-Reliance, which advises grassroots organizations, community development groups, and local government agencies. He is the coauthor, with Lawrence R. Martin, of *An Environmental Review of Incineration Technologies* (Institute for Local Self-Reliance, 1986).

JOHN SHORTSLEEVE is vice president of marketing at Foster Wheeler Power Systems, Inc.

JULIAN L. SIMON is a professor of economics at the University of Maryland. His publications include *The Ultimate Resource* (Princeton University Press, 1982), *Population and Economic Growth Theory* (Basil Blackwell, 1985), *Effort, Opportunity & Wealth* (Basil Blackwell, 1987), and *Population Matters: People, Resources, Environment, and Immigration* (Transaction Publishers, 1990).

PAUL J. STAMLER has served on the Committee on Environmental Alterations of the American Association for the Advancement of Science.

RICHARD STARNES is editor-at-large of *Outdoor Life* magazine.

WILLIAM TUCKER, a writer and social critic, has written for *Harper's* and many other publications.

ALVIN M. WEINBERG is a physicist and Distinguished Fellow of the Institute for Energy Analysis in Oak Ridge, Tennessee. He is a member of the National Academy of Sciences and the American Nuclear Society, and he is the coauthor of *The Nuclear Connection: A Reas-*

sessment of Nuclear Power and Nuclear Proliferation (Paragon House, 1985) and *The Second Nuclear Era: A New Start for Nuclear Power* (Praeger, 1985).

ELIZABETH M. WHELAN, an epidemiologist, is the executive director of the American Council on Science and Health and a member of the Environmental Protection Agency's Committee on Pesticides and Toxics. She is the author of *Preventing Cancer* (Norton, 1984).

LANGDON WINNER is a professor in the Department of Science and Technology Studies at Rensselaer Polytechnic Institute. He is the author of *Autonomous Technology: Technics-Out-Of-Control as a Theme in Political Thought* (MIT Press, 1977) and *The Whale and the Reactor: A Search for Limits in an Age of High Technology* (University of Chicago Press, 1986).

NIRA BRONER WORCMAN is a free-lance journalist from Brazil. She specializes in high technology and environmental issues. In 1988–89, she was a Knight science journalism fellow at the Massachusetts Institute of Technology.

ROBERT G. WRIGHT is the director of the asbestos and hazardous waste department of the Laborers' International Union of North America in Washington, D.C.

INDEX

acid rain: 9, 14, 67, 99, 105, 106, 267, 268, 275, 320; controversy over, 174–184
active ingredients, in pesticides, definition of, 143, 144
additives, food, 159, 165, 166
affluence, pollution and, 85, 89–91
Agency for International Development (AID), 148, 272
Agent Orange, 199
agriculture: 86, 279, 280, 288, 310, 316; effect of population growth on, 73, 74; *see also,* crop diseases; crop damage; fertilizer; herbicides; pesticides
agroforestry, 272
air pollution: 8, 11, 13, 61, 63, 64, 65, 67, 68, 76, 86, 87, 88, 89, 155, 159, 161, 164, 165, 166, 279, 307; controversy over, 98–109
Alar, 8, 160
aldrin/dieldrin, 146, 160, 161
alternative fuels, 115, 126, 277
aluminum: 220; recycling of, 219, 221; soil, and acid rain, 176, 177
Amazon, controversy over deforestation in Brazil's, 248–261
American Cancer Society (ACS), 159, 160–161
American Forestry Association (AFA), 11, 12, 14
Ames, Bruce, 161, 162
animals, endangered species of, 38–49
aquifer, 231, 234
aromatics, hydrocarbons called, 103, 104
asbestos, 8, 75
ash: incinerator, 13, 219, 220, 221, 222, 223, 225; scrubber, 106
Associated General Contractors (AGC), hazardous waste worker training program of, 206–208
aulocomnium moss, and acid rain, 176
automobiles: electric, 77–78; fuel-efficient, 13, 267–268, 269; pollution by, 14, 67, 75, 90, 91, 98, 99, 106, 178

backpackers, 29, 30, 31
back-to-the-land movements, 20, 31
batteries, pollution from 220–221
Benedick, Richard Elliot, on the Montreal Protocol, 286–293
benzene, as carcinogen, 99, 100, 102, 103, 161, 210
Bickel, Amanda, on the Montreal Protocol, 294–301
biological pest controls, 88, 142, 148
biomass, 114, 271, 294
birth defects: due to parent's hazardous occupation, 188, 193, 196, 197, 198, 199, 200; pollution and, 55, 62
blame-the-victim-theory, of cancer causation, 157, 161
bona fide occupational qualifications (BFOQ), as defense for sex discrimination against women in hazardous occupations, 191–192
Brazil, controversy over deforestation in, 248–261, 272
Briggs, Shirley A., on pesticides, 140–149
British Petroleum, air quality concerns of, 98–104
Brower, David, 28, 67
Brown, Lester, 19, 315
Bureau of Land Management, 23, 26, 27
burning, of municipal solid waste, controversy over, 218–226
business necessity, of disparate treatment of fertile women in hazardous occupations, 188, 191, 192, 194

Califano, Joseph, 167, 168
cancer: 146, 307; controversy over chemicals as cause of, 154–169; ozone depletion and increase in skin, 286, 288, 295; pollution and, 55, 56, 61, 77, 199; radiation and, 124, 125
carbon dioxide (CO_2), and greenhouse effect, 9, 76, 78, 81, 88, 99, 108, 117, 118, 119, 248, 266–282, 314
carbon monoxide (CO), as air pollutant, 8, 77, 99, 102, 103, 104, 106
carbon tetrahydrochloride, hazards of, 295, 296, 297–298, 299
carcinogen, benzene as, 99, 100, 102, 103, 161, 210
carpooling, 106, 108
carrying-capacity, of the earth, 20, 76, 257, 317
Carter, Luther J., on Yucca Mountain, Nevada, as nuclear waste repository, 231–239
cataracts, increase in eye, and ozone depletion, 288, 295
cement, pollution from production of, 85, 86, 90
characterization, of potential nuclear waste disposal sites, 231, 232, 235, 241
chemicals, controversy over, as cause of cancer, 154–169
chemotherapy, 159, 161
chlordane, 141, 145–146, 156, 160, 161, 165
chlorofluorocarbons (CFCs): 9, 14; and controversy over global warming, 266–282; *see also,* Montreal Protocol

chromosomal damage, from pollution, 62
citizens' groups, environmental, growth of, 148
Civil Rights Act of 1964, and controversy over sex discrimination toward women in hazardous occupations, 188–200
Clean Air Act, 6, 7, 55, 102, 103, 107, 123, 180, 182, 184, 315
Clean Water Act, 6, 7, 55
climate change: 18, 118, 286, 288, 314; *see also,* global warming
clustering: of cancer cases, 155; of infertility due to workplace chemicals, 197, 198
coal, as energy source, 115, 116, 123, 125, 127, 275, 276
coal industry, and controversy over acid rain, 178, 180, 308
Commoner, Barry: 75, 78; on pollution and population growth, 80–91
Comp, T. Allan, on the success of the environmental movement, 11–14
composting, 218–221, 224
computer industry, as hazardous occupation, 196, 197, 199
conservation, 7, 12, 20, 32, 34, 119, 127, 254
Conservation Era, 23, 25, 26, 28–29
consumers, and environmental movement, 5, 12, 76, 118, 307, 318
Convention on International Trade in Endangered Species of Wild Flora and Fauna (CITES), 45, 46, 47
coolants, CFCs as, 287, 298, 299, 300
cornucopian fallacies, 312–320
Corporate Average Fuel Economy standards for automobiles, 268–269
Corr, Michael, on pollution and population growth, 80–91
Council on Environmental Quality (CEQ), 5, 38–39, 40, 46, 49, 142, 164
crop damage, from acid rain, 178
crop diseases, food damage by, 133, 138

daminozide, *see* Alar
DDT, 8, 145–146, 147
Deaf Smith, Texas, as potential nuclear waste disposal site, 231, 232
deforestation, 9, 11, 14, 18, 34, 78, 248–261, 267, 271–273, 276, 277, 308, 320
Department of Energy (D.O.E.), U.S.: 222, 269, 270, 271; and Yucca Mountain nuclear waste site, 235, 236, 238, 240
desertification, 316, 320
detergents, phosphates in, 77, 78, 82, 83, 85, 86, 87
diet, cancer and, 161, 162, 166
discrimination, sexual, controversy over women and, in hazardous occupations, 188–200
disparate treatment, of fertile women in hazardous occupations, 188, 189–190, 191
Doll, Sir Richard, 161, 168
drought, 266, 278

Earth Day, and controversy over success of the environmental movement, 4–14
earthquake risk, at Yucca Mountain nuclear waste disposal site, 233, 234–235, 240, 241
economies of scale, 74, 75
Ehrlich, Paul: 27, 81–82; on the impact of population growth, 72–79
electric power production: 106, 115, 125, 126, 270; pollution and, 88–89, 90, 180
Ellerman, A. Denny, on acid rain, 181–184
emissions standards, auto, 14, 106, 107, 108
endangered species: 192, 248, 310; controversy over, 38–49
Endangered Species Act, 6, 7, 38, 44, 49
Endangered Species Scientific Authority (ESSA), 45, 46, 47, 49
energy efficiency, 13, 106, 107, 268, 270–271, 272
Environmental Defense Fund (EDF), 146, 148, 178
environmental impact statements (EISes), 5–6, 7, 142, 148, 260
environmental movement, controversy over success of, 4–14, 248–261
Environmental Protection Agency (EPA): 6, 99, 104, 176, 236; and controversy over science, risk & public policy, 54–55, 61; chemicals and 156, 165; and the Montreal Protocol, 288, 292, 296; and municipal solid waste, 220, 226, 227, 228; pesticides and, 142, 143, 144, 199, 200; *see also,* Superfund
Epstein, Samuel S., on cancer-causing chemicals, 154–163
Equal Employment Opportunity Commission, 189, 200
erosion, soil, 6, 9, 34, 78, 88, 137
ethanol, 103, 271
ethylene dibromide (EDB), 156, 160, 161, 162, 199
eutrophication, 74, 77, 178
extinction, of plant and animal species, 4, 20, 34, 39, 272, 286, 309

facts, vs. values, in risk assessment and management, 56, 64
famine, 132, 309
Fearnside, Philip M., on deforestation in Brazilian Amazonia, 255–261
Federal Insecticide, Fungicide and Rodenticide Act (FIFRA), 142, 144, 145
fertile women, controversy over sex discrimination and employment of, in hazardous occupations, 188–200
fertilizer, 74, 75, 77, 78, 86, 178, 316

fetal protection policies (F.P.P.s), in hazardous occupations, 198, 199, 200
finite resources, myth of, 307, 317–319
Finneran, Hugh M., on sex discrimination toward women in hazardous occupations, 188–193
fish, effect of acid rain on, 176, 178, 179
Fish and Wildlife Service, U.S., 45, 46, 48, 49
fission, nuclear, 114, 115, 117, 118, 122
foam, CFCs in, 298, 299
forests, controversy over effect of acid rain on, 174–184
formaldehyde, 100, 161
fossil fuels, burning of, 9, 89, 105, 108, 114, 117, 118, 156, 267, 270–271, 272, 273, 274, 276, 277
Freedom of Information Act, 146, 147
French, Hilary F., on air pollution, 105–109
frontier spirit, American, environmentalism and, 28–30
Furtick, William R., on pesticides, 132–139
fusion, nuclear, 114, 117, 118

game management, in national parks, 24–25
gametotoxins, 198–199
gas guzzler tax, for automobiles, 269
gasoline, 100, 271
genetic damage: from pollution, 62; from workplace chemicals, 198, 200
genetic susceptibility, cancer and, 154, 157
Gilbert, Bil, on the success of the environmental movement, 4–10
glaciers, melting of, 275, 281, 282
glass, incineration of, 219–220; recycling of, 245
Global Environment and Human Needs, The (President's Council on Environmental Quality), 38–39
Global Releaf, 12, 14
Global 2000 report, 38, 309, 312, 313, 314, 316, 317, 319
global warming: 11, 12, 21, 99, 105, 248, 296, 298; controversy over, 266–282
Gloucester Environmental Management Services (GEMS) landfill site, as Superfund failure, 209–214
glycol ethers, hazards of, 197, 199
government procurement, recycling and, 225, 271
Grant, Lindsey, on the cornucopian fallacies, 307–320
"grants equal to taxes" (GETT), under the Nuclear Waste Policy Act, 237, 238
greenhouse effect: 105, 118, 296; controversy over, 266–282
"greening" of America, 12, 14
groundwater, 74, 156, 212, 236, 241

habitat, destruction of, and endangered species, 40, 41, 46, 47, 48
Hair, Jay, 4, 9
halons, 287, 292, 293, 294, 296, 299
Hanford, Oregon, as potential nuclear waste disposal site, 231, 232
Hayes, Denis, on nuclear power, 122–128
hazard, vs. risk, in cost/benefit analysis, 62–63, 66
hazardous occupations, controversy over sex discrimination and women in, 188–200
hazardous waste: 106, 156, 222, 223; *see also,* Superfund
heat pump, 116, 269
heavy metals, 76, 77, 78, 198, 220
heptachlor, 146, 156, 160, 161, 165
herbicides, 78, 81, 85, 86, 138, 144, 149, 200, 316
Haldren, John P., on the impact of population growth, 72–79
hydrocarbons, air pollution from, 99, 100, 103, 106
hydrochlorofluorocarbons (HCFCs), 295, 298, 299
hydroelectric power, 117, 268, 271, 277

immune system, CFCs and, 288, 295
incineration, waste: 13; controversy over, 218–226
indigenous peoples, efforts of, to save Brazilian rain forest, 248, 253–254
inert ingredients, in pesticides, definition of, 143
infinite earth fallacy, 307, 317–319
insecticides: 142, 156; natural, 139
insects, food damage by, 133, 134, 138
integrated pest management (IPM), 148–149
integrated resource recovery, of municipal solid waste, 218–221
Inter-American Development Bank (IDB), 248, 256
interferon, 160, 168
International Agency for Research on Cancer (IARC), 161, 167

Kahn, Herman, rebuttal of views of, on improving quality of life, 312–320

Laborers International Union of North America (IUNA), hazardous waste worker training program of, 206–208
land use, and deforestation, 257–258
landfill crisis, 11, 209–214, 218–226
Lassa fever, 77
law, environmental, 7, 12, 54–55, 147–148
lead, pollution from, 83, 100, 199, 200, 210, 220
least-cost utility planning, 271
Leopold, Aldo, 24
life expectancy, increasing, 306
life-style, cancer and, 154, 157, 161

light water nuclear reactor, 120
logging, commercial, and deforestation, 259–260
Love Canal, New York, 62, 212
low-till farming methods, 149
lung cancer, smoking and, 154, 157, 158, 160, 161, 162, 165, 166, 167, 168
Luoma, Jon R., on acid rain, 174–180
Lutzenberger, Jose, 248

Makhijani, Arjun and Annie, on the Montreal Protocol, 294–301
malnutrition, 77, 135
Malone, Charles R., on Yucca Mountain, Nevada, as nuclear waste repository, 240–243
Marburgvirus, 77
Marshall, Carolyn, on sex discrimination toward women in hazardous occupations, 194–200
Marshall, Robert, 24
mass burning, of municipal solid waste, 218, 222–226
mass transit, 90
McDonald, John G., on air quality, 98–104
McDonnell Douglas Corp. v. Green, 189–190
media, portrayal of environmental movement in, 11, 21, 253, 266
men, reproductive damage to, from toxic chemicals, 199, 200
Mendes, Chico, 248, 249, 254
mercury, pollution from, 86, 221
metals, recycling of ferrous, 219
methane, as greenhouse gas, 266, 267, 275, 277, 287
methanol, 100, 101
methyl chloroform, hazards of, 294, 295, 296, 297, 299
methylene chloride, 199
microwaves, hazards of, 197
minorities, lack of, in the environmental movement, 13, 76
miscarriage: due to parent's hazardous occupation, 188, 196, 197, 199, 200; and toxic chemicals, 209, 212
monitored retrieval storage plan, for nuclear waste, 238
monofills, 223
Montreal Protocol, controversy over, 286–301
mosses, growth of, and acid rain, 176–177
Muir, John, 23, 27
multiple cropping, 137
Multiple Use and Sustained Yield Act, 25
municipal solid waste (MSW) disposal, controversy over, 218–226
mutagenic effects: 4, 124; of chemicals, 155, 162, 189, 190, 198, 199

Nader, Ralph, 67
Nashville Gas Co. v. Satty, 191
National Academy of Sciences, 19, 56, 143, 158, 267, 269, 272
National Acid Precipitation Assessment Program (NAPAP), 179, 182
National Ambient Air Quality Standards (NAAQS), 181, 184
National Cancer Institute (NCI), 159, 160–161
National Coal Association, and controversy over acid rain, 181–184
National Energy Policy Act, 267
National Forest Service, 23, 24, 25, 29
National Institute for Environmental Health Sciences (NIEHS), 206
National Institute for Occupational Safety and Health (NIOSH), 155, 159, 197, 198
National Institutes of Health, 159
National Park Service, 24, 25
national parks, 8, 76
National Resources Defense Council, 34, 106, 124, 148
National Science Foundation, 46, 49
National Wildlife Federation, 4, 7, 9, 26
natural gas, 14, 117, 118, 276
Nelson, Gaylord, 6
Netherlands fallacy, 75
Nevada, controversy over Yucca Mountain site in, as nuclear waste repository, 230–242
newspaper, recycling of, 219
Next Million Years, The (Darwin), 114
Next 200 Years, The (Kahn), 313
NIMBYism (not-in-my-backyard), and nuclear waste disposal sites, 232–233
nitrogen oxide: and acid rain, 174, 176, 178; air pollution from, 83, 88, 91, 99, 100, 103, 104, 106, 107, 223
noble savage, concept of, 30–31
noise pollution, 55
NTA (nitrilotriacetic acid), 77
nuclear power: 60, 63, 77, 78, 79, 271, 277; controversy over, 114–128
nuclear waste repository, controversy over Nevada's Yucca Mountain as, 230–242
nuclear weapons, 13, 122

occupational carcinogens, 157, 162, 167
occupational hazards, and controversy over women and sex discrimination regarding, 188–200
Occupational Safety and Health Administration (OSHA), 56, 160, 165, 168, 198, 206, 207
ocean levels, rising, 275, 280–282
oil, as energy source, 115, 116, 117, 118, 127, 268, 269, 270, 308, 318
oil industry, concerns about air quality of, 98–104
olefins, hydrocarbons called, 103
Olson, Robert K., on the intrinsic value of wilderness, 18–22

Our Nature Program, Brazil's, 249, 252, 256, 258
oxygenates, in gasoline, 103, 104
ozone: air pollution from ground-level, 99, 100, 102, 103, 104, 175, 178, 182, 267; destruction of atmospheric, 9, 11, 12, 267, 268, 275, 286–301

particulate air pollution, 76, 78, 99, 181, 307
passive solar design, 270, 271
pastoral impulse, 31
pasture, and deforestation, 257–258
Patrinos, Ari, on global warming, 274–282
Pauling, Linus, 124
perchloroethylene, 161, 297
pessimism, in the environmental movement, 9
pesticides: 55, 74, 76, 77, 78, 81, 82, 85, 86, 157, 159, 165, 166, 195, 198, 199, 249, 252, 315; controversy over, 132–149
Peto, Richard, 161, 168
petrochemicals, toxic, 141–142, 155, 156, 199
petroleum industry, 193; *see also,* fossil fuels; oil industry
petroleum products, pollution from 85, 275, 276
phenol, 87
phosphates, in detergents, 77, 78, 82, 83, 87
photovoltaic cells, 271
Pinchot, Gifford, 23, 25
plants, endangered species of, 38–49
plastics: biodegradable, 12–13; pollution by, 78, 81, 85, 86, 87; recycling of, 219
Platt, John, 78
plutonium, 124, 235, 236, 238
politics, ecology and, 18–19, 20, 23
Politics of Cancer, The (Epstein), 162
pollution: 7, 21, 55, 292, 299; *see also,* air pollution; water pollution
polyethylene terephthlate (PET) beverage bottles, recycling of, 219
polystrichum moss, and acid rain, 176
Population Bomb, The (Ehrlich), 27
population growth, world: 7, 18, 19, 28, 132, 135, 136, 259, 267, 272, 273, 277, 312, 315, 317, 318; controversy over, 72–91
Powell, John Wesley, 23
predator control, in national parks, 24–25
Pregnancy Disability Amendment, to Title VII of the Civil Rights Act, 190–191
Preservation Era, 23, 25, 26, 32
primitivism, and environmentalism, 32–34
Program for Alternative Fluorocarbon Toxicity Testing (PAFT), 299
Project Independence, 126
propellants, CFRCs as, 287, 288, 290, 294, 298, 299
Public Utilities Regulatory Policies Act (PURPA), 222

quality of life, controversy over improving, 306–320

Rachel Carson Institute, views on pesticides of, 140–149
racism, in the environmental movement, 13, 76
radioactive waste: 77, 89, 119, 123, 126; controversy over disposal of, 230–243
rain forests, destruction of, 9, 39, 40, 248–261
reactors, nuclear, controversy over, 114–128
Reagan administration, environmental policies of, 41–42, 143, 157, 159, 160, 197, 231, 291, 293
recycling, 13, 14, 108, 218–221, 223, 224–225, 297
red dye #2, 164
reformulated gasoline, 100, 101
Regenstein, Lewis, on endangered species and human survival, 38–43
Reilly, William K., 211–212
religion, and environmentalism, 33–34
renewable energy sources, 73, 108, 114, 117, 123, 270, 271, 272
reproductive capacity: 55; employee, and controversy over sex discrimination toward women in hazardous occupations, 188–200
resistive heating, 116
Resource Conservation and Recovery Act, 7
restricted pesticides, 145
restrictions, on fertile women in hazardous occupations, 188, 194–195
retrofitting, 116
"right-to-know" workplace registration, 160
risk-benefit analysis, science, and public policy, controversy over, 54–68, 138, 143, 193, 207
Roadless Area Resources Evaluation (RARE), 27
robust packaging, 235, 238
Roche, Robert, on the integrated approach to managing solid waste, 218–221
rodents, food damage by, 133
Romanticism, and environmentalism, 30, 33, 34
Roosevelt, Theodore, 23
Rousseau, Jean Jacques, 30, 31
rubber tappers, influence of, on Brazilian deforestation, 248, 253–254
Ruckelshaus, William: 6; on science, risk, and public policy, 54–59

saccharin, 164
Sagebrush Revolt, 24
salination, soil, 78
Sand County Almanac, A (Leopold), 24
Schneider, Claudine, on global warming, 266–273

science, risk, and public policy, controversy over, 54–68
scrubbers, to control nitrogen oxide emissions, 106, 183, 223
sea level, rising, 275, 280–282, 314
Seldman, Neil, on waste management, 222–226
sewage, water pollution by, 8, 74, 75, 81, 91, 178
sex discrimination, controversy over women and, in hazardous occupations, 188–200
Shortsleeve, John, on the integrated approach to managing solid waste, 218–221
Sierra Club, 21, 23, 26, 27, 28, 30, 148
Silent Spring (Carson), and pesticide use, 140–149
Simon, Julian L.: on the improving quality of life, 306–311; rebuttal of, 312–320
Sinking Ark, The (Myers), 39
siting policy, for nuclear reactors, 120
Smithsonian Institution, 46, 49
smog, 75, 76, 77, 81, 82, 99, 102, 107, 266, 288
smoking, lung cancer and, 154, 155, 157, 160, 161, 162, 165, 166, 167, 168
Snyder, Gary, 28
soil erosion, 6, 9, 34, 78, 88, 137
solar energy, 106, 114, 116, 117, 118, 119, 123, 127, 270, 271, 277
solid waste, controversy over incineration of, 218–226
solvents: 141, 143, 161; CFCs as, 294, 297, 298
Source Performance standards, for mass burning of municipal solid waste, 224
sphagnum moss, and acid rain, 176
Stamler, Paul J., on pollution and population growth, 80–91
Starnes, Richard, on the sham of endangered species, 44–49
stewardship, concept of, 32, 34, 267, 293
stillbirth, due to parent's hazardous occupation, 188, 199, 200
sulphur dioxide (SO_2): and acid rain, 174, 176, 178, 180, 181, 183, 184; as air pollutant, 8, 9, 77, 99, 106, 108, 296, 307
superconductivity, 277
Superfund: 142; controversy over, 206–214
Supreme Court, rulings of, regarding sex discrimination toward women in hazardous occupations, 189–190, 191, 192
surfactants, 143
survival groups, 31
sustainable development, 9, 13, 21, 253, 254

TCE (1,1,1 trichloroethane), 297
1080, 8
teratogenic effects, of chemicals, 155, 189, 190, 194, 195, 198
termites, use of chlordane to destroy, 156, 161
thermal pollution, 123
Thoreau, Henry David, 22, 23, 28
threatened species, 42, 47
Three Mile Island, 119, 120
Times Beach, Missouri, 212
Title VII, and controversy over sex discrimination toward women in hazardous occupations, 188–200
Tolba, Mostafa, 290, 292, 293
toluene, 210
Toronto Conference on the Changing Atmosphere, 266
tort law, and prenatal injury caused by hazardous occupations, 192–193
toxic chemicals: 8; controversy over, 154–169
Toxic Substance Control Act, 157
Toxic Substances Strategy Committee (TSSC), 164
toxic waste: 11, 55, 67, 106, 108, 127, 142, 267; *see also,* Superfund
toxins, natural, 139, 155
Transuranium Registry, 124
trichloroethylene, 161, 297
Tris, 162
tritium, 236
Tucker, William, on the intrinsic value of wilderness, 19–34

Ultimate Resource, The (Simon), 310, 315
ultraviolet radiation, increase in, 286, 295
underground testing, of nuclear weapons, 238
United Nations Environment Program (UNEP), 39, 42, 147, 288, 290, 297, 300
United Nations Food and Agriculture Organization (FAO), 133, 134, 137
uranium, 115, 116, 118, 119, 124, 235
Usery v. Tamiami Trail Tours, Inc., 192
utility planning, least-cost, 271

values, vs. facts, in risk assessment and management, 56, 64
van den Bosch, Robert, 33, 148
vector-control programs, for insects, 134, 138
Velsicol Chemical Company, as producer of chlordane, 140–141, 146
vinyl chloride, hazards of, 198, 199
volatile organic solvents (VOCs), 297
volcanic activity: 277, 294, 296; risk of, at Yucca Mountain nuclear waste disposal site, 233, 234, 235, 240, 241

waste disposal: 7, 13, 55, 119; controversy over incineration for, 218–226; *see also,* Superfund
waste-to-energy companies, 218, 220
water pollution: 8, 11, 40, 65, 78, 83, 86, 87, 155, 161, 298; *see also,* acid rain
water power, 106, 114

weeds, food crop damage by, 133
Weeks v. Southern Bell Telephone & Telegraph Co., 191
Weinberg, Alvin M., on nuclear energy, 114–121
Whelan, Elizabeth M., on cancer-causing chemicals, 164–169
wilderness, controversy over intrinsic value of, 18–34
Wilderness Act, 20, 23, 24, 29
Wilderness Society, 20, 26
wind power, 106, 118, 123, 125, 127, 277
Winner, Langdon, on science, risk, and public policy, 60–68
women: 13; sex discrimination toward, in hazardous occupations, 188–200
wood preservatives, 141
Worcman, Nira Broner, on Brazil's thriving environmental movement, 248–254
workplace hazards: employee health and, 68, 154, 156, 160, 163, 165, 298; controversy over sex discrimination regarding, 188–200
World Bank, 137, 248, 249, 250, 256, 271
World Food Conference, 133, 134
World Health Organization, 166, 167
World Wildlife Fund, 39
Worldwatch Institute, 19, 40, 315
wrongful death, prenatal injury caused by hazardous occupations as, 193

X-rays, hazards of, 200

yard waste, composting of, 220
Yucca Mountain, Nevada, controversy over, as nuclear waste repository, 230–242

Zero Population Growth, 7